Evolutionary Stasis and Change in the Dominican Republic Neogene

Aims and Scope Topics in Geobiology Book Series

Topics in Geobiology series treats geobiology – the broad discipline that covers the history of life on Earth. The series aims for high quality, scholarly volumes of original research as well as broad reviews. Recent volumes have showcased a variety of organisms including cephalopods, corals, and rodents. They discuss the biology of these organisms-their ecology, phylogeny, and mode of life – and in addition, their fossil record – their distribution in time and space.

Other volumes are more theme based such as predator-prey relationships, skeletal mineralization, paleobiogeography, and approaches to high resolution stratigraphy, that cover a broad range of organisms. One theme that is at the heart of the series is the interplay between the history of life and the changing environment. This is treated in skeletal mineralization and how such skeletons record environmental signals and animal-sediment relationships in the marine environment.

The series editors also welcome any comments or suggestions for future volumes.

Series Editors
Neil H. Landman, landman@amnh.org
Peter Harries, harries@shell.cas.usf.edu

The titles published in this series are listed at the end of this volume

Evolutionary Stasis and Change in the Dominican Republic Neogene

Edited by

Ross H. Nehm
The Ohio State University
College of Education and Human Ecology
and Department of Evolution, Ecology
and Organismal Biology
Columbus, USA

and

Ann F. Budd
University of Iowa
Department of Geoscience
Iowa City, USA

Springer

Editors
Ross H. Nehm
Ann F. Budd

ISBN: 978-1-4020-8214-6 e-ISBN: 978-1-4020-8215-3

Library of Congress Control Number: 2008920069

Cover illustration: Photo credits: Ross H. Nehm and Severino Dahint

Printed on acid-free paper

9 8 7 6 5 4 3 2 1

springer.com

Preface

Science is supposedly ultimately constrained by the nature of the physical world, meaning that changes in scientific methods and practice are supposed to be away from those with less utility and toward those that are more revealing, useful, and productive of insights into the nature of that world. In practice, however, science is no less susceptible to fads, culture shifts, and pendulum swings than any other realm of human endeavor. This is an especially important feature of science to keep in mind in the present climate of shrinking government funding (at least in proportion to the demand) and the resulting susceptibility of individual scientists and entire disciplines to being influenced by the changing priorities of funding agencies (even if, as such agencies maintain, those priorities come ultimately "from the community"). The present volume is in several important respects a testimonial to both the threats and opportunities that such scientific culture swings pose, both for the individual researcher and a wider field.

When scientific research in the Dominican Republic Neogene began more than a century ago, paleontology was an essentially descriptive discipline, focused mainly on finding, describing, and documenting the taxa represented in the fossil record, and (especially in invertebrate paleontology) on using these taxa for biostratigraphic correlation. Despite the successful integration of paleontology into the Modern Evolutionary Synthesis in the middle of the twentieth century (Simpson, 1944, 1953; Jepsen et al., 1949; Gould, 1983), the vast majority of paleontological research continued in this tradition, and most paleontological papers – including the fundamental works on the Dominican Neogene – were some version of "a new X from the Y of Z-land" (Gould, 1989:114).

The structure of paleontology, at least in the U.S., began to change in the late 1960s and early 1970s in association with at least three significant developments, each of which was to have significant influence on paleontological research in the Dominican Republic Neogene. The first was an increased interest in the ecology of fossil taxa (in addition to simply using fossils for paleoenvironmental reconstruction). There was a burst of research activity around this new slant on "paleoecology" as a new generation of paleontologists sought to interpret fossil assemblages by close comparison with living communities. Although by the early 1980s this research program had lost much of its focus, it did produce some innovative and lasting contributions, including attempts at documenting long-term patterns of biological communities in the shallow ocean (Allmon and Bottjer, 2001).

The second major development was the Deep Sea Drilling Project (DSDP) (see, e.g., Hsu, 1992; Corfield, 2001). This enormous (and well-funded) project was influential to paleontology in two significant ways. Scientifically it provided both abundant new data and a new temporal and (in many ways) intellectual framework for applying fossils to answering questions of Earth history, including climate, sea-level, temperature, and ocean circulation and nutrient status. Although it was concerned almost exclusively with microfossils, the DSDP clearly demonstrated the unique value of paleontology to reconstructing the biotic and abiotic environment in a modern high-tech scientific context. Methodologically, it also demonstrated – not least to paleontologists themselves – how paleontology could be an integral part of large-scale, multidisciplinary "big science".

The third development was the percolation of aspirations among the younger generation of paleontologists to contribute in substantive and unique ways not just to geology but to evolutionary biology. These stirrings led to what became known broadly as "paleobiology", a major subfield of which became devoted to the compilation of taxonomic data from the literature, a research program that came to be known as "quantitative" or "analytical" paleobiology (Gilinsky and Signor, 1991; Sepkoski, 2005). This and related research programs emphasized theoretic over descriptive approaches and new methods of analysis of existing systematic data from the fossil record as much or more than the acquisition of new data. It brought paleontology to the "high table" of evolutionary theory (Maynard Smith, 1984; Eldredge, 1995), and – intentionally or not – it diminished the status of traditional descriptive systematics for its own sake.

The lessons and implications of the first two of these developments – the DSDP and paleoecology – were not lost on the founders of the Dominican Republic Project (DRP). In the late 1970s this group concluded that land-based, macropaleontology could benefit from a DSDP-style, large-scale, international, multi-investigator approach to creating and compiling taxonomic, stratigraphic, and paleoecological data (Saunders et al., 1986; Jung, 1993). At the core of the new project were two main ideas. First was an emphasis on a rigorous stratigraphic and sampling protocol that would be used by all project participants. This would, the organizers thought, avoid many of the biases inherent in different investigator's styles of sampling, and would allow data from many researchers to be readily compiled and compared. Second was the decision to distribute sorted samples to systematic specialists around the world. This would, thought the project leaders, bring to bear a much more powerful set of specialists than would be possible with only one or a very few systematists.

With the benefit of almost 30 years of hindsight, several aspects of the DRP experiment are noteworthy. Most conspicuously, the common stratigraphic and sampling regimes were enormously valuable and used by almost all participants, and provided an excellent model in these respects for the subsequent Panama Paleontology Project (PPP; see Jackson et al., 1996; Collins and Coates, 1999). By comparison, the DRP systematics results were both more and less successful than one might have anticipated or hoped. Although it received significant funding provided by the Swiss National Science Foundation, the DRP never had the financial resources to support the work of the individual systematic researchers who volunteered to take on various taxonomic groups. This inevitably contributed to sometimes lengthy delays in, and sometimes total abandonment of, production of the individual systematic monographs. Although DRP coordinators and collections staff at the Natural History Museum in Basel tried to keep close track of the collections that had been sent out, some were never seen or heard from again. (This experience was not lost on the coordinators of the PPP, who explicitly chose not to distribute material to numerous independent specialists.) Finally, although the DRP organizers certainly envisioned that the data resulting from the project would almost certainly be used for research into broader paleobiological topics, they did not specify in advance what those topics would or should be. Although the DRP was enormously innovative in its approach to centralizing stratigraphy and sampling while decentralizing its systematics, it was, *as a project*, not particularly innovative in the applications of the data that resulted. It was, rather, left to individual researchers to use their or others' data to investigate whatever topic was of interest to them.

Which brings us to the third of those three critical 1970s-era developments in paleontology. As noted by Nehm and Budd in the present volume, many of the subsequent studies that used DRP data were of great significance for areas of paleobiology such as evolutionary tempo and mode and diversity, extinction and turnover. Yet these were not explicitly goals of the project at the outset. In other words, careful attention to making large, well-documented, and well-curated collections within a common, standardized, high-resolution stratigraphic framework made possible the fruitful application of the resulting data to larger theoretic questions. High-quality descriptive paleontology of the "traditional" sort permitted high-quality synthetic paleontology of the newer sort later.

Laudable though this outcome – and its copious illustration in the present volume – is, anyone who has written or reviewed an NSF proposal in the last 20 years knows that something is amiss here. It is almost impossible today to obtain funding for generation of basic systematic data without specifying beforehand to what larger (preferably pressing) theoretic use those data will be put. As an NSF program officer once put it to me, "there is an infinity of groups that need systematic revisions; we can only fund those that are interesting" because they can be used to address an "interesting" question. Thus the fundamental structure of the DRP, the success of which the present volume celebrates, would almost certainly not be fundable in this form by NSF or similar agencies today.

It has been frequently noted that paleontologists are a generally solitary lot, not especially well-suited to the large-scale collaboration and group-think often associated with "big science" projects. Historically, it is often observed, we have mostly pursued research that required relatively little infrastructure, aside from space to store our collections, a library, a microscope, and a means of travel. These attributes have been bemoaned as keeping paleontology out of the "big science" scene. We have, it is said, never "gotten our act together" and "gotten our share of the pie" the way the physicists, astronomers, or genomicists have. The difficulty of getting paleontologists to collaborate on one or a small number of larger topics or problems is highlighted by the multiplicity of national and international meetings and reports, most supported at least in part by NSF, that have attempted over the past couple of decades to chart a common, collaborative, "big-science" research agenda for paleontology (e.g., "Geobiology of Critical Intervals", Stanley et al., 1997; "Paleontology in the 21st Century", Lane et al., 2000; and most recently "Future Research Directions in Paleontology", FRDP, Bottjer, 2007).

It is noteworthy that the big collaborative projects in paleontology that *have* succeeded have been, in large part, not question-based, but (literally) data-based, such as the *Treatise on Invertebrate Paleontology* and the Paleobiology Database. In this context it is interesting that the recent FRDP report (Bottjer, 2007) includes as one of its five major objectives "Database and Museum Collection Development and Integration". The authors of the FRDP write: "Museum collections, databases and informatics are an integral part of the infrastructure of paleontology at present, and will continue to be so into the future. In order to be dynamic and useful resources, both require long-term support. Further, these two infrastructural resources are quite naturally complementary and interlinked. … Databases and museums undergird integrative multiuser research initiatives as well as individual projects. Being able to combine different datasets provides opportunities to ask new and more widely ranging questions in deep time studies. … Thus, both require long-term support and stability."

The present volume supports this objective and demonstrates the profound utility of well-coordinated data supported by carefully-collected and well-curated collections, and the editors have gone to considerable lengths to emphasize these themes. I suggest, however, that we might take this lesson even more seriously. As a discipline, paleontology might recognize, reiterate, and celebrate that "big paleontology" cannot be successfully modeled closely after "big physics" or "big astronomy" or "big molecular biology". Our major collaborations appear to be most fruitful in the coordination and assembling of large data sets, not necessarily in their interpretation around a narrow predetermined set of large or "important" questions. The actual generation of much of our data, especially systematics, and its application to questions about the history of the Earth and its life appear to require the dedicated attention of one or a very small number of individual researchers.

This does not make our science less than physics, astronomy, or genomics; it makes it different. It means that more projects like the DRP are needed – applied to both new field collections and existing museum collections (Jackson and Johnson, 2001; Allmon, 2005) – in order to generate and make available large quantities of new, high-quality systematic, stratigraphic, and paleoecologic data. It may be that the precise questions to which these data can be applied cannot now be specified. But that does not mean that the data are and will not be valuable. Indeed, many questions will not occur to us until the data are generated.

Finally, it should be noted that the DRP was and is a truly international, multi-institutional effort, involving museums, universities, and numerous individual researchers, including a number of Ph.D. students. The project was begun by Swiss paleontologists, and soon involved scientists from Tulane University, and eventually from dozens of other institutions around the world. In this context, I cannot help but note with pride (albeit more of the kind felt by the fan on the sidelines than of the player in the game) the prominent role that the Paleontological Research Institution (PRI) has played in this story since the early twentieth century. PRI's founder, Cornell professor Gilbert Harris, was the major advisor of Carlotta Maury, who conducted the first comprehensive overview of the macrofauna of the Cibao Valley, and published it in her landmark monograph (Maury, 1917a,b). Her collections remain today at PRI. When the DRP was started in the late 1970s, its architects chose PRI as the publisher of its systematic monographs in its journal, *Bulletins of American Paleontology*. To date, 22 such contributions have appeared, and more are in press and in preparation. With the retirement of Emily Vokes from Tulane in 1995 the large collections of Dominican fossils that she had assembled with her late husband Harold over more than three decades came to PRI. The involvement of a small museum in upstate New York in a project organized by a major European museum and a husband-and-wife academic team at a private university in Louisiana – now taken over by a new generation of researchers at an even more far-flung spectrum of institutions – is perhaps a fitting testament to how paleontology at its best (big, small, or otherwise) works.

References

Allmon, W.D., 2005, The importance of museum collections in paleobiology, *Paleobiology*, **31**(1):1–5.

Allmon, W.D. and Bottjer, D.J., 2001, Evolutionary paleoecology: the maturation of a discipline, in: *Evolutionary Paleoecology. The Ecological Context of Macroevolutionary Change* (W.D. Allmon and D.J. Bottjer, eds.), Columbia University Press, New York, pp. 1–8.

Bottjer, D.J. (ed.), 2007, Future Research Directions in Paleontology: report of a Workshop held April 8–9, 2006. The Paleontological Society, Knoxville, Tennessee.

Collins, L.S. and Coates, A.G. (eds.), 1999, A paleobiotic survey of Caribbean faunas from the Neogene of the Isthmus of Panama. *Bull. Am. Paleontol.*, **357**:351.

Corfield, R., 2001, *Architects of Eternity. The New Science of Fossils*. Headline Book Publishing, London, 338 p.

Eldredge, N., 1995, *Reinventing Darwin. The Great Debate at the High Table of Evolutionary theory*. Wiley, New York, 244 p.

Gilinsky, N.L. and Signor, P.W. (eds.), 1991, *Analytical Paleobiology. Short Courses in Paleontology*, No. 4. The Paleontological Society, Knoxville, Tennessee, 216 p.

Gould, S.J., 1983, Irrelevance, submission, and partnership: the changing role of palaeontology in Darwin's three centennials, and a modest proposal for macroevolution, in: *Evolution from Molecules to Men* (D.S. Bendall, ed.), Cambridge University Press, Cambridge, pp. 347–366.

Gould, S.J., 1989, *Wonderful Life: The Burgess Shale and the Nature of History*. Norton, New York, 347 p.

Hsu, K.J., 1992, *Challenger at Sea: A Ship That Revolutionized Earth Science*. Princeton University Press, Princeton, NJ, 454 p.

Jackson, J.B.C., and Johnson, K.G., 2001, Measuring past biodiversity, *Science*, **293**:2401–2404.

Jackson, J.B.C., Budd A.F., and Coates A.G. (eds.), 1996, *Evolution and Environment in Tropical America*. University of Chicago Press, Chicago, IL, 425 p.

Jepsen, G., Simpson G.G., and Mayr E. (eds.), 1949, *Genetics, Paleontology, and Evolution*. Princeton University Press, Princeton, NJ, 474 p.
Jung, P., 1993, The Dominican Republic project. *Am. Paleontol.*, **1**(5):1–3.
Lane, H.R., Lipps, J., Steininger, F.F., Kaesler, R.L., Ziegler, W., and Lipps, J. (eds.), 2000, Fossils and the future. Paleontology in the 21st century. Senckenberg-Buch Nr. 74, Frankfurt, 290 p.
Maury, C.J., 1917a, Santo Domingo type sections and fossils. Part 1, *Bull. Am. Paleontol.*, **5**(29):1–251.
Maury, C.J., 1917b, Santo Domingo type sections and fossils. Part 2, *Bull. Am. Paleontol.*, **5**(30):1–43.
Maynard Smith, J., 1984, Palaeontology at the high table, *Nature*, **309**:401–402.
Saunders, J.B., Jung, P., and Biju-Duval, B., 1986, Neogene Paleontology in the northern Dominican Republic. Part 1. Field surveys, lithology, environment, and age, *Bull. Am. Paleontol.*, **89**(323):1–79.
Sepkoski, D., 2005, Stephen Jay Gould, Jack Sepkoski, and the 'Quantitative Revolution' in American paleobiology, *J. Hist. Biol.*, **38**(2):209–237.
Simpson, G.G., 1944, *Tempo and Mode in Evolution*. Columbia University Press, New York, 237 p.
Simpson, G.G., 1953, *The Major Features of Evolution*. Columbia University Press, New York, 434 p.
Stanley, S.M. (Steering Committee Chair) et al., 1997, Geobiology of Critical Intervals (GOCI). A proposal for an initiative by the National Science Foundation. Sponsored by the Paleontological Society, Knoxville, TN, 82 p.

Ithaca, NY — Warren D. Allmon

Contributors

Tiffany S. Adrain
Department of Geoscience, 121 TH, The University of Iowa, Iowa City, IA 52242, USA, tiffany-adrain@uiowa.edu

Warren D. Allmon
Paleontological Research Institution and Department of Earth and Atmospheric Sciences, Cornell University, 1259 Trumansburg Road, Ithaca, NY 14850, USA, allmon@museumoftheearth.org

Brian R. Beck
Centre for Marine Studies, University of Queensland, Brisbane, Queensland 4072, Australia, b.beck@uq.edu.au

Ann F. Budd
Department of Geoscience, 121 TH, The University of Iowa, Iowa City, IA 52242, USA, ann-budd@uiowa.edu

Rhawn F. Denniston
Department of Geology, Cornell College, Mount Vernon, IA 52314, USA, RDenniston@cornellcollege.edu

Charles C. Evans
Evans Environmental and Geosciences, 14505 Commerce Way, Miami Lakes, FL 33016, USA

Maria Harvey
Department of Biology, The City College, C.U.N.Y., Convent Avenue at 138th Street, New York, NY 10031, USA

Carole S. Hickman
Department of Integrative Biology and Museum of Palaeontology, University of California, Berkeley, CA 94720, USA, caroleh@berkeley.edu

Kenneth G. Johnson
Department of Palaeontology, Natural History Museum, Cromwell Road, London SW7 5BD, UK, K.Johnson@nhm.ac.uk

James S. Klaus
Department of Geological Sciences, University of Miami, 43 Cox Science Building, Coral Gables, FL 3133, USA, jsklaus@rsmas.miami.edu

Jermaine Lawson
Department of Biology, The City College, C.U.N.Y., Convent Avenue at 138th Street, New York, NY 10031, USA

Jupiter Luna
School of Education, The City College, C.U.N.Y., Convent Avenue at 138th Street, New York, NY 10031, USA

Donald F. McNeill
Marine Geology and Geophysics, Rosenstiel School of Marine and Atmospheric Science, University of Miami, 4600 Rickenbacker Causeway Miami, FL 33149, USA, dmcneill@rsmas.miami.edu

Ross H. Nehm
College of Education and Human Ecology and Department of Evolution, Ecology, and Organismal Biology, 333 Arps Hall, The Ohio State University, Columbus, OH 43210, USA, nehm.1@osu.edu

Stephanie C. Penn
Department Of Geology, Cornell College, Mount Vernon, IA 52314, USA

Rysanek Rivera
Department of Biology, The City College, C.U.N.Y., Convent Avenue at 138th Street, New York, NY 10031, USA

Holly A. Schultz
Department of Geology, University of California, Davis, One Shields Avenue Davis, CA 95616, USA, schultz@geology.ucdavis.edu

Juw Won Park
Department of Computer Science, Information Technology Services, 2860-65 UCC, The University of Iowa, Iowa City, IA 52242, USA
jwpark@cs.uiowa.edu

Contents

Chapter 1
Palaeobiological Research in the Cibao Valley of the Northern Dominican Republic

Ross H. Nehm[1] and Ann F. Budd[2]

Contents

1.1 Introduction

The Cibao Valley of the northern Dominican Republic has been of great interest to geoscientists for more than a century because its rich fossil fauna, temporally long-ranging sections, and geographically widespread exposures collectively provide an excellent system for innovative palaeobiological research. In order to provide context for the research studies presented in this volume, we begin with a brief overview of the history of palaeobiological research in the Cibao Valley of the Dominican Republic from the 1800s to the present. Subsequently, we summarize new research on the DR Neogene in this volume as well as new educational efforts and infrastructure that have been developed to strengthen and support the evolution of this international research effort.

[1]The Ohio State University, Columbus, OH, USA. Email: nehm.1@osu.edu

[2]Department of Geoscience, University of Iowa, Iowa City, IA, USA. Email: ann-budd@uiowa.edu

R.H. Nehm, A.F. Budd (eds.) *Evolutionary Stasis and Change in the Dominican Republic Neogene*,

1.2 Overview of Past Palaeobiological Research in the Cibao Valley

The Tainos, the indigenous inhabitants of Hispaniola, used the word "Cibao" to describe the rocky lands of the island's central mountain range. Today "Cibao" is used primarily to describe the fertile valley bordered on the north by the Cordillera Septentrional and on the south by the Cordillera Central. The Río Yaque del Norte bisects the valley along its east-west axis and drains westward towards Monte Cristi and into the Caribbean Sea. A series of north-south trending rivers (e.g., the Río Cana, Río Gurabo, and Río Mao) connect to the Río Yaque del Norte. It is these rivers that have collectively exposed the several thousand meters of fossil-rich sedimentary rock that have been the focus of palaeobiological inquiry for more than a century (Fig. 1.1).

Research in the Cibao Valley by European and North American scientists began in the mid-1800s. The first studies were very small in scope and involved single scientists

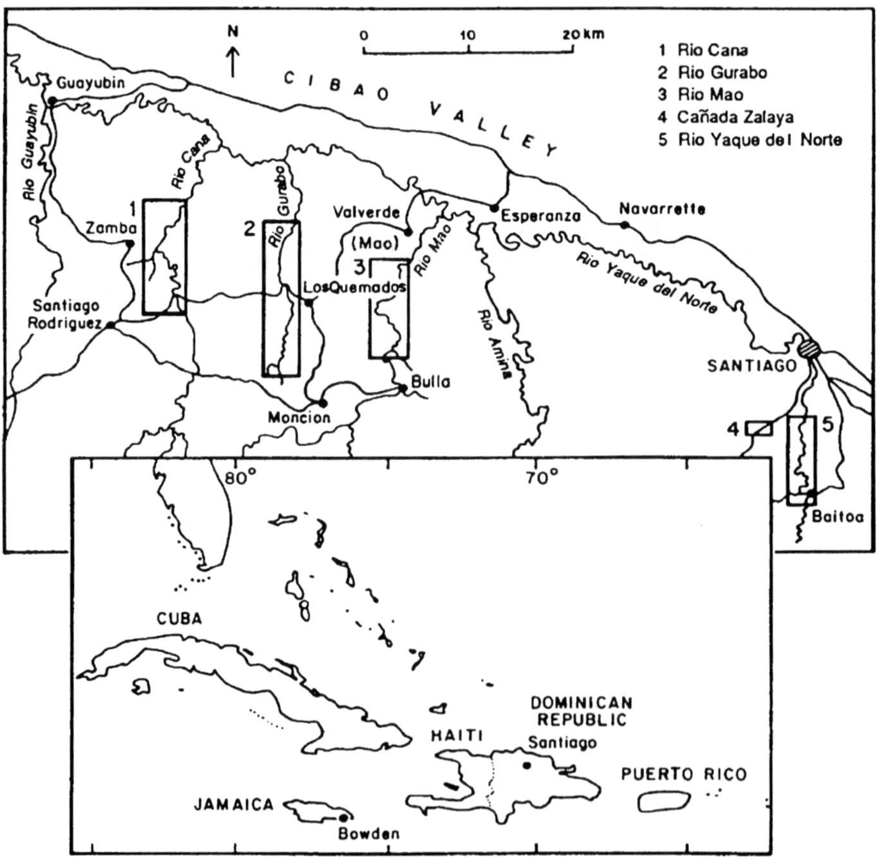

Fig. 1.1 Map of the Cibao Valley of the Northern Dominican Republic, with major river sections encompassed by boxes (Modified from Nehm and Geary, 1994)

rather than collaborative research teams. In the 1850s a series of papers by G.B. Sowerby II (1850), Lonsdale (1853), and Heneken (1853) described some of the first localities and fossil invertebrate species from the region. Duncan (1864) described 27 species of corals in the Heneken collections from the Dominican Republic, 20 of which were new. He also described two new genera: *Antillia* and *Teleiophyllia* (=*Manicina*). Type specimens were deposited in the Natural History Museum in London, UK. Eighteen of the new species are zooxanthellate corals, including three species of *Placocyathus*, two of *Stylophora*, one of *Dichocoenia*, three of *Antillia* (one of which is currently *Trachyphyllia bilobata*), two of *Teleiophyllia* (=*Manicina*), one of *Meandrina*, four of *Plesiastrea* (including two currently assigned to *Solenastrea*, one to *Stephanocoenia*, and one to *Montastraea*), one of *Siderastrea*, and one of *Pocillopora*. Three additional species were described in Duncan (1868). Vaughan (1919) later revised Duncan's names, finding a total of 28 species. Work on molluscs continued with Gabb (1873), Pilsbury and Johnson (1917), and Pilsbury (1922). But the most comprehensive work on the geology and fossils of the Cibao Valley in the early 20th century was conducted by Carlotta Joaquin Maury (Fig. 1.2).

Fig. 1.2 Major scientists instrumental in the development of the Dominican Republic Neogene as a palaeobiological research system. Top row, left to right: Carlotta Maury, T.W. Vaughan, and Peter Jung. Bottom row, left to right: John Saunders, Harold Vokes, and Emily Vokes

Maury was the first scientist to conduct a comprehensive overview of the fossils that occur in the layers of rock exposed by rivers in the Cibao Valley. Her tumultuous expedition of 1916 (during the Dominican revolution and American military invasion) involved collecting and identifying hundreds of new species of molluscs and many other invertebrates. Maury also revised estimates of the geological age of the sedimentary rocks and redefined geological formations. Dr. Maury is also noteworthy in being one of few women from the turn of the century to complete a doctorate in the sciences (at Cornell University) and be employed as a professional scientist. Her 1917 study is a classic reference that is still used today by mollusc researchers. Vaughan (1919) studied the corals in her collections and provided a chart listing the occurrences of coral species in each of Maury's zonal units.

Following the US invasion, T.W. Vaughan and his associates from the US Geological Survey conducted a major reconnaissance study of the general geology, stratigraphy, and economic geology of the Dominican Republic, including the regions of Cordillera Septentrional, Samaná Peninsula, Cibao Valley, Cordillera Central, as well as additional regions in the southern part of the country. Their work resulted in a 268 page memoir (Vaughan et al., 1921), two chapters of which have been particularly relevant to subsequent palaeontological studies of the Cibao Valley (chapter 4 by Wythe Cooke on geology and geologic history, and chapter 6 by T.W. Vaughan and W.P. Woodring on stratigraphic palaeontology). Chapter 6 of the memoir provides detailed descriptions of localities and faunal lists, including foraminifera, corals, bryozoans, molluscs, crustaceans, and echinoids, thereby setting the stage for the studies of systematics and palaeoecology in the present volume. Vaughan and Hoffmeister (1925) later described nine coral species based on the Gabb collections, all of which were new.

A series of other revisions of the ages and nomenclature of the Cibao sections were made by Maury (1929, 1931), Weyl (1940, 1966), Bermudez (1949), Butterlin (1954), Ramirez (1956), Van den Bold (1968, 1969), Bowin (1966), and Seiglie (1978) (see also McNeill et al., this volume). In 1961, Pflug illustrated and updated the scientific names of many of the species descriptions of Sowerby's Dominican fossil molluscs.

By the 1970s, Harold and Emily Vokes of Tulane University were working on the living and extinct molluscs of the Caribbean region (Vokes, 1979). Their field research efforts produced major new collections of mollusc material from the Cibao Valley (and elsewhere in the Caribbean) that remain of considerable importance (now housed at the Palaeontological Research Institution in Ithaca, New York). Unbeknownst to the Vokeses, a group of European scientists were planning a large-scale research project to resample, map, and study the fossil rich sedimentary rocks of the Cibao Valley. The two groups joined forces in the late 1970s and established the Dominican Republic Project (DRP), which moved our understanding of the system forward considerably.

To understand how and why the Dominican Republic Project progressed in the way it has, it is important to note how scientific research itself has changed over the past several decades. In some respects, the DRP was a harbinger of future geoscience research efforts. Today, large-scale, interdisciplinary, and international scientific research projects such as the Deep Sea Drilling Project in oceanography or the Human Genome Project in molecular biology are becoming increasingly common. By the 1970s, scientists in many fields were beginning to recognize that the amount of information, number of research methods, and range of specialties had increased to such a degree that it was difficult for a single scientist to possess the methodological tools and conceptual knowledge necessary for addressing many research questions. The recognition emerged that collaborative teams focused on the same research questions, but specializing in different subfields, could collectively test scientific hypotheses more accurately, efficiently, and economically. The DRP was one of the first multidisciplinary and international research projects in the field of palaeobiology.

In the mid-1970s, a group of European scientists (Peter Jung, macropalaeontologist, Switzerland; John Saunders, geologist and micropalaeontologist, England; and Bernard Biju-Duval, geologist, France) began planning a large-scale research project to resample the invertebrates of the Cibao Valley, re-map the region, and measure the stratigraphic sections with greater precision. The founders of the DRP embraced a collaborative approach to doing science. In order to understand the Cibao Valley system, many specialists were clearly necessary, including field geologists, geochronologists, stratigraphers, palaeobiologists, systematists, and evolutionary biologists. It is difficult, if not impossible, for any single researcher to have the breadth of knowledge to accomplish all of these goals.

The European team planned to precisely determine the ages of the sections, employ more appropriate sampling methods, and record locality information in greater detail by relying on different specialists. Each year from 1977 to 1980 Saunders, Jung, and Biju-Duval were joined in the field by several other scientists and Dominicans from nearby communities (see Saunders, Jung, and Biju-Duval, 1986, p. 9). A total of about 50 people were involved in the collection of fossil samples. Many of the river exposures that were studied are very remote and could only be reached on foot or on horseback. (Even today, burros are needed to help carry samples out of the river valleys).

The DRP field teams collected approximately 300 samples for macrofossil study and 500 samples for microfossil study. Overall, these samples contained millions of invertebrate specimens from several tons of material. These samples were sent to the Naturhistorishes Museum Basel (NMB) Switzerland for processing, sorting, identification, and curation. The results of many years of field research were published in the "Red Book" (Saunders, Jung, and Biju-Duval, 1986). It contains a series of detailed maps of collecting localities throughout the Cibao Valley, many of which are referenced throughout this volume.

Because the DRP field team collected considerably more material than Maury or any of the other scientists who had worked in the Cibao Valley previously, many new species of invertebrates (especially corals and molluscs) were discovered. In addition, larger sample sizes were now available to (1) explore morphological variability within and among species, (2) examine variation in relation to palaeoenvironmental and lithological variables, and (3) subsequently refine species definitions made by previous workers (e.g., Sowerby, Gabb, Pflug, etc.). The extensive sampling by the DRP team also produced specimens from previously unsampled times and locations, producing more accurate and precise spatial and temporal distributions of taxa.

The Swiss team recognized that in order to identify taxa accurately, and diagnose new species appropriately, it was necessary to send collections of sorted specimens to biologists or palaeobiologists who were specialists in particular invertebrate groups. When scientists and staff at the Naturhistorisches Museum

Basel finished processing the field samples, the specimens were sent to experts from around the world. Unfortunately, there are not enough trained systematists with knowledge about invertebrate biodiversity, so many groups remain unstudied and unknown to science. Nevertheless, those systematists who participated in the project spent many years working on the samples, comparing them to other living and fossil species, and visiting museums around the world to ensure that the scientific names assigned to the specimens were correct. Once the experts identified the specimens to the species level, and performed systematic revisions, these data could be used in geographic and temporal analyses of taxonomic distributions in the Cibao Valley and elsewhere. This information was then combined with data from other studies in order to determine where else Dominican species lived in the past and if these species are living in the Caribbean Sea today.

Many systematists have published these results in the journal *Bulletins of American Palaeontology*. Currently, 22 systematic monographs on Dominican taxa have been completed (Table 1.1). After publication, the fossil material used in the

Table 1.1 Monographs in the *Bulletins of American Palaeontology* series "Neogene Palaeontology in the northern Dominican Republic"

Series #	Year	Topic	Authors
1	1986	Field surveys, lithology, environment, and age	Saunders, J., Jung P. and Biju-Duval B.
2	1986	Genus *Strombina*	Jung, P.
3	1986	Family Poritidae	Foster, A.B.
4	1987	Genus *Stephanocoenia*	Foster, A.B.
5	1987	Suborders Caryophylliina and Dendrophylliina	Caims, S.D. and Wells J.W.
6	1987	Phylum Brachipoda	Logan, A.
7	1988	Subclass Ostracoda	van den Bold, W.A.
8	1989	Family Muricidae	Vokes, E.H.
9	1989	Family Cardiidae	Vokes, H.E.
10	1990	Family Cancellaridae	Jung, P. and Petit R.E.
11	1991	Family Faviidae (Part I)	Budd, A.F.
12	1992	Genus *Spondylus*	Vokes, H.E. and Vokes E.H.
13	1992	Class Echinoidea	Kier, P.M.
14	1992	Otoliths of teleostean fishes	Nolf, D. and Stringer G.L.
15	1994	Genera *Columbella, Eurypyrene, Parametaria, Conella, Nitidella* and *Metulella.*	Jung, P
16	1996	Family Corbulidae	Anderson, L.C.
17	1996	Families Cuspidariidae and Verticordiidae	Jung, P.
18	1998	Superfamily Volutacea	Vokes, E.H.
19	1999	Family Faviidae (Part II)	Budd, A.F. and Johnson K.G.
20	2000	The Family Agariciidae	Stemann, T.A.
21	2001a	Genus *Prunum*	Nehm, R.H.
22	2001	Family Neritidae	Costa, F.H.A., Nehm R.H. and Hickman C.S.

studies was returned to the Naturhistorisches Museum Basel in Switzerland. To date, more than 300 Dominican invertebrate species have been studied in great detail (taxonomically, stratigraphically, and ecologically) by systematists who are experts on their respective biological groups. We know of no other geological research system that offers *species-level* data of this quality. These data form the raw material for many other scientific research questions, as discussed below and in other chapters of this volume.

In addition to basic research on the age, lithology, and environment of the Cibao Valley sections and particular taxonomic groups, additional effort has focused on evolutionary questions (e.g., Cheetham, 1987; Cheetham et al., 2001; Nehm, 2001a, b, c, d, 2005; Budd et al., 1996; Budd, 2000; Costa et al., 2001). For example, Dominican invertebrate groups have been used in several detailed quantitative analyses of evolutionary change (e.g., Cheetham, 1986, 1987; Nehm and Geary, 1994; Anderson, 1994; Nehm, 2001a, b, c, d; Cheetham and Jackson, 1996; Marshall, 1995). Some of these studies (e.g., Cheetham, 1986, 1987) figure prominently in evolutionary biology textbooks as benchmark cases of punctuated equilibrium (for example, see Futuyma, 1998). Additionally, speciation research in the Dominican Republic is important because the DRP is one of only a few research systems in the world where several unambiguous cases of morphological stasis and punctuated speciation in multiple lineages of invertebrate animals are known to occur (Cheetham, 1986, 1987; Nehm and Geary, 1994). Finally, the Dominican Republic Neogene provides an important window into the biodiversity of the Caribbean region prior to the Plio-Pleistocene mass extinction (Allmon et al., 1993) (see Table 1.2 for a list of major studies).

The first major attempt at synthesizing DRP research was a symposium organized by Nehm and Budd and held at the 2001 North American Palaeontological Convention (NAPC) in Berkeley, California. This symposium (Species-level and Community-level Stability: Case Studies from the Dominican Republic Neogene) brought together researchers from around the world, reviewed what we had learned in the past 20 years, and included examples of how the DRP research system could be used to address new questions in ecology and evolution. The present edited volume is an outgrowth of that symposium, and summarizes ongoing collaborative research that is currently being conducted as part of a new phase of the DRP.

1.3 Review of Chapters in this Volume

1.3.1 Geology, Paleoenvironment and Taphonomy

The first set of chapters, by McNeill et al., Dennison et al., and Nehm and Hickman, explore geological, palaeoenvironmental, and taphonomic issues relating to the Cibao Valley sections. Of particular importance is McNeill et al.'s revised temporal framework for the Río Cana and Río Gurabo sections, which has been incorporated

Table 1.2 A summary of published palaeobiological research studies employing the Dominican Republic Neogene

Topic	Taxon	Year	Authors
Evolutionary stasis and change	Bryozoa	1986	Cheetham, A.H.
	Coral	1986	Foster, A.B.
	Coral	1987	Foster, A.B.
	Bryozoa	1987	Cheetham, A.H.
	Coral	1988	Budd, A.F.
	Bryozoa	1999	Jackson, J.B.C. and Cheetham, A.H.
	Coral	1990	Budd, A.F.
	Coral	1991	Budd, A.F.
	Bryozoa	1994	Jackson, J.B.C. and Cheetham, A.H.
	Gastropoda	1994	Nehm, R.H. and Geary, D.H.
	Bryozoa	1995	Cheetham, A.H. and Jackson, J.B.C.
	Gastropoda	2001a	Nehm, R.H.
	Gastropoda	2005	Nehm, R.H.
	Bryozoa	2007	Cheetham, A.H. et al.
Environment and evolution	Coral	1990	Budd, A.F.
	Gastropoda	1991	Budd, A.F. and Johnson, K.G.
	Coral	1993	Budd, A.F.
	Bivalvia	1994	Anderson, L.C.
Diversity, extinction, and turnover	Coral	1995	Johnson, K.G. et al.
	Coral	1996	Budd, A.F., Johnson, K.G. and Stemann, T.A.
	Coral	1996	Budd, A.F. et al.
	Bryozoa	1996	Cheetham, A.H. and Jackson, J.B.C.
	Coral	1997	Budd, A.F. and Johnson, K.G.
	Bryozoa	1998	Cheetham, A.H. and Jackson, J.B.C.
	Coral	1999	Budd, A.F. and Johnson, K.G.
	Bryozoa	1999	Cheetham, A.H. et al.
	Coral	2000	Jackson, J.B.C. and Johnson, K.G.
	Coral	2000	Budd, A.F.
	Coral	2001	Budd, A.F. and Johnson, K.G.
	Coral	2003	Klaus, J.S., and Budd, A.F.
Development and evolution	Coral	1983	Foster, A.B.
	Coral	1988	Foster, A.B. et al.
	Bryozoa	2001	Cheetham, A.H. et al.
	Gastropoda	2001b	Nehm, R.H.
	Gastropoda	2001c	Nehm, R.H.
Community evolution	Coral	2003	Klaus, J. and Budd, A.F.
	Coral	1996	Jackson, J.B.C. et al.
Biogeography	Coral	1989	Budd, A.F.
	Coral	1994	Budd, A.F. and Guzman, H.
Phylogeny reconstruction	Coral	1993	Potts, D.C. et al.
	Bryozoa	1994	Jackson, J.B.C. and Cheetham, A.H.
	Coral	2001	Budd, A.F. and Klaus, J.

in all subsequent chapters. The newly reported age dates are not only more tightly constrained but they also suggest that the lower portions of the Río Gurabo and Río Cana sections are considerably younger than previously interpreted (see also Johnson et al., this volume). McNeill et al. review basic background information about the geologic setting and regional stratigraphy of the Cibao Valley and provide a historical overview of past stratigraphic research. They describe how the observed patterns are related to changes in climate and sea level as well as closure of the Central American isthmus. Their interpretations of water depth in the Mao Formation differ significantly from previous work in that a shallowing upward trend is detected which corresponds with the onset of Northern Hemisphere glaciation.

As a first step toward better understanding the link between changing environmental conditions and shallow marine species diversity, Denniston et al. construct carbon and oxygen isotope and Sr/Ca profiles from an exceptionally well-preserved coral collected along Rio Gurabo in the Gurabo Formation. Stable isotope ratios reveal well-behaved sinusoids, indicating primary isotopic signals, but their attempts to deconvolve subannual salinity and sea surface temperature ranges were hampered by the poor fit of modern Sr/Ca-SST relationships to their Miocene coral. The oxygen isotope ratios, if assumed to represent water temperature alone, suggest a seasonal range of approximately 2°C.

Despite growing interest in the effects of taphonomic processes on palaeobiological patterns (Kidwell, 2001), little work has investigated these relationships in the DR Neogene. Nehm and Hickman use the unique morphological attributes of turbinid gastropod species—each animal possesses two skeletal hardparts (shell and operculum) with different preservation potentials—to investigate and compare palaeobiological signals using the two structures in the Río Cana and Gurabo sections. They reject the hypothesis that shells and opercula from the same species produce similar measures of diversity, abundance, and stratigraphic range. If turbinid shells alone had been studied, abundance would have been underestimated by 75% and species richness would have been underestimated by 60%. Although they found that significantly fewer shells were preserved and/or sampled than opercula, studies of size patterns in shells and opercula were similar. Their broadest finding is that "taphonomic extrapolation" between morphologically similar objects may be problematic: they find that unique biological and ecological factors likely influence palaeobiological signals to an equal or greater extent than physical biostratinomic processes. Clearly, much greater consideration of taphonomic processes is necessary in the DR and perhaps other regions.

1.3.2 Species-Level Patterns of Evolutionary Stasis and Change

Four chapters in this volume focus on patterns of evolutionary stasis and change in coral and mollusc species: Budd and Klaus, Schultz and Budd, Beck and Budd, and Nehm.

Budd and Klaus examine evolutionary patterns within an ecologically dominant species complex of reef corals, the *Montastaea "annularis"* complex, which is widely distributed across the Caribbean today. Using both a geometric morphometric dataset and a dataset consisting of traditional linear measurements, they perform a series of canonical discriminant analyses to recognize species, trace their stratigraphic distributions, and examine morphologic change within the complex as a whole and within individual species. The results show that a total of eight species existed in the northern Dominican Republic during the Mio-Pliocene, and that diversity remained roughly constant (3–5 species per formation) through the three Yaque Group formations (Cercado, Gurabo, Mao), covering a time span of approximately 3 million years. This diversity is comparable to that previously observed in the complex both during the Plio-Pleistocene and today. Speciation and extinction rates were approximately 1–2 species per million years through the DR sequence, and the complex as a whole exhibited morphologic stasis. However, morphologic disparity (differences among species) was higher in the Mio-Pliocene than it is today. In contrast, careful examination of one relatively long-ranging species within the complex revealed directional change in some, but not all, species diagnostic morphologic features.

Schultz and Budd expand previous work on the less common *Montastraea "cavernosa"* complex by using larger sample sizes and employing geometric morphometrics in concert with traditional distance measurements. Their study reveals significantly more variation within the complex, three new species, and several very short-lived species. Thus, some of the species delineated by Budd (1991) are likely more than one species. Schultz and Budd's work underscores how systematic work dramatically affects interpretations of stasis and change, and corroborates Jackson and Cheetham's (1999) findings that rigorous taxonomy and splitting morphospecies as finely as possible are essential for testing the theory of punctuated equilibrium.

Beck and Budd's chapter explores evolutionary patterns in the reef coral *Siderastrea* using geometric morphometric and traditional techniques. Unlike the previous two chapters, the four species that are distinguished are discrete and do not overlap, and have relatively long stratigraphic ranges. They find that several species display evolutionary stasis over a period of approximately >5 million years. Methodologically they demonstrate that traditional measures, if used exclusively, may cause the misidentification of colonies and that 2D geometric morphometrics are the most accurate method for species diagnosis.

Nehm focuses on evolutionary stasis and change within species of the abundant and widely distributed *Prunum maoense* group. Because *Prunum* species possess clear morphological markers of adulthood, it was possible to compare equivalent ontogenetic stages through time and space. Morphometric analyses using traditional distance measurements and geometric landmarks produced generally similar evolutionary patterns, with no net morphological change characterizing adults through time. Perhaps the most interesting problem raised by the chapter is the meaning and significance of rare "*P. latissimum*" phenotypes throughout the spatial and temporal range of *P. maoense*. Are these individuals iteratively produced parallelisms arising from the *P. maoense* lineage, or persisting holdouts of the ancestral *P. latissimum* lineage? Nehm discusses the significance of each interpretation for models of species-level change in the fossil record and goes on to argue that such outliers may be crucial for understanding evolutionary stasis.

Previous studies of species-level change in DR invertebrates have indicated that, in general, no long-term directional evolutionary trends occur (Foster, 1986; Cheetham, 1986, 1987; Anderson, 1994; Nehm and Geary, 1994; Nehm, 2001a; Nehm 2005; Cheetham et al., 2007). Overall, the four new studies of species-level stasis and change in this volume generally corroborate these previous findings. More detailed comparisons are problematic, however, due to the different methodological approaches used in these studies. Additionally, reef corals tend to be restricted to a narrow range of environmental conditions and their species are widely distributed across the Caribbean region. They therefore have low numbers of stratigraphic occurrences relative to other taxonomic groups. The question remains as to whether similar processes are responsible for patterns of stasis in corals, mollusks, and bryozoans. One important factor that has received increasing attention in recent years is community-level processes, which are addressed in the next section.

1.3.3 Stability and Change in Coral and Mollusc Assemblages

Coordinated stasis is an observed pattern in which faunal assemblages and their constituent species appear to stay stable for millions of years prior to experiencing rapid faunal turnover. This pattern has generated considerable interest in the palaeontological community, and has been used to hypothesize that community stability and species-level morphological stability may be associated over long time spans (Brett and Baird, 1995; Ivany and Schopf, 1996, and references therein). Considering that species-level stasis characterizes many of the species studied from the DR Neogene (see above), do their associated communities also display stability in time and space?

Reef corals represent one of the best studied faunal groups in the Dominican Republic Neogene. Klaus et al. examine changes in coral communities through the sequence using three different approaches: (1) Persistence of individual species from one formation to the next, (2) quantitative analysis of presence/absence data within 21 lithostratigraphic units, and (3) quantitative analysis of relative abundance data obtained from line transects. The results indicate that 61% of species persist from the oldest to youngest formation in the sequence, and that presence/absences of species do not change through the sequence, suggesting community stasis. However, statistical analyses show that the relative abundances of species and the ecological dominance structures of reef communities (grouped into massive and branching subsets) do in fact change. The abundances of two *Goniopora* species, *Gardineroseris*, and *Montastraea endothecata* decrease through geologic time; whereas the abundances of massive *Porites* and *Montastraea cavernosa* increase. *Pocillopora* decreases in branching coral communities. The observed changes appear to be related to a combination of environmental (both local and regional) and evolutionary factors, leading up to the closure of the Central American Isthmus.

Although the Río Gurabo has figured prominently in studies of evolutionary stasis and change within coral, bryozoan, and mollusc species, little work has explored the associated mollusc assemblages. Rivera et al., like Klaus et al., investigate faunal-level patterns in the Cibao valley sections. Rivera et al. specifically investigate faunal change in mollusc-rich assemblages from the Río Gurabo section and find that the assemblages display considerable variability in composition, relative abundance, species richness, and trophic distributions through time. As the authors note, their study of >16,000 individuals from more than 300 species encompasses only a small portion of the exposed section, and consequently it will be necessary to study other portions of the section before a complete understanding of faunal change within Río Gurabo is attained. Nevertheless, Rivera et al. do demonstrate that significant faunal differences characterize the lower and upper regions of the section. Expansion of their work should be able to provide a more precise analysis of the relationships between species-level and community-level change throughout this important section.

Johnson et al.'s study is of the broadest scale in this volume, and tests: (1) the effects of revised age dates (based on McNeill et al., this volume) on the timing and magnitude of origination and extinction events in the Caribbean reef coral fauna as a whole, and its patterns of diversity through time, and (2) the importance of the DR fossil reef coral occurrences in general in understanding origination and extinction events in Neogene Caribbean reef corals, as well as their patterns of diversity through time. Comparisons of first occurrences in the DR based on old and new age dates reveal a shift in regional first occurrences from 7–9 Ma to 5–7 Ma using new age dates, and an unrecognized sampling gap across the Caribbean during the late Miocene. These patterns are further accentuated when the DR reef coral occurrences are excluded altogether from the database. In contrast, excluding occurrences from the Plio-Pleistocene Limon Basin of Costa Rica resulted in only minor change in the timing of origination and extinction events, although they do affect estimates of the magnitude of Plio-Pleistocene extinction. The study attests to the importance of incorporating multiple taxonomic and stratigraphic interpretations into palaeontological databases, and comparing analyses using data based on different interpretations.

1.3.4 Education and Infrastructure

The final two chapters of this volume focus on the importance of education and infrastructure in international multidisciplinary research and development. In a departure from the science research focus of other chapters, Nehm, Luna, and Budd discuss two science education projects closely tied to the Dominican Republic Project: (1) Science education in US schools with predominantly

Dominican American students, and (2) international outreach and development work with Dominican undergraduates. Both projects were spurred by the recognition that the persistent lack of Dominican and Dominican American involvement in the DR project over the past 30 years would require new approaches and greater attention to outreach. The chapter begins with a review of four interrelated DRP science education efforts with Dominican American students: (1) 'Funds of knowledge' research relating to 'sense of place' in immigrant secondary students; (2) development of curricula and resources relating to the DRP; (3) science teacher professional development; and (4) involvement of Dominican American middle school, high school, and college students and teachers in DRP research projects. The chapter continues with an overview of two workshops for Dominican undergraduates. The goal of the first workshop (based in Santo Domingo) was to demonstrate how studies of fossil reef systems, thousands to millions of years old, are relevant to addressing modern-day issues in reef conservation. The second workshop (based in Mao) trained students and researchers in collection care and management, including preventive conservation, collection organization, and data preservation and management. Overall, the chapter highlights the importance of science education in the development and maintenance of successful international science research efforts.

A final goal of this edited volume is to demonstrate the importance of specimen-based research in palaeontology to the study of evolution. As described in McNeill et al. (this volume), one of the chief goals of a new phase of the DR project is to expand collections so that patterns of evolutionary stasis and change can be analyzed within individual lineages, as well as in benthic marine communities. Two important infrastructural components are essential to specimen-based research: (1) museum collections, and (2) databases.

Government agencies and administrators of natural history museums must be made aware of the central importance that care and maintenance of collections play in quality studies of species and communities through geological time. Collections are made during fieldwork and are costly to collect or recollect. They therefore should be maintained and developed for future research. Many collecting sites become inaccessible over time because of building development, access and collecting restrictions, or are collected out (e.g., the NMB localities in the Baitoa Formation along Río Yaque del Norte south of Santiago). As collections develop and grow they contain more material and information than could be collected in a single fieldwork project. This mass of information provides a valuable contribution to, and forms the basis of, many large scale database initiatives and literature-based research, as well as primary research. Collections often contain material that is later recognized or rediscovered as being scientifically important according to research developments. As new research techniques are discovered, collections continue to be important sources of information. In the case of modern endangered species (e.g., corals), use of museum collections reduces the necessity to collect threatened populations and lessens the negative impacts of scientific collecting.

The collections on which the research in this book is based are deposited primarily at the Natural History Museum in Basel, Switzerland (NMB;

http://www.nmb.bs.ch), the Paleontology Repository of the University of Iowa (SUI; http://www.uiow.edu/~geology/paleo), and the Paleontological Research Institution in Ithaca, New York (PRI; http://ww.pri.org). Lists of studied specimens are provided in the appendices to chapters by Budd and Klaus, Schultz and Budd, and Beck and Budd.

Another, equally important infrastructural component of specimen-based research involves databases containing specimen, locality, and taxonomic information and facilitating taxonomic standardization (described by Budd et al., this volume, in the chapter on the NMITA database). Today, databases not only contain the information traditionally held in museum specimen catalogues and locality registers, but they also allow this information to be searchable in many different ways, and make it readily available online to the scientific community. In addition, databases contain the information traditionally assembled by systematists to make specimen identifications, recognize new species, evaluate the status of existing species, and revise higher level classification. They provide a mechanism for standardizing taxonomy so that palaeontological occurrence data can be used to perform spatial and temporal analyses of biodiversity. Moreover, as described in Budd et al. (this volume), taxonomic databases facilitate gathering, organizing, and sorting information that is routinely assembled when preparing a taxonomic monograph. Finally, as demonstrated in Johnson et al. (this volume), modern databases can be designed to allow for multiple interpretations (e.g., multiple alternative identifications for any given specimen, age interpretations for any given stratigraphic unit, and classification systems for higher level taxa). Databases for reef corals (described in Budd et al., this volume) have been developed for: (1) specimen and locality data, and stratigraphic interpretations (Cenozoic Coral Database, 'CCD', in Microsoft Access and available to project members), (2) taxonomic data (Neogene Marine Biota of Tropical America, 'NMITA', in Oracle and available online), and (3) palaeontological occurrence data (Statistical Analysis of Palaeontological Occurrence Data, 'STATPOD', originally in R and available to project members). Queries of the first database provide the foundation for the second and third databases. Information in the second database are shared with other community databases in palaeontology.

1.4 Goals of this Book

The past decade has witnessed the gradual departure of the scientists most instrumental to the development of the Dominican Republic Neogene into a modern palaeobiological research system. The chief architects of the Dominican Republic Project (Peter Jung and John Saunders) have retired, Harold Vokes has passed away, and Emily Vokes has retired. Additionally, the Naturhistorisches Museum Basel, which served as a locus for DR research for the past 30 years, has directed its scientific focus to other topics. In light of these changes, we view this edited volume as a transitional effort that attempts to build an empirical, conceptual, and historical bridge between the accomplishments of past DR project workers and future students, scientists, and research questions.

While past work on the Dominican Republic Neogene has explored a diverse array of palaeobiological questions, this volume demonstrates that many revisions need to be made to our understanding of the geological and palaeoenvironmental framework, and many significant macroevolutionary questions remain to be answered. The expansion of geoscience research to include educational outreach has also fostered the development of two science education projects. We hope that this volume serves as a vehicle for moving research on the Dominican Republic Neogene forward, and provides a useful starting point for the next generation of students and researchers.

Acknowledgments This volume is dedicated to Peter Jung and John Saunders for their field-based and specimen-based approach to palaeontology and their careful work with collections. AFB would also like to thank Jörn Geister for introducing her to the DR Project. We thank the National Science Foundation for support. We greatly appreciate comments and reviews from Emily Vokes.

References

Allmon, W.D., Rosenberg, G., Portell, R., and Schindler, K.S., 1993, Diversity of Atlantic Coastal Plain mollusks since the Pliocene, *Science*, **260**, 5114:1626–1629.

Anderson, L.C., 1994, Palaeoenvironmental control of species distributions and intraspecific variability in Neogene Corbulidae (Bivalvia: Myacea) of the Dominican Republic, *J. Palaeontol.*, **68**:460–473.

Anderson, L.C., 1996, Neogene Paleontology in the northern Dominican Republic, 16, The Family Corbulidae (Mollusca: Bivalvia), *Bull. Am. Paleontol.*, **110**:351, 1–34.

Bermudez, P.J., 1949, Tertiary smaller foraminifera of the Dominican Republic, *Cushman Lab. Foram. Res., Spec. Publ.*, **25**:322.

Bold, W.A. and Van Den, 1968, Ostracoda of the Yague Group (Neogene), Dominican Republic, *Bull. Am. Paleontol.*, **94**:329.

Bold, W.A. and Van Den, 1969, Neogene Ostracoda from southern Puerto Rico, *Carib. J. Sci.*, **9**, 3–4:117–125.

Bold, W.A. and Van Den, 1988, Neogene palaeontology of the northern Dominican Republic. 7. The subclass Ostracoda (Arthropoda: Crustacea), *Bull. Am. Paleontol.*, **94**:1–105.

Bowin, C., 1966, Geology of central Dominican Republic: a case history of part of an island arc, *Geol. Soc. Am. Mem.*, **98**:11–84.

Brett, C.E. and Baird, G.C., 1995, Coordinated stasis and evolutionary ecology of Silurian to Middle Devonian faunas in the Appalachian Basin, in: *New Approaches to Speciation in the Fossil Record* (Erwin, D.H. and Anstey, R.L., eds.), Columbia University Press, New York, pp. 285–315.

Budd, A.F., 1988, Large-scale evolutionary patterns in the reef coral *Montastraea*: the role of phenotypic plasticity, *Proceedings of the Sixth International Coral Reef Symposium*, Townsville **3**:393–402.

Budd, A.F., 1989, Biogeography of Neogene Caribbean reef-corals and its implications for the ancestry of eastern Pacific reef-corals, *Mem. Assoc. Australas. Palaeontol.*, **8**:219–230.

Budd, A.F., 1990, Long-term patterns of morphological variation within and among species of reef-corals and their relationship to sexual reproduction, *Syst. Bot.*, **15**:150–165.

Budd, A.F., 1991, Neogene Paleontology in the Northern Dominican Republic, 11, The Family Faviidae (Anthozoa: Scleractinia), Part I, The Genera *Montastraea* and *Solenastrea*, *Bull. Am. Paleontol.*, **101**, 338:5–83.

Budd, A.F., 1993, Variation within and among morphospecies of *Montastraea*, *Courier Forschungs-institut Senckenberg*, **164**:241–254.

Budd, A.F., 2000, Diversity and extinction in the Cenozoic history of Caribbean reefs, *Coral Reefs*, **19**:25–35.

Budd, A.F. and Guzman, H., 1994, *Siderastrea glynni,* a new species of scleractinian coral (Cnidaria: Anthozoa) from the eastern Pacific, *Proc. Biol. Soc. Washington*, **107**: 591–599.

Budd, A.F. and Johnson, K.G., 1991, Size-related evolutionary patterns among species and subgenera in the *Strombina*-group (Gastropoda: Columbellidae), *J. Paleontol.*, **65**:417–434.

Budd, A.F. and Johnson, K.G., 1997, Coral reef community dynamics over 8 myr of evolutionary time: stasis and turnover, *Proc. 8th Int. Coral Reef Symp.*, **1**:423–428.

Budd, A.F. and Johnson, K.G., 1999a, Origination preceding extinction during late Cenozoic turnover of Caribbean reefs, *Paleobiology*, **25**:188–200.

Budd, A.F. and Johnson, K.G., 1999b, Neogene Paleontology in the Northern Dominican Republic, The Family Faviidae (Anthozoa: Scleractinia), Part II, The Genera *Caulastraea, Favia, Diploria, Manicina, Hadrophyllia, Thysanus,* and *Colpophyllia, Bull. Am. Paleontol.*, **113**:356.

Budd, A.F. and Johnson, K.G., 2001, Contrasting evolutionary patterns in rare and abundant species during Plio-Pleistocene turnover of Caribbean reef corals, in: *Evolutionary Patterns: Growth, Form, and Tempo in the Fossil Record* (Jackson, J.B.C., Lidgard, S., and McKinney, F.K., eds.), University of Chicago Press, Chicago, IL, pp. 295–325.

Budd, A.F. and Klaus, J.S., 2001, The origin and early evolution of the *Montastraea "annularis"* species complex (Anthozoa: Scleractinia), *J. Paleontol.*, **75**, 3:527–545.

Budd, A.F., Stemann, T.A., and Johnson, K.G., 1994, Stratigraphic distributions of genera and species of Neogene to Recent Caribbean reef corals, *J. Paleontol.*, **68**:951–977.

Budd, A.F., Johnson, K.G., and Stemann, T.A., 1996, Plio-Pleistocene turnover and extinctions in the Caribbean reef coral fauna, in: *Evolution and Environment in Tropical America* (Jackson, J.B.C., Budd, A.F., and A.G. Coates, eds.), University of Chicago Press, Chicago, IL, pp. 168–204.

Butterlin, J., 1954, La géologie de la République d'Haiti et ses rapports avec celle des régions voisines, *Mémoires de l'Institut Français d'Haiti*, 406–407.

Cairns, S.D. and Wells, J.W., 1987, Neogene Paleontology in the Northern Dominican Republic, 5, The Suborders Caryophylliina and Dendrophylliina (Anthozoa: Scleractinia), *Bull. Am. Paleontol.*, **93**, 328:23–43.

Cheetham, A.H., 1986, Tempo of evolution in a Neogene Bryozoan: rates of morphologic change within and across species boundaries, *Palaeobiology,* **12**:190–202.

Cheetham, A.H., 1987, Tempo of evolution in a Neogene Bryozoan: are trends in single morphologic characters misleading?, *Palaeobiology*, **13**:286–296.

Cheetham, A.H. and Jackson, J.B.C., 1995, Process from pattern: tests for selection versus random change in punctuated bryozoan speciation, in: *New Approaches to Speciation in the Fossil Record* (Erwin, D. and Anstey, R., eds.), Columbia University Press, New York.

Cheetham, A.H. and Jackson, J.B.C., 1996, Speciation, extinction, and the decline of arborescent growth in Neogene and Quaternary Cheilostome Bryozoa of Tropical America, in: *Evolution and Environment in Tropical America* (Jackson, J.B.C., Budd, A.F., and Coates, A.G., eds.), University of Chicago Press, Chicago, IL, pp. 205–233.

Cheetham, A.H. and Jackson, J.B.C., 1998, The fossil record of cheilostome bryozoa in the Neogene and Quaternary of Tropical America: adequacy for phylogenetic and evolutionary studies, in: *The Adequacy of the Fossil Record* (Donovan, S.K. and Paul, C.R.C., eds.), Wiley, Chichester, England, pp. 227–242.

Cheetham, A.H., Jackson, J.B.C., Sanner, J., and Ventocillo, Y., 1999, Neogene cheilostome bryozoa in Tropical America: comparison and contrast between the Central American isthmus (Panama, Costa Rica) and the north-central Caribbean (Dominican Republic), in: A Paleobiotic

Survey of the Caribbean Faunas from the Neogene of the Isthmus of Panama (Collins, L.S. and Coates, A.G., eds.), *Bull. Am. Paleontol.*, **357**:159–192.

Cheetham, A.H., Jackson, J.B.C., and Sanner, J., 2001, Evolutionary significance of sexual and asexual modes of propagation in Neogene species of the bryozoan *Metrarabdotos* in Tropical America, *J. Palaeontol.*, **75**:564–577.

Cheetham, A.H., Sanner, J., and Jackson, J.B.C., 2007, *Metrarabdotos* and related genera (Bryozoa: Cheilostomata) in the late Paleogene and Neogene of Tropical America, *Paleontol. Soc. Mem.*, **67**:1–96.

Costa, F.H.A., Nehm, R.H., and Hickman, C., 2001, Neogene palaeontology in the Northern Dominican Republic, 22, The Family Neritidae, *Bull. Am. Paleontol.*, **359**:47–71.

Duncan, P.M., 1864, On the fossil corals of the West Indian Islands, Part II, *Quart. J. Geol. Soc. Lond.*, **20**:20–44.

Duncan, P.M., 1868, On the fossil corals of the West Indian Islands, Part IV, *Quart. J. Geol. Soc. Lond.*, **24**:9–33.

Foster, A.B., 1983, The relationship between corallite morphology and colony shape in some massive reef-corals, *Coral Reefs*, **2**:19–25.

Foster, A.B., 1986, Neogene palaeontology in the Northern Dominican Republic, 3, The Family Poritidae (Anthozoa: Scleractinia), *Bull. Am. Paleontol.*, **90**, 325:47–123.

Foster, A.B., 1987, Neogene palaeontology in the northern Dominican Republic, 4, The genus *Stephanocoenia* (Anthozoa: Scleractinia: Astrocoeniidae), *Bull. Am. Paleontol.*, **93**:328, 5–22.

Foster, A.B., Johnson, K.G., and Schultz, L.L., 1988, Allometric shape change and heterochrony in the free-living coral *Trachyphyllia bilobata* (Duncan), *Coral Reefs*, **7**:37–44.

Futuyma, D.J., 1998, *Evolutionary biology*, Sinauer, Sunderland, MA.

Gabb, W.M., 1873, On the topography and geology of Santo Domingo, *Am. Philos. Soc., Trans.*, **15**:49–259

Heneken, T.S., 1853, On some Tertiary deposits in San Domingo: With Notes on the Fossil Shells, by J.C. Moore, Esq., F.G.S.; and on the Fossil Corals by W. Lonsdale, Esq., F.G.S. *Quart. J. Geol. Soc. Lond.*, **9**:115–134.

Ivany, L.C. and Schopf, K.M. (eds.), 1996, New perspectives on faunal stability in the fossil record, *Palaeogr. Palaeoclim. Palaeoecol.*, **127**:1–359.

Jackson, J.B.C., 1994, Community unity?, *Science*, **264**:1412–1413.

Jackson, J.B.C. and Cheetham, A.H., 1994, Phylogeny Reconstruction and the Tempo of Speciation in Cheilostome Bryozoa, *Paleobiology*, **20**:407–423.

Jackson, J.B.C. and Cheetham, A.H., 1999, Tempo and mode of speciation in the sea, *Trends Ecol. Evol.*, **14**:72–77.

Jackson, J.B.C. and Johnson, K.G., 2000, Life in the last few million years, in: *Deep time: Paleobiology's Perspective* (Erwin, D.H., and Wing, S.L., eds.), *Paleobiology* (Suppl.), **26**:221–235.

Jackson, J.B.C., Budd, A.F., and Pandolfi, J.M., 1996, The shifting balance of natural communities?, in: *Evolutionary Paleobiology: Essays in Honor of James W. Valentine* (Erwin, D., Jablonski, D., and Lipps, J., eds.), University of Chicago Press, Chicago, IL, pp. 89–122.

Johnson, K.G., Budd, A.F., and Stemann, T.A., 1995 Extinction selectivity and ecology of Neogene Caribbean reef corals, *Paleobiology*, **21**:52–73.

Jung, P., 1986, Neogene Paleontology in the Northern Dominican Republic, 2, The genus *Strombina* (Gastropoda: Columbellidae), *Bull. Am. Paleontol.*, **90**:324, 1–42.

Jung, P., 1994, Neogene Paleontology in the Northern Dominican Republic, 15, The genera *Columbella, Eurypyrene, Parametaria, Conella, Nitidella* and *Metulella* (Gastropoda: Columbellidae), *Bull. Am. Paleontol.*, **106**:344, 1–45.

Jung, P., 1996, Neogene Paleontology in the Northern Dominican Republic, 17, The Families Cuspidariidae and Vertcordiidae (Mollusca: Bivalvia), *Bull. Am. Paleontol.*, **110**, 351:1–45.

Jung, P. and Petit, R.E., 1990, Neogene Paleontology in the Northern Dominican Republic, 10, The Family Cancellaridae (Mollusca: Gastropoda), *Bull. Am. Paleontol.*, **98**, 334:1–62.

Kidwell, S.M., 2001, Preservation of species abundance in marine death assemblages, *Science*, **294**:1091–1094.

Kier, P.M., 1992, Neogene Paleontology in the Northern Dominican Republic. 13. The Class Echinoidea (Echinodermata), *Bull. Am. Paleontol.,* **102**, 339:13–23.

Klaus, J.S. and Budd, A.F., 2003, Comparison of Caribbean coral reef communities before and after Plio-Pleistocene faunal turnover: analyses of two Dominican Republic reef sequences, *Palaios*, **18**:3–21.

Logan, A., 1987, Neogene Paleontology in the Northern Dominican Republic. 6. Phylum Brachipoda, *Bull. Am. Paleontol.*, **93**:328, 44–52.

Lonsdale, W., 1853, Notes on the fossil corals of San Domingo, in: On some Tertiary deposits in San Domingo (Heneken, J.S.), *Quart. J. Geol. Soc. Lond.,* **9**:132–134.

Marshall, C., 1995, Stratigraphy, the true order of species originations and extinctions, and testing ancestor-descendent hypotheses among Caribbean Neogene bryozoans, in: *New Approaches to Speciation in the Fossil Record* (Erwin, D.H. and Anstey, R.L., eds.), Columbia University Press, New York, pp. 208–235.

Maury, C.J., 1917a, Santo Domingo type sections and fossils, Part 1, *Bull. Am. Palaeontol.,* **5**, 29:1–251.

Maury, C.J., 1917b, Santo Domingo type sections and fossils, Part 2, *Bull. Am. Palaeontol.,* **5**, 30:1–43.

Maury, C.J., 1929, Porto Rican and Dominican stratigraphy, *Science*, **70**, 1825:609.

Maury, C.J., 1931, Two new Dominican formational names, *Science*, **73**, 1880:42–43.

Morris, P.J., 1996, Testing patterns and causes of faunal stability in the fossil record, with an example from the Pliocene Lusso Beds of Zaire, *Palaeogr. Palaeoclim. Palaeoecol.*, 127.

Nehm, R.H., 2001a, Neogene Paleontology in the northern Dominican Republic, 21, The genus *Prunum*, *Bull. Am. Paleontol.*, **359**:1–46.

Nehm, R.H., 2001b, Linking macroevolutionary pattern and developmental process in marginellid gastropods, in: *Evolutionary Patterns: Growth, Form, and Tempo in the Fossil Record* (Jackson, J.B.C., Lidgard, S., and McKinney, F.K., eds.), University of Chicago Press, Chicago, IL, pp. 159–195.

Nehm, R.H., 2001c, The developmental basis of morphological disarmament in *Prunum* (Neogastropoda; Marginellidae), in: *Beyond Heterochrony* (Zelditch, M.L., ed.), Wiley, New York, pp. 1–26.

Nehm, R.H., 2001d, Species-level morphological stability in Neogene marginellids from the Dominican Republic, in: Program and Abstracts, North American Paleontological Convention, *PaleoBios*, **21**:2, 97.

Nehm, R.H., 2005, Patterns and processes of evolutionary stasis and change in *Eratoidea* (Gastropoda: Marginellidae) from the Dominican Republic Neogene, *Carib. J. Sci.,* **41**:189–214.

Nehm, R.H. and Geary, D., 1994, A gradual morphologic transition during a rapid speciation event in marginellid gastropods (Neogene; Dominican Republic), *J. Paleontol.*, **68**:787–795.

Nolf, D. and Stringer, G.L., 1992, Neogene Paleontology in the Northern Dominican Republic, 14, Otoliths of teleostean fishes, *Bull. Am. Paleontol.,* **102**, 340:41–81.

Pflug, H.D., 1969, Molluscen aus dem Tertiar von St. Domingo, *Acta Humboltiana*, Series Geologica et Palaeontologica, NR **1**:1–107.

Pilsbury, H.J., 1922, Revision of W.M. Gabb's Tertiary Mollusca of Santo Domingo, *Proc. Acad. Nat. Sci. Phil.*, **73**:305–435.

Pilsbury, H.J., and Johnson, C.W., 1917, New Mollusca of the Santo Domingo Miocene, *Proc. Acad. Nat. Sci. Phil.*, **69**:150–202.

Potts, D.C., Budd, A.F., and Garthwaite, R.L., 1993, Soft tissue vs skeletal approaches to species recognition and phylogeny reconstruction in corals, *Courier Forschungs-institut Senckenberg*, **164**:221–231.

Ramirez, N., 1956, *Paleontologia Dominicana*, Publications De la Universidad de Santo Domingo, Ser. 4, **103**, 1:1–26.

Saunders, J.B., Jung, P., and Biju-Duval, B., 1986, Neogene palaeontology of the northern Dominican Republic, 1, Field surveys, lithology, environment, and age, *Bull. Am. Paleontol.,* **89**, 323:1–79.

Seiglie, G.A., 1978, Comments on the Miocene-Pliocene boundary in the Caribbean Region, *Ann. de Centre, Univ. Savoi, Sc. Naturel.*, **3**: 71–86.

Sowerby, G.B., II, 1850, Descriptions of new species of fossil shells found by J. S. Heniker, Esq., pp. 44–53, in: J.C. Moore, on some Tertiary beds in the island of San Domingo; from notes by J.S. Heniker, Esq. with remarks on the fossils, *Quart. J. Geol. Soc. Lond.*, **6**:39–53.

Vaughan, T.W., 1919, Fossil corals from Central America, Cuba, and Porto Rico with an account of the American Tertiary, pleistocene, and recent coral reefs, *US Nat. Hist. Mus. Bull.*, **130**:189–524.

Vaughan, T.W. and Hoffmeister, J.E., 1925, New species of fossil corals from the Dominican Republic, *Bull. Mus. Comp. Zool., Harvard*, **67**:315–326.

Vaughan, T.W., Cooke, W., Condit, D.D., Ross, C.P., Woodring, W.P., and Calkins, F.C., 1921, A geological reconnaissance of the Dominican Republic, *Geol. Surv. Dom. Rep., Mem.*, **1**:1–268.

Vokes, E.H., 1979, The age of the Baitoa Formation, Dominican Republic, using mollusca for correlation, *Tulane Stud. Geol. Paleontol.*, **15**:105–116.

Vokes, E.H., 1998, Neogene Paleontology in the Northern Dominican Republic, 18, The super Family Volutacea (in part), *Bull. Am. Paleontol.*, **113**, 354:1–54.

Vokes, E.H., 1989a, Neogene Paleontology in the Northern Dominican Republic, 8, The Family Muricidae (Mollusca: Gastropoda), *Bull. Am. Paleontol.*, **97**, 332:1–94.

Vokes, H.E., 1989b, Neogene Paleontology in the Northern Dominican Republic, 9, The Family Cardiidae (Mollusca: Bivalvia), *Bull. Am. Paleontol.*, **97**, 332:87–141.

Vokes, H.E. and Vokes, E.H., 1992, Neogene Paleontology in the Northern Dominican Republic, 12, The genus *Spondylus* (Bivalvia: Spondylidae). *Bull. Am. Paleontol.*, **102**, 339:1–13.

Weyl, R., 1940, Blockmeere in the Cordillera Central von Santo Domingo (LVestindien), *Zeitschnyt der Deutschen Geologischen Gesellschaf*, **92**:175–179.

Weyl, R., 1966, *Geologie der Antillen.* Gebruder Borntraeger, Berlin, 410 pp.

Chapter 2
An Overview of the Regional Geology and Stratigraphy of the Neogene Deposits of the Cibao Valley, Dominican Republic

Donald F. McNeill[1], James S. Klaus[1], Charles C. Evans[2], and Ann F. Budd[3]

Contents

2.1 Introduction

This chapter sets the tectonic, geologic, and stratigraphic stage for many of the chapters that follow—it is largely a review, but some new results and reinterpretations are included. Our new results include refined age dates for the key Miocene

[1]Department of Geological Sciences, University of Miami, 43 Cox Science Building, Coral Gables, FL, 3133. Email: jsklaus@rsmas.miami.edu

[2]Evans Environmental and Geosciences, 14505 Commerce Way, Miami Lakes, FL 33016, USA. Email: CEvans@eeandg.com

[3]Department of Geoscience, University of Iowa, 121 Trowbridge Hall, Iowa City, IA 52242, USA. Email: ann-budd@uiowa.edu

R.H. Nehm, A.F. Budd (eds.) *Evolutionary Stasis and Change in the Dominican Republic Neogene*,

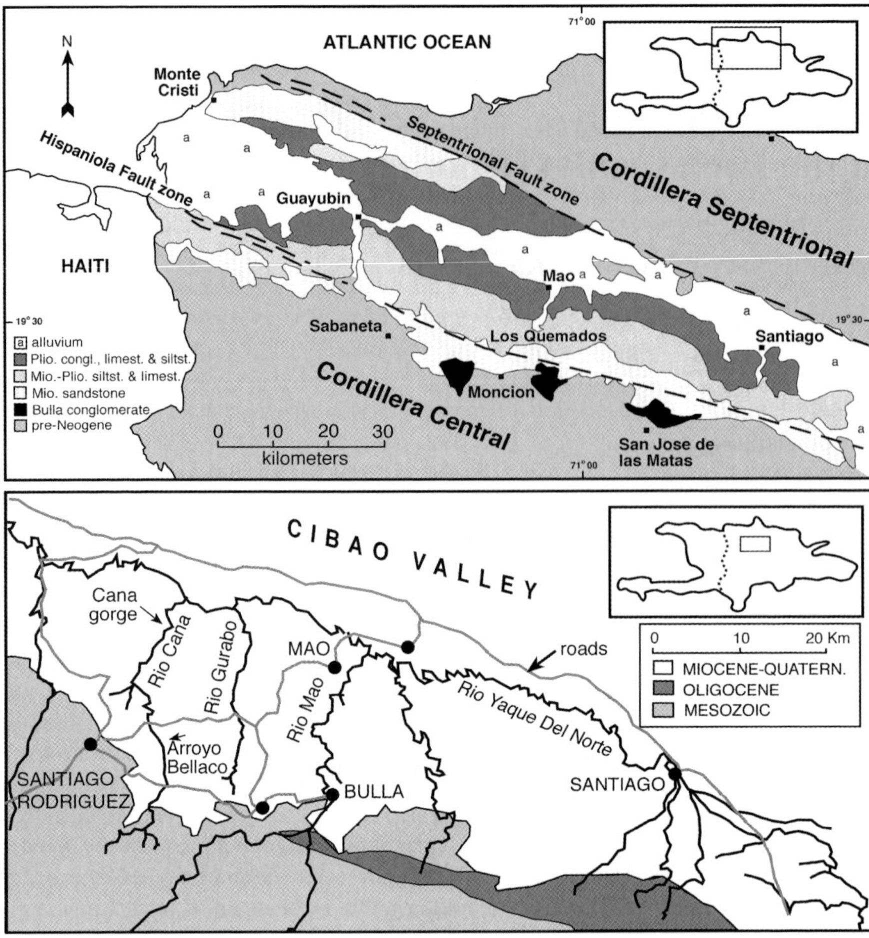

Fig. 2.1 Location map for the Cibao Valley (upper) with generalized geology of the basin from Blesch (1966). The rivers (black lines) that dissect the southern flank of the basin and provide access to the key sections along the Rio Gurabo and Rio Cana are shown in the lower figure. The roads are shown in gray (Figure from Saunders et al. 1986)

and Pliocene sections along the Río Gurabo and the Río Cana (Fig. 2.1). This project has focused on four main objectives. These include: improving the age model of the key sections, updating the stratigraphic framework of the basin, improving the depositional interpretation using benthic foraminifera, and expanding collections of key fauna to assess evolutionary changes during a period of major turnover in both coral and mollusc fauna. We build on previous studies, especially those of Saunders et al. (1986), Evans (1986a), and Vokes (1989).

The Cibao Valley was at one time an open shelf and seaway along the northern margin of Hispaniola. Subsequent uplift of the island, associated with nearby plate boundary interactions, has exposed a relatively undeformed Miocene and

Pliocene section. The uplifted sequence consists of a wedge-shaped deposit of Neogene marine sediment. Both siliciclastic and carbonate facies occur in the Cibao Basin, the siliciclastics shed from the adjacent Cordillera Central, and the carbonates mainly from *in-situ* skeletal precipitation. The section is especially rich in corals (colonial, solitary), molluscs (bivalves, gastropods), and bryozoans, as well as various microfossils (Budd and Johnson, 1999; Nehm and Geary, 1994; Saunders et al., 1986; van den Bold, 1988; Vokes, 1979).

2.2 Tectonic Setting of the Cibao Valley, Northern Dominican Republic

2.2.1 Geology of the Caribbean Region

The Caribbean region is geologically complex. The relatively small Caribbean Plate (Fig. 2.2) interacts with the surrounding North American, Cocos, Nazca and South American plates through a variety of plate boundary interactions (Draper et al., 1994). There is active subduction along the Lesser Antilles and Central America,

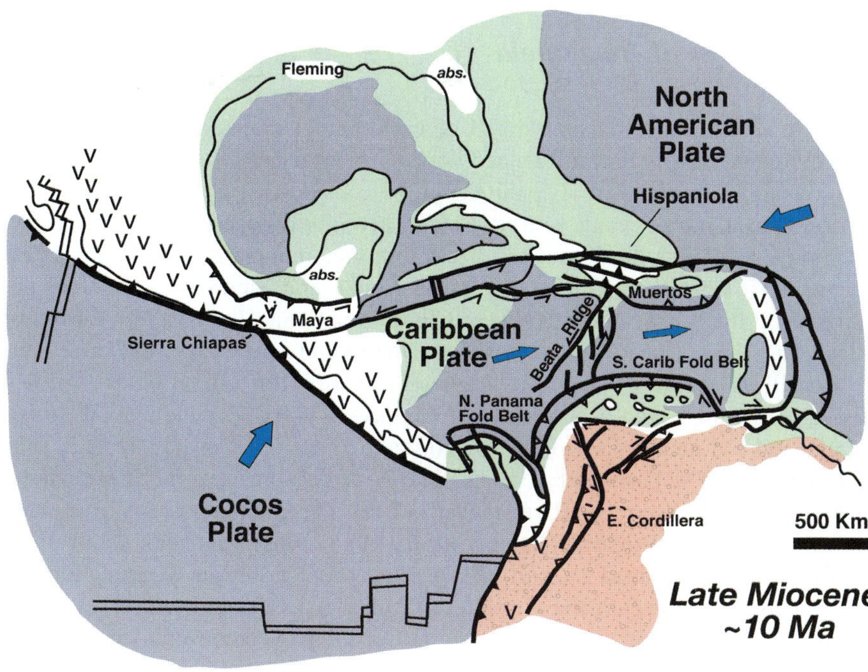

Fig. 2.2 Tectonic setting of the Caribbean plate from Pindell (1994). The Cibao Valley is a result of transpressional forces as the Caribbean plate moves east relative to the North American plate

strike-slip motions on the northern and southern boundaries, and sea floor spreading in the Cayman Trough. The Caribbean Plate is moving eastwards with respect to both North and South America at a rate of about 1 to 2 cm $year^{-1}$ (Mann et al., 1990). The eastward movement of the Caribbean Plate has resulted in subduction of the Atlantic Ocean crust under the eastern margin of the Caribbean, producing the Lesser Antilles island arc system. Eastward motions of the Pacific and Cocos Plates with respect to the Caribbean and North America have resulted in subduction of these plates beneath the western margin of the Caribbean Plate in Central America.

Pindell (1994) provides a summary of an evolved plate tectonic model for the Caribbean. One hundred and sixty million years ago, the super continent Pangea had just recently started to separate, and the Proto-Caribbean Seaway was beginning to form. By 90 Ma the Proto-Caribbean Seaway was being subducted along the western side and formed an island-arc. Between 90 and 70 Ma this island arc migrated eastward and became the eastern margin of the Caribbean plate. By 70 Ma the Costa Rica—Panama island arc had emerged to form the western margin of the plate. By 35 Ma the northern Caribbean had collided with the Bahamas (Eocene) and Central America had spanned the gap between North and South America. Strike-slip motion characterized the northern and southern boundaries of the Caribbean. By 10 Ma the Caribbean region had nearly assumed its modern configuration (Fig. 2.2).

2.2.2 Geology of Hispaniola

Strike-slip motion and transpressional conditions that affected the Cibao Valley were the major factors that influenced development of the basin over the past 10 million years. The island of Hispaniola can be considered part of a mature island arc formed in an intra-oceanic setting (Bowin, 1966). Physiographically, Hispaniola is comprised of four northwest-southeast trending mountain ranges (Cordillera Septentrional, Cordillera Central, Sierra de Neiba, and Sierra de Bahoruco) and separated by three lower lying valleys (Cibao Valley, San Juan Valley, Enriquillo Valley) (Fig. 2.3). The twin peaks of Pico Duarte and La Pelona (3087 m) within the Cordillera Central mark the highest elevation of the Greater Antilles. Lithologically, the island is composed of Cretaceous-Early Eocene igneous, metamorphic and sedimentary substrate that forms the basement for late Tertiary sedimentary basins. The basement of Hispaniola is made up of several fault-bounded blocks, or geological terranes (Draper et al., 1994). The geologic history of adjacent terranes is often quite distinct (Draper et al., 1994). Basement rock south of the Cordillera Central formed as part of a Cretaceous Caribbean Oceanic Plateau. Basement rock underneath the Cordillera Central is associated with Cretaceous-Eocene volcanic arcs. Rocks underlying the Cordillera Septentrional are additionally associated with a Cretaceous-Eocene forearc.

The island remains tectonically active today, with reports of major earthquakes (magnitude 6.5) occurring as recently as 2003 and smaller earthquakes occurring

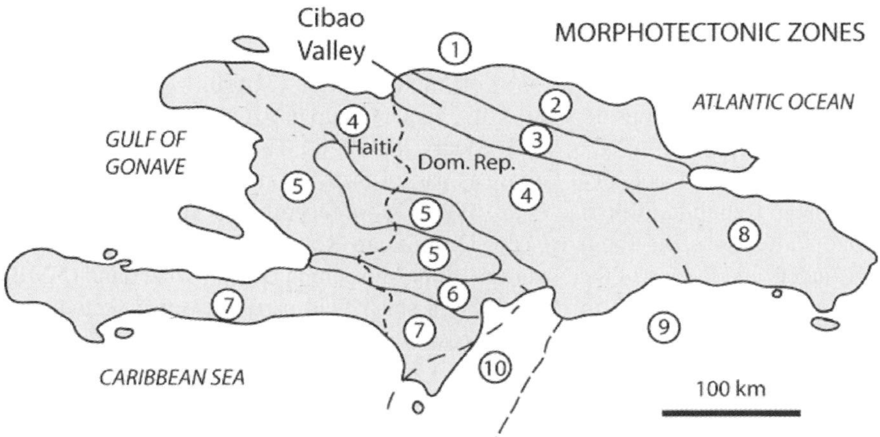

Fig. 2.3 Morphotectonic zones from Lewis and Draper (1990) and Draper et al. (1994). The Cibao Valley (Zone 3) is flanked by mountain belts to the north (Zone 2, Cordillera Septentrional) and to the south (Cordillera Central, Zone 4). The remaining zones include: Zone 1 = Old Bahama Trench; Zone 5 = Northwestern-south-central zone which includes the following features: Plateau Central-San Juan Valley-Azua Plain; Sierra el Numero; Presqu'ile de Nord-Ouest; Montaignes Noires; Chaines de Matheux-Sierra de Neiba; and Sierra de Martin Garcia; Zone 6 = Ile de la Gonave-Plaine de Cul-de-Sac-Enriquillo Valley; Zone 7 = southern peninsula which includes Massif de la Selle-Massif de la Hotte-Sierra de Bahoruco; Zone 8 = eastern peninsula which includes Cordillera Oriental and the Seibo coastal plain; Zone 9 = San Pedro basin and north slope of the Muertos Trough; and Zone 10 = Beata Ridge and southern peninsula of Barahona

quite frequently. Episodes of intense shaking along the northern coast occurred in 1946, 1887, and 1842 (McCann and Pennington, 1990). The 1842 and 1887 events were likely related to high-angle faults in the Cibao Valley. The 1946 earthquake was located further east and associated with the thrust zone off Puerto Rico (McCann and Pennington, 1990).

2.3 Neogene Geology and Paleontology of the Cibao Valley

2.3.1 The Cibao Valley Basin

The Cibao Basin lies between the Cordillera Central and the Cordillera Septentrional in the northern Dominican Republic (DR). Together the Cibao Basin and eastern and central Cordillera Septentrional define a large synclinal structure with its axis approximately parallel to that of the Cibao Basin (Mann et al., 1991; Mann, 1999). The Cibao Basin is traversed by the Río Yaque del Norte, which along with four smaller streams (Río Gurabo, Río Cana, Río Mao, Río Amina), exposes Eocene to Pliocene mixed siliciclastics and carbonates. The bulk of the thick (~5,000 m) and well-preserved sequence (Mann et al., 1991) is composed of Miocene-Pliocene deposits of the Cercado, Gurabo, and Mao Formations. Collectively these formations make up the Yaque Group, a remarkably continuous northward (seaward)-prograding wedge of sediments shed off the Cordillera Central.

The richly fossiliferous deposits of the Yaque Group have attracted scientific attention for over a century (Gabb, 1873; Maury, 1917a, b). The first major group effort was led by the US Geological Survey as part of their reconnaissance geology of the Dominican Republic (Vaughan et al., 1921). More recently the Yaque Group of the Cibao Valley was the focus of "the Dominican Republic (DR) project" led by J. Saunders and P. Jung of the Natural History Museum in Basel, Switzerland (NMB). During three field seasons in 1978–1980, a small field party measured sections and collected samples of micro- and macrofossils at closely spaced intervals along nine river sections in the Cibao Valley. Age dates for the sections were determined through study of planktic foraminifera and nannofossils (Saunders et al., 1986).

Several geologic maps exist for the Cibao Valley. Blesch (1966) showed the main lithologies and estimated ages in the valley, but generally avoided using formation names (except Bulla Conglomerate). Antonini (1968, 1979) published a detailed geologic map of the southern flank of the Cibao Valley and divided the different rock types into temporal units. His main groups included: Belted Metamorphic Deposits (pre-Miocene); Piedmont Upland Deposits (lower Miocene); Piedmont Plateau Deposits (lower to upper Miocene); Piedmont Plains Deposits (upper Miocene to Pliocene); and River Valley Floor Deposits (Pleistocene to Recent). A reproduction of the Antonini map, with a few minor changes we have noted from our fieldwork, is shown in Fig. 2.4. The most prominent feature of the geologic map is the northwest-southeast orientation of the main lithologic units. This orientation parallels the structural alignment of the valley and the pre-Neogene belt of metamorphic rocks (Antonini's pre-Miocene Belted Metamorphic Deposits). Within the Neogene sedimentary units, the alternation of limestone and siliciclastic-rich units, and their differential erosion, has produced the linear (and topographic) trends distinctly evident on the map of Antonini (1979).

2.3.2 Historical Foundations to the Regional Stratigraphy of the Cibao Basin

Geologic study of the Cibao Valley started in the mid-1800s with a basic description of selected sites and key fossils. The first discussion of the stratigraphy of the Cibao Basin was by Gabb (1873), who considered a single, unnamed formation of late Miocene age. Following the pioneering studies of Maury (1917a, b; 1919) and Cooke (1920), the US Geological Survey published the benchmark study on the strata of the Cibao Basin (Vaughan et al., 1921). This study contained key sections on the geology and stratigraphy (chapter by W. Cooke) and on the stratigraphy and paleontology (chapter by T.W. Vaughan and W.P. Woodring). The Vaughan and Woodring chapter contained detailed information on the molluscs, corals, and some earlier published information on formaminifera (Cushman, 1919). The first

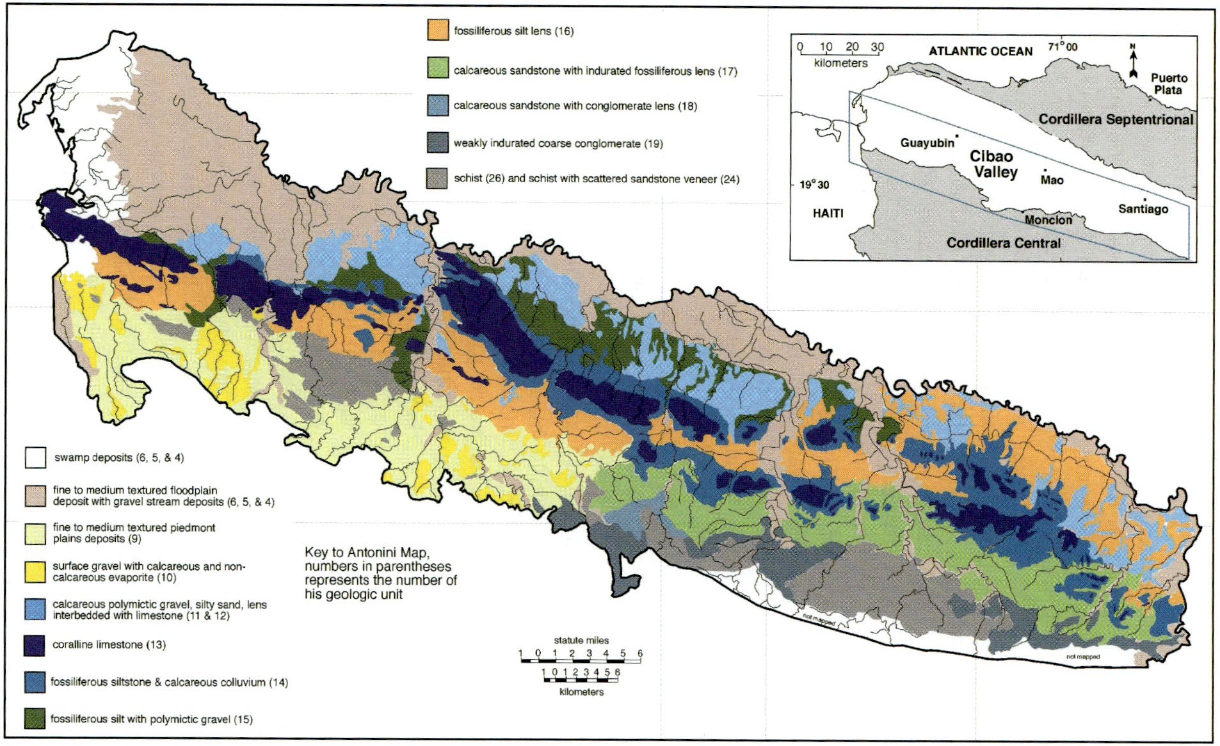

Fig. 2.4 Geologic map of the southern flank of the Cibao Valley based on the map of Antonini (1979). Note the coralline-rich units shown in dark blue that form several sub-parallel ridges

formalization of the stratigraphic units in the Cibao Basin was with the naming of formations within the Yaque Group. Maury (1919) named the Cercado and Gurabo Formations based on two index fossil zones recognized in her 1917 studies. Several years later, Cooke (1920) expanded the formations within the Yaque Group to include the Bulla Conglomerate, Baitoa Formation, Cercado Formation, Gurabo Formation, Mao Adentro Limestone, and the Mao Clay. The next major advancement in refining the stratigraphy of the Cibao Basin came with several projects in the 1970s and 1980s. Vokes (1979) summarized microfossil age dates from several sites in the Cercado, Gurabo, and Mao Formations. A stratigraphic study by J.B. Saunders, P. Jung, and B. Biju-Duval published in *Bulletins of American Paleontology* (1986) provided the most comprehensive evaluation of the Cibao Basin deposits. This study (Saunders et al., 1986) examined the lithology, stratigraphy, geologic history, and biostratigraphy of two main sections (Río Gurabo and Río Cana) and several other subsidiary river sections. Since the 1920s at least sixteen modifications to the stratigraphic nomenclature have been proposed (Saunders et al., 1986, their Table 1). The most significant modification has been the relegation of the Bulla Conglomerate to member status within the Cercado Formation because of its discontinuous lateral nature.

Over the last 35 years, the usage of four Neogene formations has been relatively consistent—Baitoa Formation (early/middle Miocene), Cercado Formation (middle/late Miocene), Gurabo Formation (late Miocene to early Pliocene), and the Mao Formation (early to late Pliocene). In this chapter, we are mainly concerned with the three younger formations, the Cercado, Gurabo, and Mao (Fig. 2.5). *Cercado Formation*—in the Río Gurabo and Río Cana sections this unit is predominantly sandstone but contains variations in lithology ranging from pebble stringers, conglomerate lenses, lignite beds, and reef limestone (Saunders et al., 1986; Evans, 1986a). The depositional setting is interpreted to be a relatively shallow shelf. *Gurabo Formation*—in the study area predominantly a siltstone: lithologic variations are considerable, ranging from interbedded silts and coral debris, to plain massive siltstone with occasional sands and gravel lenses (Saunders et al., 1986; Evans, 1986a). The depositional setting is thought to include a transition from a relatively shallow shelf setting to a steep upper-slope setting. *Mao Formation*—in the Río Cana and Río Gurabo sections the lithology of this unit is again highly variable. Evans (1986a) recognized four different lithofacies: bedded siltstone, conglomerate, interbedded coral boundstone-siltstone, and a clean siltstone. The depositional setting is likewise variable, for example, the conglomerates and bedded limestone a prograding shelf margin (Evans, 1986a). The uppermost part of the formation is also likely middle shelf facies (see uplift discussion below).

2.4 Ages of the Cibao Basin Deposits

We present here our preliminary results of the age synthesis. The effort to refine the ages is especially important because it allows intra-basin correlation of the key sections, it solidifies the interpretation of intrabasin paleoenvironment, it allows better temporal evaluation of species changes, it allows calculation of an average sedimentation rate, it allows correlation to changes in eustatic sea level, and it provides the opportunity to correlate to other fossil-rich sites around the Caribbean. As part of an effort to unify the stratigraphy of the Cibao Basin we have begun a study to refine the ages of the main sections along the Río Gurabo and the Río Cana.

2.4.1 Existing Biostratigraphy and Age Models

Over the past 87 years, nineteen different age models have been proposed for the sequences of the Cibao Valley. Several of the recent biostratigraphic studies (Saunders et al., 1986; Vokes 1979, 1989) provide slightly different age models. The Saunders et al. (1986) age model employs the use of planktic foraminifers and calcareous nannofossils. The Vokes (1979) ages are based on molluscs and their known ranges in other Caribbean sections and some spot ages using planktic foraminifers. The two models are shown in Fig. 2.5. This uncertainty in age was

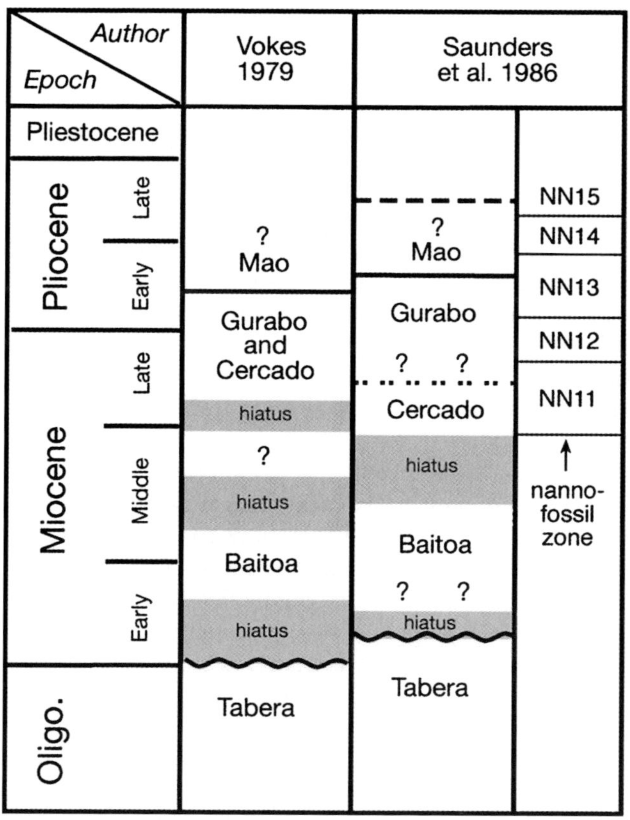

Fig. 2.5 Summary of the stratigraphic schemes of Vokes (1979) and Saunders et al. (1986). The right side of the figure shows the calcareous nannofossil zones for the key formations

partly the motivation for our ongoing chronostratigraphic reevaluation of the main Cibao Basin sections. The Vokes model indicates a relatively short period of deposition for the Cercado and Gurabo Formations, ranging from late Miocene to early Pliocene. The Saunders et al. study indicates a late Miocene age for the Cercado Formation, a late Miocene to early Pliocene age for the Gurabo Formation, and an early-middle Pliocene age for the Mao Formation.

The more robust of the two data sets is the microfossil biostratigraphy developed by Saunders et al. (1986), and tied to faunal datums of the Berggren et al. (1985) timescale (which we have converted to the Berggren et al., 1995b timescale). The Cercado Formation is assigned an NN11 age (Late Miocene; Tortonian and Messinian, ~8.6–5.6 Ma in Berggren et al., 1995b timescale). The Gurabo Formation spans the upper part of NN11, NN12, and lower part of NN13 (Late Miocene to Early Pliocene, Messinian and Zanclean, 5.6–~4.5 Ma). The Mao Formation ranges from the upper part of NN13, NN14, to within NN15 (Early Pliocene, Zanclean, ~4.5–~3.6 Ma). Saunders et al. (1986) also use the ostracode *Radimella confragosa* datum (van den Bold, 1975) to help constrain the Miocene-Pliocene boundary. Unfortunately, individual foraminiferal or nannofossil datums are not given in the Saunders et al. (1986) publication, so subzone age resolution is not possible.

2.4.2 New Stratigraphic and Age Data (2003–2006)

The complete and detailed chronostratigraphy will be published upon completion of our study, but we include provisional age-depth curves for Río Gurabo and Río Cana (Fig. 2.6). These refined age models will introduce our updates and refinements to those paleontologists currently working on the Cibao sections and interested in the timing of faunal change. Our ages are based on an integration of existing biostratigraphy (Saunders et al., 1982, 1986), paleomagnetic stratigraphy, and strontium-isotope stratigraphy. Our ages are tied to the time scale of Berggren et al. (1995a, b).

Río Gurabo Section—this is where most of our efforts have been focused during the first part of the current project. We have taken available microfossil datums and updated them to the modern age ranges. We have combined these data for the first time with paleomagnetic reversal information (mostly from the Gurabo Formation) and a set of new strontium-isotope stratigraphy tied to the McArthur et al. (2001) reference database. The strontium-isotopic sample collection is currently in progress, but it will span all three formations when complete. We plot the ages as an age depth curve (Fig. 2.6) to better constrain ages along the Saunders et al. (1986) stratigraphic column. The age model indicates a latest Miocene (Messinian) age for the base of the Gurabo section (0–~150 m) within the Cercado Formation. The base of the Gurabo Formation from ~150 m to about 380 m, mainly siliciclastics with admixed coral debris, spans an age from near the Miocene-Pliocene boundary to middle early Pliocene. This unit is overlain by a limestone unit ~20 m

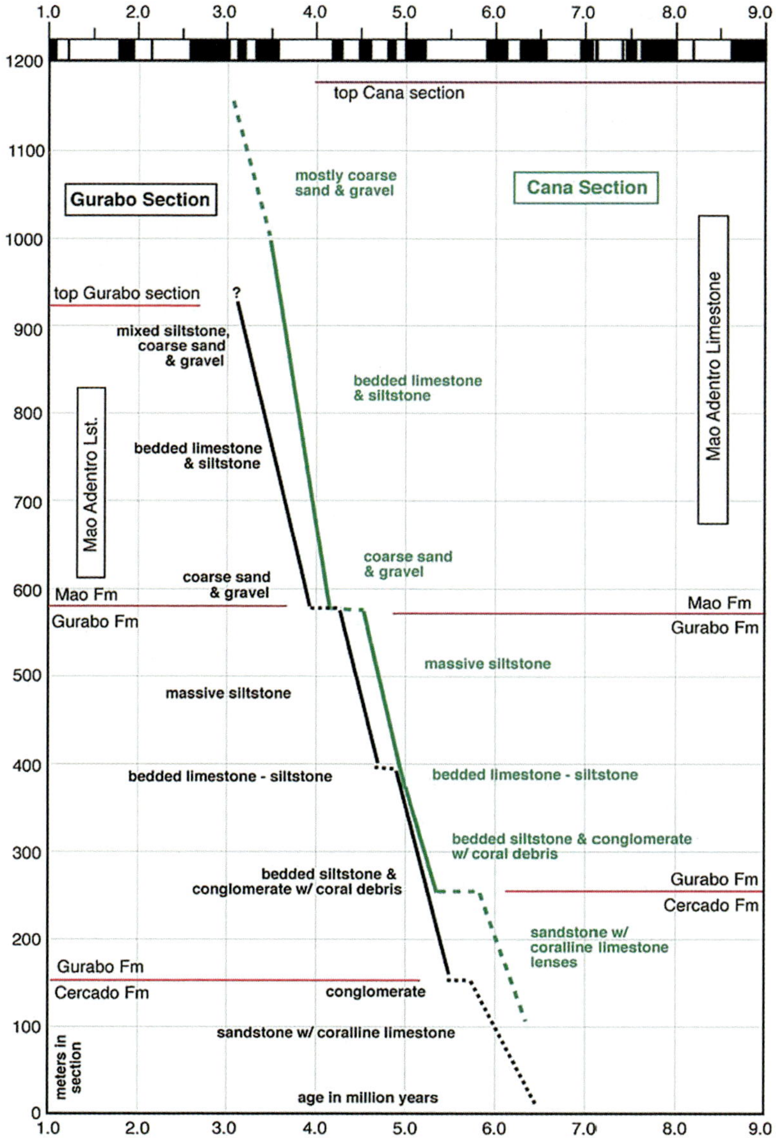

Fig. 2.6 Age-depth plots for the Rio Gurabo and Rio Cana sections based on existing biostratigraphic data of Saunders et al. (1986) and new data (strontium-isotope ages and paleomagnetic reversal stratigraphy). This is a preliminary summary of our best age estimates at this time and does not include details on our revised age estimates

thick, and constrained in age to mid-early Pliocene. From ~400 m to the top of the Gurabo Formation around 580 m, the unit is fine-grained siliciclastics of late early Pliocene age. The Gurabo-Mao contact likely represents an erosional hiatus (of unknown duration) within the deeper-water siliciclastic slope deposits and probable shallowing due to a change of sea level. The Mao Formation consists of a basal sandstone and conglomerate, middle limestone, and upper siliciclastic unit. The lower and middle Mao Formation in this section, from ~580 to 800 m, was deposited relatively rapidly and is early late Pliocene in age. The overlying sands and conglomerates that comprise the uppermost Mao Formation are poorly constrained, but appear to be latest Pliocene.

Río Cana Section—as part of our ongoing project we have made an effort to refine the age of the Río Cana section (Fig. 2.6). We utilize the existing biostratigraphy from Saunders et al. (1986), the few available microfossil "spot" samples (Vokes, 1979) and some newly acquired strontium-isotope age ranges. No paleomagnetic data yet exist for this section, although samples have been collected in 2005 and 2006. As with the Rio Gurabo section we tie our age-depth curve to the stratigraphic section of Saunders et al. (1986). The lower Río Cana section, within the Cercado Formation is dated using strontium isotopes and microfossils to be latest Miocene (Messinian) (Maier et al., 2007). The Cercado-Gurabo contact (~250 m) is likely erosional, with an age very near the Miocene-Pliocene boundary. The base of the Gurabo Formation, siliciclastics from ~250 to ~360 m, is earliest Pliocene. An intermediate limestone (~360–400 m), exposed near the confluence of Cañada Zamba with the Río Cana is middle early Pliocene. From ~400 m to the top of the Gurabo Formation at ~575 m, planktic foraminifera-rich siliciclastics, an indicator of deeper water, are also middle early Pliocene. The Mao Formation is marked by a series of coarse sand and gravel units from the underlying contact (~575 m) to ~675 m. We estimate the age of these deposits to be latest early Pliocene to about the early/late Pliocene boundary. The transition to limestone at ~675 marks a period of relatively rapid accumulation very near the early/late Pliocene boundary (~3.6 Ma). Coarse siliciclastics overlying the Mao limestone at Cana Gorge are poorly constrained in age, but appear to be late Pliocene.

A comparison of the tentative age assignments between the two sections shows some interesting similarities. The age of the Cercado-Gurabo Formation contact at both sections appears to be very close to the Miocene-Pliocene boundary at about 5.3 Ma. The Gurabo-Mao Formation contact ranges from ~3.9 to 4.2 Ma in Río Gurabo and ~4–4.2 Ma in Río Cana. Surprisingly, the formation contacts appear to be generally age consistent (within our resolution). Some lithostratigraphic similarities are also found. The Cercado Formation at Arroyo Bellaco (Evans, 1986a), a tributary to the Río Cana, has a fairly well developed reef unit, and we tentatively correlate this "patch" reef to small coral build-ups in the similar age sediments in the Río Gurabo section. These intermittent, Messinian-age reef patches mark the earliest development of Neogene reefs in the Cibao Basin (the Baitoa Formation contains some isolated corals) (Maier et al., 2007). Further up section, both river sections show an earliest Pliocene shallow-water siliciclastic unit overlain by a reefal limestone. This limestone unit is distinguished by alternating beds of coral debris and mud-rich siliciclastic sediment. It is interesting to note that these bedded limestones may be correlative to the coral-bearing section at Angostura Gorge described in Saunders et al. (1986), although the age assignment at Angostura remains uncertain and the possibility exists that the Angostura reef may be Messinian. Above this bedded limestone (and above Angostura Gorge), a major deepening occurs based on planktic foraminifera (upper part of Gurabo Formation) and the lithologies are dominated by mostly fine-grained siliciclastics. Differing from previous interpretations, as explained below, our reevaluation indicates that the overlying Mao Formation may transition to shallower conditions in both sections, with coarse near shoreface sands and gravel overlain by a thick shallow-water limestone. Marine muds, sands and conglomerates cap the sequence.

2.5 Depositional Setting Along the Basin Margin

2.5.1 Cibao Basin Morphology

The Cibao Basin is bounded in the south by the Hispaniola Fault Zone (along the northern edge of the Cordillera Central) and in the north by the Septentrional Fault Zone (Fig. 2.1). Saunders et al. (1986) prepared detailed maps and stratigraphic sections of the river sections that incise into the sedimentary wedge on the south side of the basin. They also provided bed-by-bed descriptions of lithologies, sedimentary features, and fossil content. Evans (1986a, b) performed a detailed study of the sedimentology of the sequence and interpreted patterns within the context of depositional models. The configuration of the basin during the Neogene is uncertain, but Evans (1986a) provided evidence that the basin likely had a central shallow marine high in the area of Guayubin (Fig. 2.1). He pointed out that based on the existing distribution of Pliocene marine strata a ridge probably continued across the basin and surrounded (on the south, west, and north) a deep trough that opened to the northeast and east. This interpretation was supported by an eastward shift of 25–55 in the depositional strike and dip in the Río Cana deposits which correlated with a northwest deflection in the Zamba Hill as it followed the north-south oriented ridge (Evans, 1986a). The northern margin of the basin was likely a shallow ridge or possibly emergent with development of the Cordillera Septentrional. Nagle (1979) mapped Pliocene limestones on the southern side of the Cordillera Septentrional, indicating a shallow foundation that probably defined the northern boundary of the Cibao Basin during the Pliocene.

Saunders et al. (1986) calculated the formation thicknesses for the sections along the major rivers with the assumption that dip was post-depositional. In the Río Gurabo section, thicknesses of 158, 423, and 339 m were estimated for the Cercado, Gurabo, and Mao Formations, respectively. The Río Cana section had thicknesses of 276, 299, and 615 m for the three formations. Evans (1986a) working on the sedimentology of the same sections recognized that the beds were best viewed as large-scale cross-beds (clinothems) and recommended thickness be calculated perpendicular to bedding. Evans (1986a) revised thicknesses of the formations, and on average: Cercado = 230 m; Gurabo = 230 m; and Mao = 200 m.

2.5.2 Sedimentology of the Cibao Basin

Evans (1986a) recognized the main sedimentary successions and plotted them relative to the sections constructed by Saunders et al. (1986). Evans also described in detail the lithofacies and made sedimentologic interpretations as to the depositional setting. We briefly summarize that work below, but the reader is encouraged to review the dissertation of C.C. Evans for a more thorough discussion. Evans (1986a) recognized several lithofacies within each of the formations.

The Cercado Formation has two facies: the laterally discontinuous (Bulla) conglomerate; and the more common sandstone facies. The conglomerate is likely a series of four, isolated fan-shaped deposits of a shoreline fan-conglomerate (Antonini, 1979). The sandstone facies is a clean, fine- to medium-sand with pebble stringers, conglomerate lenses, patches of lignite, and reef limestone (Fig. 2.7a). The unit is generally horizontally laminated or trough cross-laminated with coarse grains and skeletal debris at the base of the troughs. The depositional environment is thought to be a nearshore setting with well-developed coral reefs and areas of protected lagoon.

The Gurabo Formation has three facies: bedded siltstone A; massive siltstone; and bedded siltstone B. The bedded siltstone A consists of 0.3–1.5 m thick beds of soft siltstone interlayered with thin (<25 cm), well-cemented beds of coarse biogenic calcareous material (coralline debris and upright, in place corals) (Fig. 2.7b). Dip is less than 3.5°. Evans (1986a) interprets these to represent shallow, above storm wave base shelf deposits. The bedding is a result of the background silt accumulation with mud tolerant fauna being interrupted by events that introduce shallow-water debris or locally concentrate coarse material. The massive siltstones also have a dip less than 3.5° and as the name implies have beds that are 2 m or greater (Fig. 2.7c). These massive siltstones are interpreted by Evans to have been deposited below fair weather wave base, but above storm wave base in an outer shelf setting. The bedded siltstone B is similar to bedded siltstone A except that the bedding planes dip at 6° or greater, channels and discontinuity surfaces are encountered, and matrix-supported, coarse channel fills with erosional base occur. The depositional setting for this facies indicates a relatively steep (>6°) depositional slope, part of the deeper shelf.

The Mao Formation has three members comprised of four lithofacies: the bedded siltstone B (as described in the Gurabo Fm.), conglomerate facies, interbedded boundstone-siltstone, and siltstone. The conglomerate facies is comprised of interbedded sandstone and conglomerates with evidence of syndepositional slumping, all dips are >6° and beds are 10 cm to 3 m thick (Evans, 1986a). This facies is interpreted to occur in a shelf setting of ~100 m water depth. The interbedded limestone-siltstone facies consists of beds of coral floatstone and small coral mounds (~2 m thick) (Fig. 2.7d). This unit has dips up to 20° (average = 13°). This facies contains numerous coral species, some allochtonous and some probably *in situ* (mounds). The bedded limestone probably represents middle to outer shelf deposition in relatively clear water (within the photic zone); both transported and in-place corals occur. The siltstone facies occurs locally within the three facies described above.

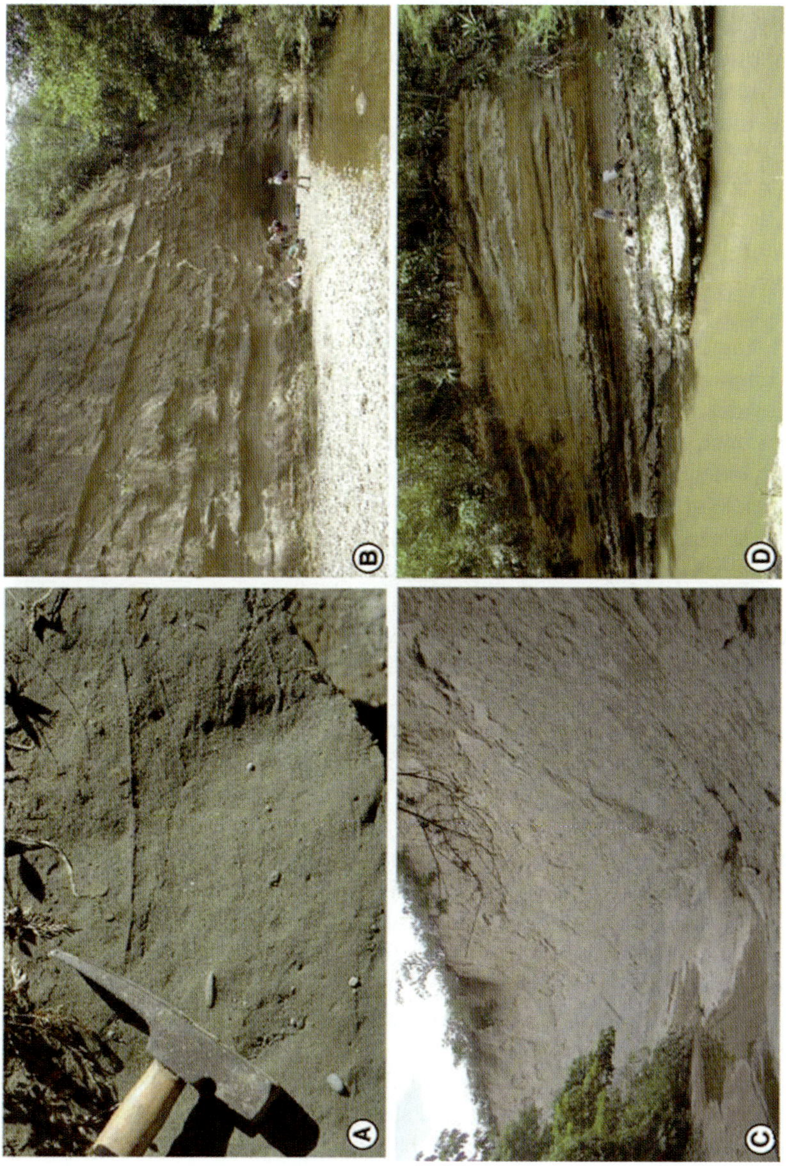

Fig. 2.7 Photographs of the key lithologic units in the Cibao Basin sections. A—Late Miocene cross-bedded sands from the late Cercado Formation along the Rio Gurabo. B—Bedded siltstone and limestone from the Pliocene section along the Rio Gurabo. C—Early Pliocene massive siltstone from the Rio Gurabo section. D—Bedded limestone and siltstone from the Cana Gorge on the Rio Cana

2.5.3 Mao Formation and Uplift of the Cibao Basin

Uplift history of the Cibao Basin is generally constrained by the termination of open marine deposition. The youngest unit in the Yaque Group is the Mao Formation. That formation has been divided into three members: the basal Mao Adentro Limestone, the intermediate Mao Clay Member, and the upper, unnamed member (Bermudez, 1949).

The depositional history of the Mao Formation, as summarized from Saunders et al. (1982, 1986, pp. 16–17) indicated deeper-water (>100 m) deposition near the contact with the underlying Gurabo Formation. This interpretation is based on channeling, load features, flame structures and silt casts (600–700 m on their Rio Gurabo column). A suite of molluscs that range from 650–750 m on the same column support this relatively deep water interpretation. Up-section these deposits transition to fairly shallow-water limestone with corals and oysters (Mao Adentro Limestone member); followed by a mud unit (Mao Clay at Río Cana); and finally sands and conglomerates of the youngest unit (unnamed member).

In our reevaluation of the Mao Formation, the sequence does show a shallowing from the erosional, channelized base (Rio Gurabo section) upward to conglomeritic fluvial or fluviodeltaic deposits (500 to ~630 m in Rio Cana section). These coarse gravels deepen slightly as the shelf prograded seaward and the delta front became more carbonate rich. Water depths continued to deepen to ~20 m or more as prograding clinothems developed and deposited the Mao Adentro Limestone. The Mao Adentro Limestones were likely deposited in the photic zone based on in-place coral accumulations (Evans, 1986a). Saunders et al. (1986) make water depth estimates on the order of 100 m for the uppermost limestones and mud interbeds. Water depth, however, likely varied greatly depending on location on the clinothem, either on the foreslope or on the topset. Based on visual estimates of the height of the clinothems water depth could range from ~20 to 50 m. The uppermost part of the section, including the Mao Clay Member and the unnamed member are interpreted to have been deposited in considerably deeper water. Near the base of the Mao Clay, Saunders et al. (1986) estimate depth in excess of 100 m based on a rich planktic foraminiferal fauna. Furthermore, the overlying unnamed member is purported (Vokes, 1989) to make an abrupt change to water depths of >350 m.

The unnamed member, uppermost of the formation, consists of coarse sands, gravels, and some large allochtonous blocks (Saunders et al., 1986). Lithologically, this would be grouped with Evan's (1986a) conglomerate lithofacies (although he described this lithofacies from the base of the formation below the Mao Adentro Limestone). These deposits also exhibit small channels, cross-beds, and slumps that were interpreted to indicate "…turbidity and slump influxes into an environment of indeterminate water depth." (Saunders et al., 1986). Vokes (1989) interprets the upper, unnamed member of the Mao Formation to be at least 350 m deep based on several species of molluscs (p. 41, muricids that are extant; *Chicoreus (Siratus) articulatus* and *Chicoreus (Siratus) formosus)*. Other shallow-water molluscs were thought to be admixed with the deepwater species through gravity flows.

The transition from relatively shallow water clinothem deposition to water depths in excess of 350 m is intriguing. With uplift of the island reported to occur around 3 Ma, this proposed sudden deepening in the unnamed member is somewhat unexpected. As such, we tentatively reexamine some of the water depth interpretations for the unnamed member. First, the swale-type bedding and cross-bedding features can also be found in relatively shallow (<100 m) settings. For example, the transition zone from shallow subtidal (~5 m depth) to shelf (up to 100 m depth) environments can have cross-bedding, mixed sand and gravel deposits, channels, and contain shell layers, mud pebble intraclasts, plant debris, and wood pieces (Reineck and Singh, 1980; Howard and Reineck, 1981). Second, the muricids that are used to indicate depth have modern ranges of 12 to 16 m down to several hundreds of meters (database of Academy of Natural Sciences of Philadelphia; http://data.acnatsci.org). Third, the planktic foraminifers are found in silt clasts, and these clasts could very likely be a result of reworking of the Gurabo Formation deep-water facies (massive siltstone). As noted above, mud clasts are noted from studies of modern shore to shelf transects. Fourth, the sequence is in conflict with the sea-level record from the Bahamas over that time period where the late Pliocene (~3–2 Ma) deposits show consistent shallowing and subaerial exposure due to lowered sea level. Last, the ostracode data of van den Bold (1988) reflects a mixing of shallow water and deep-water ostracodes in the deposits of the Mao Formation. van den Bold (1988) proposed downslope transport to explain this mixing. We feel an alternative interpretation that could explain this faunal mixing is that of water being upwelled from the deeper parts of the basin and open-ocean to the east.

The key deep-water ostracode species that were used to infer water depth have recently (Borne et al., 1999) been shown to occur admixed with shallow-water ostracodes in reef deposits and shallow forereef sediments (due to upwelling) in the Limon Basin of Costa Rica. We thus propose that the upper, unnamed member of the Mao Formation may actually record a shallowing of the basin due to both sea level change and perhaps the onset of regional uplift. The channelization and slumps may indeed be an indication of earthquakes associated with tectonic uplift (Vokes, 1989), but slumping may have occurred shallower than previously thought. These upper Mao Formation slumps are exceptionally well exposed in the upper part of the Rio Gurabo section (site NMB 16118–15833 on Saunders et al., 1986, their Fig. 2.4). Further basinward, the alluvium of the Río Yaque del Norte unfortunately covers the transition from final marine deposition to subaerial exposure and fluvial conditions. Also in support of this alternative interpretation, pollen data (Saunders et al., 1986, p. 17) indicate that there is a gradual change in pollen and spores up section to a slightly more temperate flora, either a result of uplift or a cooler climate.

If confirmed, the progressive shallowing of the Mao Formation combined with an age of middle late Pliocene for the top of the formation (see discussion below), would suggest a transition from marine conditions to emergence sometime around 3 Ma. This age is coincident with a general lowering of sea level with the onset of Northern Hemisphere glaciation (Shackleton et al., 1984). Support for a ~3 Ma age for uplift also comes from studies of the tectonism of the region. McCann and

Pennington (1990) studying the seismicity and marine geophysical data (Ladd, 1976) suggested a post-Miocene timing of strike-slip and vertical motion along northern Hispaniola. Lewis and Draper (1990) concluded that a main deformation phase that included the uplift of the Cordillera Central must have occurred about 3–4 Ma, and the uplift of the whole island has continued in the Quaternary.

2.6 Regional Paleoceanographic Setting of the Cibao Basin

2.6.1 Late Miocene-Pliocene Paleoceanography and Sea Level Change

Several significant changes occurred in the Caribbean Sea during Neogene deposition in the Cibao Basin. The overriding controls on sedimentation were likely changes in eustatic sea level and the tectonic movement (gradual uplift) of the basin. We briefly summarize four intervals over the period from about 7 Ma to 2 Ma (Fig. 2.8). The four intervals include the late Miocene (Messinian); the early Pliocene global transgression; the middle Pliocene highstand and forced regression; and the late Pliocene closure of the Central American Seaway (CAS) and sea level lowstand.

Late Miocene—glacial cooling started about 10 Ma and glacial conditions were established at about 7 Ma in both the northern and southern hemisphere (Larsen et al., 1994). This late Miocene global cooling was manifest as lowstands of sea level, generating hiatuses on many of the shallow water shelves around the Caribbean basin and elsewhere (Keller and Barron, 1983; Haq et al., 1988). In the Cibao Valley the Messinian deposits of the Cercado Formation record a change from exclusively shallow-water siliciclastics to isolated coral-bearing deposits, some of which form well developed reefs. In the Río Cana section, the exposed reef at Arroyo Bellaco shows progressive development from isolated corals, to small reef mounds, to large head corals, to pocilloporid-dominated (*Pocillopora, Stylophora*) reefs (Evans, 1986a; Klaus and Budd, 2003). On the Río Gurabo, a small, fairly diverse coral community is developed just below the contact with the overlying Gurabo Formation. The initiation of reefs in the late Miocene may be combination of several factors. The sea level during the later parts of the Messinian were low due to glacial events (MMi-2 event at 5.7 Ma, Abreu and Anderson, 1998) as glacial conditions developed in the North Atlantic (Larsen et al., 1994) and West Antarctic ice sheet (Zachos et al., 2001). The progressive restriction of water through the CAS (Coates et al., 1992) may have affected circulation throughout the Caribbean, and one effect is a consequence may have been decreased upwelling intensity. This decrease in upwelling may have produced conditions (less nutrients?) that were favorable for the establishment of coral reefs throughout the region (Maier et al., 2007).

Early Pliocene—the Miocene/Pliocene boundary and the early Pliocene marked a period of sea level lowstand and a major sea level rise, respectively. The Cercado-Gurabo Formation contact is close to the Miocene-Pliocene boundary and may

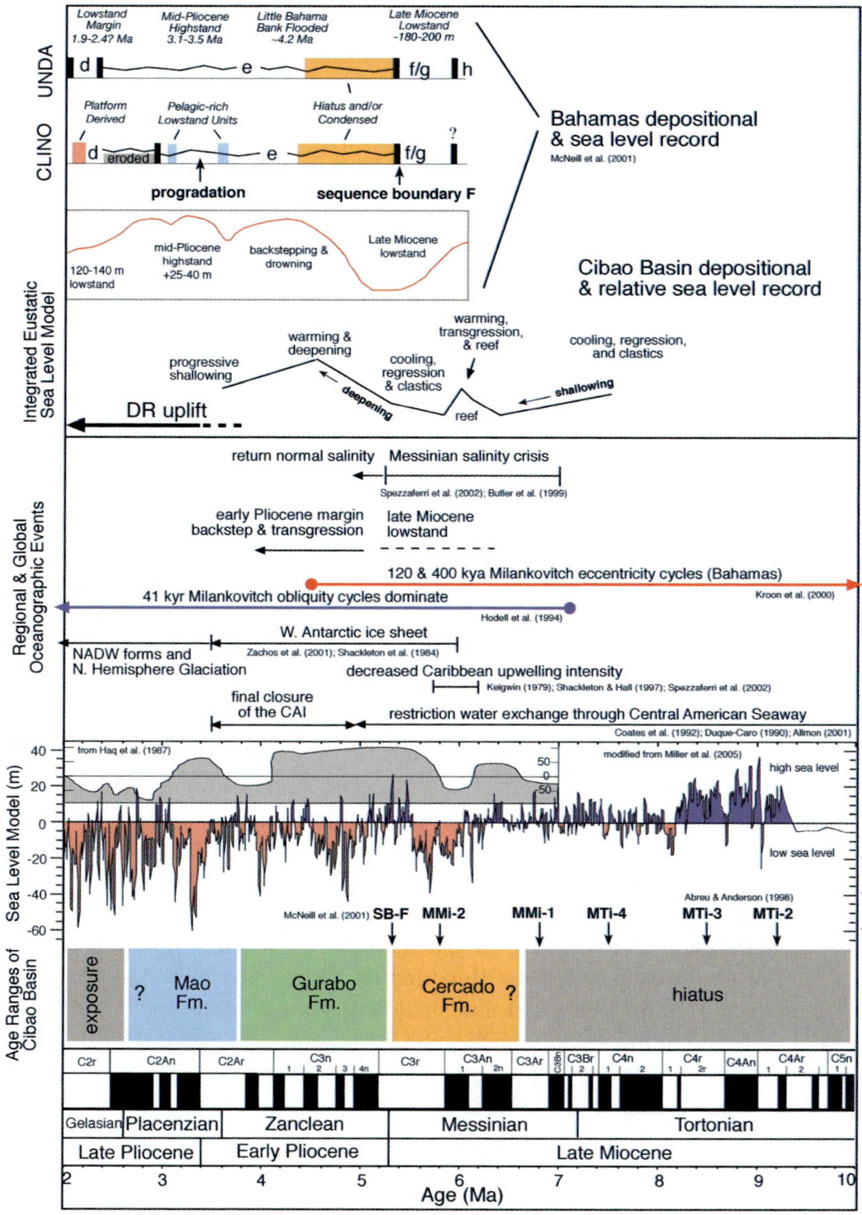

Fig. 2.8 Summary of regional and global paleoceanographic events that may have influenced deposition within the Cibao Basin. A late Messinian reef is coincident with a drop in sea level and a reduction in regional upwelling. The early Pliocene transgression, seen globally, likely correlates to the abrupt deepening recorded in the massive siltstone of the Gurabo Formation. The prograding limestones of the Mao Formation correlate temporally with margin progradation along western Great Bahama Bank. Uplift of the basin is recorded in the shallowing at the top of the Mao Formation that occurred simultaneously with the late Pliocene fall in sea level due to onset of Northern Hemisphere glaciation

represent a depositional hiatus. The early Pliocene period around the Caribbean records a period of transgression as sea level rose and flooded shelf areas. In the Bahamas, this highstand event is marked by backstepping of the platform margin and deposition of deeper water facies on top of the late Miocene shallow-water deposits (McNeill et al., 2000). In the Cibao Basin, the Gurabo Formation records this global transgression. The basal part of the formation records shelf deposits including bedded limestones (bedded siltstone A of Evans, 1986a). The bedded limestones give way to massive siltstones, rich in planktic foraminifers, that represent the rapid deepening and drowning of the basin. This time period is one of transition from eccentricity-dominated cycles (120 and 400 ky) to higher frequency 41 ky cycles (Hodell et al., 1994; Kroon et al., 2000). The bedded siltstones and limestones commonly found in the Gurabo Formation, as well as the Mao Formation, may be a manifestation of these orbital-driven climate and sea level changes. The later part of the early Pliocene likely saw the progradation of the massive siltstone facies that marks the upper part of the Gurabo Formation, sea level likely reached a maximum, and began to fall.

Mid Pliocene—a fall in sea level between ~4 and 3 Ma provided the transition from a deeper water setting to progressively shallower facies. The lower Mao Formation records marine channeling with coarse sand and gravel to near shoreface fluviodeltaic deposition, both likely the result of lowered sea level near the early/late Pliocene boundary (3.5–3 Ma, Dowsett and Cronin, 1990). As conditions stabilized, coralline facies were established as a series of prograding clinothems. These deposits both aggraded and prograded to form the Mao Adentro Limestone. The interbedded silt and coral debris may have prograded basinward as part of a forced regression sequence as sea level started to fall. A forced regression may have resulted from the combination of falling sea level from development of Northern Hemisphere glaciation and regrowth of the Antarctic ice sheet, coupled with the onset of uplift of the Cordillera Central and Cibao Basin (Lewis and Draper, 1990). It is interesting to note that a major pulse of progradation in the Bahamas platform also occurred at this same time, but in that example, the driving force was entirely a lowering of sea level. It is also worth noting that deposits of the Mao Formation show a mixture of both deep water and shallow water ostracode species, more so than what is found in the underlying Gurabo Formation (van den Bold, 1988). We propose that during Mao deposition that upwelling affected the basin, either bringing in deep water species or producing ecologic conditions that supported deeper water assemblages. The question remains as to the mechanics of the upwelling during this interval—is it related to initial uplift, a new basin configuration, or regional ocean circulation? The regional shift to carbonate deposition may also be related to climate and circulation changes associated with closure of the CAS. Prior to closure, the Caribbean and western Atlantic show evidence for significant upwelling (Allmon, 2001). After closure, however, primary productivity declined within the Caribbean and salinity, temperature, and carbonate deposition increased (Keigwin, 1982; Haug and Tiedemann, 1998; Haug et al., 2001; Kameo, 2002).

Late Pliocene—from ~3 Ma onward the Cibao Basin became progressively more restricted and normal marine deposition appears to have ceased at about this time.

The sequence above the Mao Adentro Limestone consists of coarse-grained siliciclastics at the Río Cana, and coarse sands and gravels with slump blocks, cross beds, and channels at the Río Gurabo section. As we discussed above, we interpret these deposits to be deposits of shallower water than previously thought, and that they represent relative shallowing of the basin through uplift and sea level lowstand. The shallow water platforms around the Bahamas record falls in sea level and subaerial exposure starting at ~2.5 to 3 Ma (McNeill et al., 1998, 2000). The siliciclastics in the Cibao Basin at the top of the two main sections, those that form the "unnamed" member of the Mao Formation, are relatively thick in the basin (>4,000 m in northern part of basin, Bowin, 1966) and record the successive regional uplift.

2.7 Ongoing Research and Future Plans

This paper is a summary of existing stratigraphic data, sedimentology, and age information on the Río Gurabo and Río Cana sections of the Cibao Valley. We have integrated new data collected over the past two years as part of our project supported by the US National Science Foundation. The initial results now allow us to start to answer ecological-based questions posed in this project: (1) How did the overall diversity of reef corals change across the Caribbean region over time? When were rates of origination and extinction accelerated across the region, and what were their regional environmental correlates? (2) How did diversity within DR reef communities change over time? Similarly, how did their taxonomic composition and dominance structure change? Are these changes correlated with changes in the regional and/or local environment? (3) How is community change related to speciation and extinction events in the fauna as a whole? How is it related to speciation and extinction events within a diverse, highly resolved clade of ecologically dominant corals (i.e., *Stylophora*)? How do abundances change as species are added and removed from communities? Once complete, the new, preliminary information presented here will promote a dataset available to the general paleontology community through several websites. The ongoing and future research objectives include:

1. Refined chronostratigraphy—we will continue to improve the resolution of the age model throughout both main outcrop sections. Currently, the Río Gurabo section has more and better age constraints, but significant improvement in the age of the Río Cana section is already apparent.
2. Intra-basin correlation between sections—the improvement has allowed for a much-improved correlation between the Río Cana and the Río Gurabo. This new correlation allows us to test and confirm many of the previously proposed paleoenvironment and facies similarities. Already, the refined age model allows a temporal comparison to different lithofacies in the southern margin of the Cibao Basin.

3. Regional temporal correlation in the Caribbean—another advantage of a robust age model is the ability to correlate outside the basin. We anticipate having the ability to generate a regional synthesis of key events. In this paper, we already use a correlation to the Bahamas platform record to assess sea level changes and the regional manifestation of sea level events.
4. Timing of turnover transition, community change, diversity—the Cibao Basin record provides a key interval in the early transition phase of the shallow-water faunal turnover (Budd et al., 1996). The reefal sections provide a well-preserved picture of the community structure and diversity prior to turnover, as well as in the transition period where old and new species are intermixed.
5. Calculation of evolutionary rates—the refined age model in the Cibao Basin provides further data for assessing the internal processes responsible for major speciation and extinction events. The Dominican Republic sections provide a key central Caribbean record of extinctions and originations.
6. Community access to stratigraphic data and database—data from this project will be entered in the Neogene Marine Biota of Tropical America (NMITA) taxonomic database and other databases. This central database allows wide access to integrating different aspects of evolutionary change and earth history. The paleontologic, geologic and stratigraphic data will be available for shared community use.

Acknowledgments We acknowledge the support of the US National Science Foundation for this project (EAR-0446768). This project is made possible by the strong geological and paleontological foundation established by the comprehensive studies of John Saunders, Peter Jung, and Bernard Biju-Duval. We gratefully thank reviewer Scott Ishman and editor Ross Nehm.

References

Abreu, V.S. and Anderson, J.B., 1998, Glacial eustasy during the Cenozoic: sequence stratigraphic implications, *Am. Assoc. Petrol. Geol. Bull.*, **82**:1385–1400.

Academy of Natural Sciences of Philadelphia, http://data.acnatsci.org

Allmon, W.D., 2001, Nutrients, temperature, disturbance, and evolution: a model for the late Cenozoic marine record of the western Atlantic, *Palaeogeogr. Palaeoclim. Palaeoecol.*, **166**:9–26.

Antonini, G.A., 1968, *Processes and patterns of landscape change in Linea Noroeste, Dominican Republic*, Unpublished Ph.D. dissertation, Columbia University, New York.

Antonini, G.A., 1979, Physical geography of the northwest Dominican Republic, in: *Hispaniola: Tectonic Focal Point of the Caribbean—Three Geologic Studies in the Dominican Republic* (Lidz, B. and F. Nagle, eds.), Miami Geological Society, Miami, pp. 96.

Berggren, W.A., Kent, D.V., Flynn, J.J., and Van Couvering, J.A., 1985, Cenozoic geochronology, *Geol. Soc. Am. Bull.*, **96**:1407–1418.

Berggren, W.A., Hilgen, F.J., Langereis, C.G., Kent, D.V., Obradovich, J.D., Raffi, I., and Raymo, M.E., 1995a, Late Neogene chronology: new perspectives in high-resolution stratigraphy, *Geol. Soc. Am. Bull.*, **107**:1272–1287.

Berggren, W.A., Kent, D.V., Swisher III, C.C., and Aubry, M.-P., 1995b, A revised Cenozoic geochronology and chronostratigraphy, in: *Geochronology Time Scales and Global Stratigraphic*

Correlation (Berggren, W.A. ed.), SEPM (Society for Sedimentary Geology), Tulsa, pp. 130–212.

Bermudez, P.J., 1949, Tertiary smaller foraminifera of the Dominican Republic, *Cushman Lab. Foram. Res., Spec. Publ.*, **25**:322.

Blesch, R.R., 1966, Mapa geológico preliminar, República Dominicana, in: *Mapas, vol. II, Reconocimiento y Evaluación de Los Recursos Naturales de la Repúblic Dominicana*, Washington, DC, Pan American Union.

Borne, P.F., Cronin, T.M., and Hazel, J.E., 1999, Neogene-Quaternary ostracoda and paleoenvironments of the Limon Basin, Costa Rica, and Bocas del Toro Basin, Panama, in: *A Paleobiotic Survey of Caribbean Faunas from the Neogene of the Isthmus of Panama* (Collins, L.S. and A.G. Coates, eds.), *Bul. Am. Paleontol.*, **357**:231–250.

Bowin, C., 1966, Geology of central Dominican Republic: a case history of part of an island arc, *Geol. Soc. Am. Mem.*, **98**:11–84.

Budd, A.F., Johnson, K.G., and Stemann, T.A., 1996, Plio-Pleistocene turnover and extinctions in the Caribbean reef coral fauna, in: *Evolution and Environment in Tropical America* (Jackson, J.B.C., A.F. Budd and A.G. Coates, eds.), University of Chicago Press, Chicago, IL, pp. 168–204.

Budd, A.F. and Johnson, KG., 1999, Neogene paleontology in the northern Dominican Republic. 19. The family Faviidae (Anthozoa: Scleractinia). Part II. The genera *Caulastraea, Favia, Diploria, Thysanus, Hadrophyllia, Manicina,* and *Colpophyllia. Bull. Am. Paleontol.*, **109**:5–83.

Butler, R.W.H., McClelland, E., and Jones, R.E., 1999, Calibrating the duration and timing of the Messinian salinity crisis in the Mediterranean: linked tectonoclimatic signals in thrust-top basin of Sicily, *J. Geol. Soc. Lond.*, **156**:827–835.

Coates, A.G., Jackson, J.B.C., Collins, L.S., Cronin, T.M., Dowsett, H.J., Bybell, L.M., Jung, P., and Obando, J.A., 1992, Closure of the Isthmus of Panama: the near-shore marine record of Costa Rica and western Panama, *Geol. Soc. Am. Bull.*, **104**:814–828.

Cooke, C.W., 1920, Geologic reconnaissance in Santo Domingo, *Geol. Soc. Am. Bull.*, **31**:217–219.

Cushman, J.A., 1919, Fossil foraminifera from the West Indies, in: *Contributions to the Geology and Paleontology of the West Indies* (Vaughan, T.W., ed.), Carnegie Institution of Washington, Washington, DC.

Dowsett, H.J. and Cronin, T.M., 1990, High eustatic sea level during the middle Pliocene; evidence from the Southeastern US Atlantic Coastal Plain, *Geology*, **18**:435–438.

Draper, G., Mann, P., and Lewis, J.F., 1994, Hispaniola, in: *Caribbean Geology* (Donovan, S.K. and T.A. Jackson, eds.), The University of the West Indies Publishers' Association, Kingston, pp. 129–150.

Duque-Caro, H., 1990, Neogene stratigraphy, paleoceanography and paleobiogeography in northwest South America and the evolution of the Panama seaway, *Palaeogeogr. Palaeoclim. Palaeoecol.*, **77**:203–234.

Evans, C.C., 1986a, *Facies Evolution in a Neogene Transpressional Basin: Cibao Valley, Dominican Republic*, Unpublished Ph.D. dissertation, University of Miami, Coral Gables.

Evans, C.C., 1986b, *A Field Guide to the Mixed Reefs and Siliciclastics of the Neogene Yaque Group, Cibao Valley, Dominican Republic*, University of Miami Comparative Sedimentology Laboratory, Rosenstiel School of Marine and Atmospheric Science, Miami, pp. 98.

Gabb, W.M., 1873, On the topography and geology of Santo Domingo, *Am. Philos. Soc., Trans.*, **15**:49–259.

Haq, B.U., Hardenbol, J., and Vail, P.R., 1987, Chronology of fluctuating sea levels since the Triassic, *Science*, **235**:1156–1167.

Haq, B.U., Hardenbol, J., and Vail, P.R., 1988, Mesozoic and Cenozoic chronostratigraphy and eustatic cycles, in: *Sea-Level Changes-an Integrated Approach* (Wilgus, C.K., G.S.C. Hastings, B.S. Kendall, H.W. Posamentier, C.A. Ross and J.C. Van Wagoner, eds.), SEPM (Society for Sedimentary Geology), Tulsa, pp. 71–108.

Haug, G.H. and Tiedemann, R., 1998, Effect of the formation of the Isthmus of Panama on Atlantic Ocean thermohaline circulation, *Nature*, **393**:673–676.

Haug, G.H., Tiedemann, R., Zahn, R., and Ravelo, A.C., 2001, Role of Panama uplift on oceanic freshwater balance, *Geology*, **29**:207–210.

Hodell, D.A., Benson, R.H., Kent, D.V., Boersma, A., and Bied, K.R.-E., 1994, Magnetostratigraphic, biostratigraphic, and stable isotope stratigraphy of an upper Miocene drill core from the Sale Briqueterie (northwestern Morocco): a high-resolution chronology for the Messinian stage, *Paleoceanography*, **9**:835–855.

Howard, J.D. and Reineck, H.-E., 1981, Depositional facies of high-energy beach-to-offshore sequence: comparison with low-energy sequence, *Am. Assoc. Petrol. Geol, Bull.*, **65**:807–830.

Kameo, K., 2002, Late Pliocene Caribbean surface water dynamics and climatic changes based on calcareous nannofossil records, *Palaeogeogr. Palaeoclim. Palaeoecol.*, **179**:211–226.

Keigwin, L., 1979, Late Cenozoic stable isotopic stratigraphy and paleoceanography of DSDP sites from the east equatorial and central Pacific Ocean, *Earth Planet. Sci. Lett.*, **45**:361–382

Keigwin, L., 1982, Isotopic paleoceanography of the Caribbean and east Pacific: role of Panama uplift in late Neogene time, *Science*, **217**:350–353.

Keller, G. and Barron, J.A., 1983, Paleoceanographic implications of Miocene deep-sea hiatuses, *Geol. Soc. Am. Bull.*, **94**:590–613.

Klaus, J.S. and Budd, A.F., 2003, Comparison of Caribbean coral reef communities before and after Plio-Pleistocene faunal turnover: analyses of two Dominican Republic reef sequences, *Palaios*, **18**:3–21.

Kroon, D., Williams, T., Primez, C., Spezzaferri, S., Sato, T., and Wright, J.D., 2000, Coupled early Pliocene-middle Miocene bio-cyclostratigraphy of Site 1006 reveals orbitally induced patterns of Great Bahama Bank carbonate production, *Proc. Ocean Dril. Prog.*, **166**:155–166.

Ladd, J.W., 1976, Relative motion of South America with respect to North America and Caribbean tectonics, *Geol. Soc. Am. Bull.*, **94**:590–613.

Larsen, H.C., Saunders, A.D., Clift, P.D., Beget, J., Wei, W., Spezzaferri, S., and ODP Leg 152 Scientific Party, 1994, Seven million years of glaciation in Greenland, *Science*, **264**:952–955.

Lewis, J.F. and Draper, G., 1990, Geology and tectonic evolution of the northern Caribbean margin, in: *The Caribbean Region* (Dengo, G. and J.E. Case, eds.), Geological Society of America, Boulder, pp. 77–139.

Maier, K.L., Klaus, J.S., McNeill, D.F., and Budd, A.F., 2007, A late Miocene low-nutrient window for Caribbean reef formation? *Coral Reefs* in press (doi:10.1007/s00338–007–0254–6).

Mann, P., 1999, Caribbean sedimentary basins: classification and tectonic setting from Jurassic to present, in: *Caribbean Basins* (Mann, P., ed.), Elsevier, Amsterdam, pp. 3–31.

Mann, P., Schubert, C., and Burke, K., 1990, Review of Caribbean neotectonics, in: *The Geology of North America* (Dengo, G. and J.E. Case, eds.), Geological Society of America, Boulder, pp. 307–338.

Mann, P., Grenville, D., and Lewis, J.F., 1991, An overview of the geologic and tectonic development of Hispaniola, in: *Geologic and Tectonic Development of the North American-Caribbean Plate Boundary in Hispaniola* (Mann, P., D. Grenville and J.F. Lewis, eds.), Geological Society of America, Boulder, CO, pp. 152–176.

Maury, C.J., 1917a, Santo Domingo type sections and fossils: mollusca, *Bull. Am. Paleontol.*, **5**:165–415.

Maury, C.J., 1917b, Santo Domingo type sections and fossils: stratigraphy, *Bull. Am. Paleontol.*, **5**:419–459.

Maury, C.J., 1919, A proposal of two new Miocene formational names, *Science*, **50**:591

McArthur, J.M., Howarth, R.J., and Bailey, T.R., 2001, Strontium isotope stratigraphy: LOWESS Version 3: best fit to the marine Sr-isotope curve for 0–509 Ma and accompanying look-up table for deriving numerical ages, *J. Geol.*, **109**:155–170.

McCann, W.R. and Pennington, W.D., 1990, Seismicity, large earthquakes, and the margin of the Caribbean Plate, in: *The Caribbean Region, vol. H, The Geology of North America* (Dengo, G. and Case, J.E., eds.), Geological Society of America: Boulder, CO.

McNeill, D.F., Grammer, G.M., and Williams, S.C., 1998, A 5 MY chronology of carbonate platform margin aggradation, southwestern Little Bahama Bank, Bahamas, *J. Sed. Res.*, **68**:603–614.

McNeill, D.F., Coates, A.G., Budd, A.F., and Borne, P.F., 2000, Integrated paleontologic and paleomagnetic stratigraphy of the upper Neogene deposits around Limon, Costa Rica: a coastal emergence record of the Central American Isthmus, *Geol. Soc. Am. Bull.*, **112**:963–981.

McNeill, D.F., Eberli, G.P., Lidz, B.H., Swart, P.K., and Kenter, J.A.M., 2001, Chronostratigraphy of prograding carbonate platform margins: a record of sea level changes and dynamic slope sedimentation, western Great Bahama Bank, in: *Subsurface Geology of a Prograding Carbonate Platform Margin, Great Bahama Bank: Results of the Bahamas Drilling Project* (Ginsburg, R.N., ed.) SEPM Spec. Publ., **70**:101–134.

Miller, K.G., Kominz, M.A., Browning, J.V., Wright, J.D., Mountain, G.S., Katz, M.E., Sugarman, P.J., Cramer, B.S., Christie-Blick, N., and Pekar, S.F., 2005, The Phanerozoic record of global sea-level change, *Science*, **310**:1293–1298.

Nagle, F., 1979, Geology of the Puerto Plata area, Dominican Republic, in: *Hispaniola: Tectonic Focal Point of the Caribbean - Three Geologic Studies in the Dominican Republic* (Lidz, B. and F. Nagle, eds.), Miami Geological Society, Miami, pp. 29–68.

Nehm, R.H. and Geary, D.H., 1994, A gradual morphologic transition during a rapid speciation event in marginellid gastropods (Neogene: Dominican Republic), *J. Paleontol.*, **68**:787–795.

Pindell, J.L., 1994, Evolution of the Gulf of Mexico and the Caribbean, in: *Caribbean Geology* (Donovan, S.K. and T.A. Jackson, eds.), The University of the West Indies Publishers' Association, Kingston, pp. 13–39.

Reineck, H.-E. and Singh, I.B., 1980, *Depositional Sedimentary Environments with Reference to Terrigenous Clastics*, Springer, New York, pp. 551.

Saunders, J.B., Jung, P., Geister, J., and Biju-Duval, B., 1982, The Neogene of the south flank of the Cibao Valley, Dominican Republic: a stratigraphic study, *Trans. 9th Carib. Geol. Conf.*, **2**:151–160.

Saunders, J.B., Jung, P., and Biju-Duval, B., 1986, Neogene paleontology in the northern Dominican Republic 1. Field surveys, lithology, environment, and age, *Bull. Am. Paleontol* **89**:1–79.

Shackleton, N.J., Backman, J., Zimmerman, H., Kent, D.V., Hall, M.A., Roberts, D.G., Schnitker, D., Baldauf, J.G., Desprairies, A., Homrighausen, R., Huddlestun, P., Keene, J.B., Kaltenback, A.J., Krumsiek, K., Morton, A.C., Murray, J.W., and Westberg-Smith, J., 1984. Oxygen isotope calibration of the onset of ice-rafting and history of glaciation in the North Atlantic region, *Nature*, **307**:620–623.

Shackleton, N.J. and Hall, M.A., 1997, The late Miocene stable isotope record, Site 926, *Proc. Ocean Drill. Prog., Sci. Results* **154**:367–373.

Spezzaferri, S., McKenzie, J.A., and Isern, A., 2002, Linking the oxygen isotope record of late Neogene eustasy to sequence stratigraphic patterns along the Bahamas margin: results from a paleoceanographic study of ODP Leg 166, Site 1006 sediments, *Mar. Geol.*, **185**:95–120.

van den Bold, W.A., 1975, Distribution of the Radimella confragosa group (Ostracoda, Hemicytherinae) in the Late Neogene of the Caribbean, *J. Paleontol.*, **49**:692–701.

van den Bold, W.A., 1988, Neogene paleontology in the northern Dominican Republic. 7. The subclass Ostracoda (Arthropoda: Crustacea), *Bull. Am. Paleontol.*, **94**:1–105.

Vaughan, T.W., Cooke, W., Condit, D.D., Ross, C.P., Woodring, W.P., and Calkins, F.C., 1921, *A Geological Reconnaissance of the Dominican Republic: Geological Survey of the Dominican Republic*, US Geological Survey, Washington, DC, pp. 268.

Vokes, E.H., 1979, The age of the Baitoa Formation, Dominican Republic, using mollusca for correlation, *Tulane Stud. Geol. Paleontol.*, **15**:105–116.

Vokes, E.H., 1989, Neogene paleontology in the northern Dominican Republic. 8. The family Muricidae (Mollusca: Gastropoda), *Bull. Am. Paleontol.*, **97**:5–94.

Zachos, J., Pagani, m., Sloan, L., Thomas, E., and Billups, K., 2001, Trends, rhythms, and aberrations in global climate 65 Ma to present, *Science*, **262**:686–693.

Chapter 3
Constraints on Late Miocene Shallow Marine Seasonality for the Central Caribbean Using Oxygen Isotope and Sr/Ca Ratios in a Fossil Coral

Rhawn F. Denniston[1], Stephanie C. Penn[1], and Ann F. Budd[2]

Contents

3.1 Introduction

Isolation of the Pacific and Caribbean basins by closure of the Central American Seaway (CAS) in the Miocene and Pliocene produced changes in the secular physical and chemical properties of Caribbean surface waters, one possible result of which was an increase in extinction and speciation of marine biota on both sides of the Isthmus of Panama (Jackson et al., 1996; Collins and Coates, 1999). Closure of the CAS was a gradual process spanning approximately 13–2 Ma, but Caribbean environmental conditions changed significantly once water depths reached < 100 m by 4.6 million years ago (Keigwin, 1978; Coates et al., 1992, 1996, 2003; Haug and Tiedemann, 1998; Lear et al., 2003; Gussone et al., 2004). Average Caribbean surface water temperatures increased as movement of cool Pacific waters was restricted through the CAS and Caribbean waters became restricted to their own basin (Romine, 1982). Water clarity and calcium carbonate saturation may have also increased (Vermeij and Petuch, 1986), and high Caribbean evaporation rates,

[1] Department of Geology, Cornell College, Mount Vernon, IA 52314. Email: rdenniston@cornellcollege.edu

[2] Department of Geoscience, University of Iowa, Iowa City, IA 52242, USA.
Email: ann-budd@uiowa.edu

R.H. Nehm, A.F. Budd (eds.) *Evolutionary Stasis and Change in the Dominican Republic Neogene*,

coupled with westward transport of moisture-laden air to the Pacific, increased salinity in the Caribbean (Romine, 1982).

Changes in the amount of rainfall also may have been affected by closure of the CAS. Gussone et al. (2004) attributed shifts in the $\delta^{18}O$, Mg/Ca, and $\delta^{44/40}Ca$ values of Caribbean planktonic foraminifera between 4.6–4.2 million years ago to shoaling in the CAS and possibly to concomitant shifts in the position of the Intertropical Convergence Zone (ITCZ). The latter would have changed the intensity of meteoric precipitation, and thus Caribbean sea surface salinities and sea surface $\delta^{18}O$ values. Latitudinal shifts in the ITCZ have also been proposed for 4.4 million years ago (Billups et al., 1999). Such changes would have implications for ocean salinity in the southern and central Caribbean at both decadal and seasonal scales given the influence of ITCZ position over rainfall in the Orinoco River basin (Hellweger and Gordon, 2002; Watanabe et al., 2002; Estevez et al., 2003).

We lack a clear understanding of how shallow Caribbean environments changed in response to closure of the CAS. This is particularly true for changes at the seasonal scale, a time frame that plays an important role in determining the composition of shallow marine communities (McClanahan et al., 2001; Swart et al., 2001; Ateweberhan et al., 2006). For example, Teranes et al. (1996) investigated temporal changes in the seasonal ranges of venerid bivalve $\delta^{18}O$ values from both sides of the Isthmus of Panama through final closure of the CAS. Their study suggests greater seasonal temperature variability in the Caribbean in the late Miocene relative to the modern. Isotopic analysis of a 3 million-year-old coral from Florida by Roulier and Quinn (1995) supports significantly decreased seasonal temperature swings in the middle Pliocene Caribbean. Temperature reconstructions in both studies, however, were limited by the coupled influence of water temperature and salinity on oxygen isotopic ratios.

Here we present a 21-year-long, high-resolution stable isotope and trace element record of a late Miocene (~5 million-year-old) coral from the Dominican Republic (Fig. 3.1) that records shallow marine paleoenvironmental conditions prior to final closure of the CAS. This coral serves as one snapshot for the late Miocene central Caribbean that, when integrated with other records, may allow a better understanding of the role played by environmental variables in forcing Neogene faunal turnover in the Caribbean.

3.2 Geological and Environmental Setting

An area with a rich fossil fauna that has been the focus of studies aimed at better understanding the timing, rate, and mode of marine faunal speciation associated with closure of the CAS is the Cibao Valley in the northern Dominican Republic (Fig. 3.1) (Cheetham, 1986, 1987; Nehm and Geary, 1994; Nehm, 2001; Johnson and Perez, 2006). During the late Miocene and early Pliocene, the Cibao Valley was part of a tectonically active graben that was generally subsiding with time (Saunders et al., 1986). One Miocene/Pliocene unit of interest in the Cibao Valley is the

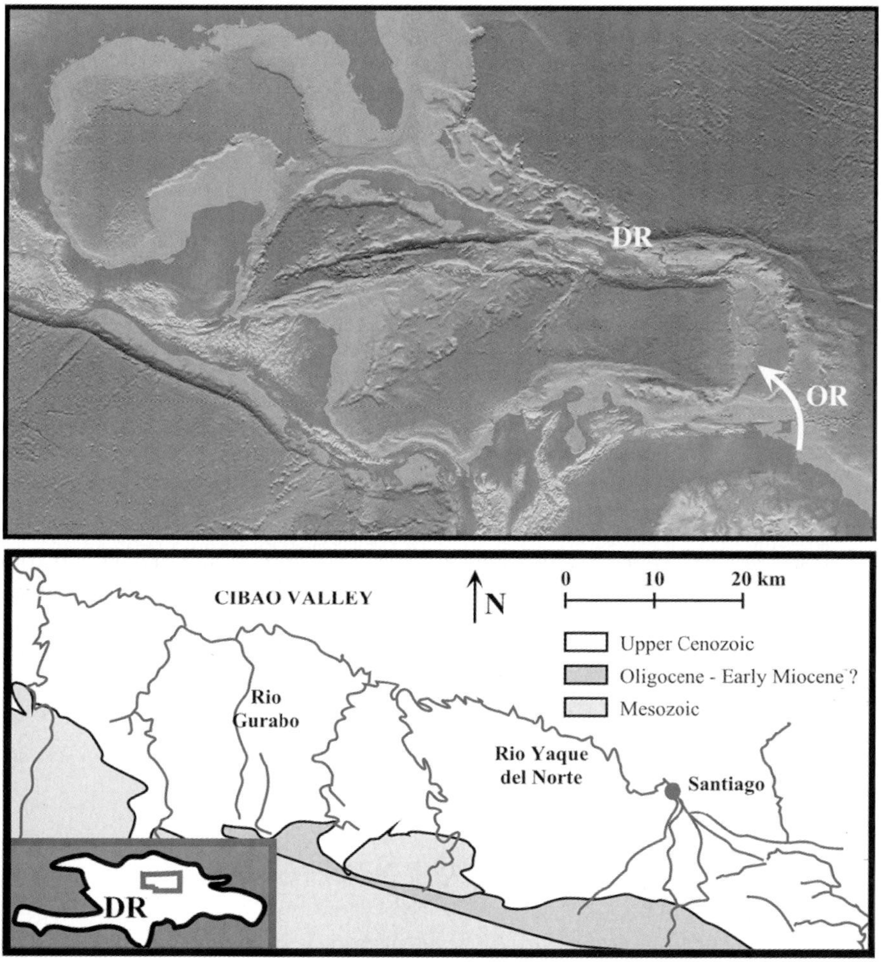

Fig. 3.1 Top: Map of the Circum-Caribbean region. DR = Dominican Republic; OR = Orinoco River discharge. Bottom: coarse geologic map of Cibao Valley

Gurabo Formation, a >400 m thick package of gently dipping, weakly indurated, coral-rich sediments. Gurabo Formation corals grew at depths of <30 m (Goreau and Wells, 1967; Graus and Macintyre, 1989) and were transported down slope as slump deposits and rapidly buried in densely-packed, fine-grained siliciclastics and clayey calcareous sediments (Evans, 1986; Saunders et al., 1986) (Fig. 3.2). Based on the excellent preservation of some of these corals, it appears that this low permeability matrix severely limited flow of marine and groundwater through the Gurabo Formation. Today, the Gurabo Formation is exposed across the Cibao Valley in a network of streams including the Rio Gurabo.

Fig. 3.2 Photograph of a coral (not the *G. hilli* discussed here) after excavation from the hillside at NMB location #15855. Clay-rich nature of the Gurabo Formation matrix at this interval is evident in the streaks left by the rock hammer

3.3 Sampling and Analytical Methods

3.3.1 Field Collection and Sample Screening Procedures

An intact coral head of *Goniopora hilli* was excavated from the Gurabo Formation on the bank of the Rio Gurabo at NMB locality 15855 (Saunders et al., 1986), located approximately 275 m in the section, about 115 m above the Cercado-Gurabo Formation contact and ~305 m below the contact with the overlying Mao Formation (Fig. 3.2). The coral was slabbed using a water-cooled trim saw and inspected macroscopically, in thin section, and with scanning electron microscopy (SEM) for signs of meteoric cements (Quinn and Taylor, 2006) and dissolution or recrystallization of the coral skeleton. SEM images obtained at the University of New Mexico Department of Earth and Planetary Sciences reveal extensive primary porosity and preservation of septal ornamentation (Fig. 3.3). A second round of SEM images was obtained at the University of Iowa Department of Geosciences on samples that had been leached by 3% acetic acid for 4 minutes in order to clarify coral microstructure. Visible in these images are fibrous aragonite crystals of the coral skeleton and clusters of aragonite crystals radiating from delicate calcification centers, structures that are readily altered by meteoric diagenesis and demonstrate the excellent preservation of this sample (Reuter et al., 2005) (Fig. 3.3).

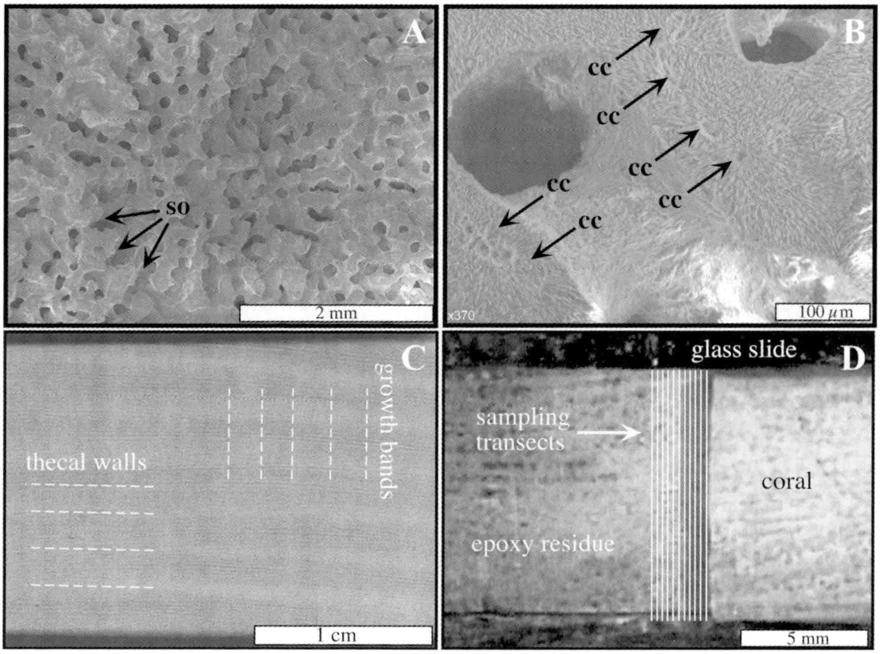

Fig. 3.3 (A) SEM image of unetched *G. hilli* with view perpendicular to maximum growth direction. Note high primary porosity and septal ornaments (so). (B) SEM image of etched *G. hilli* showing fibrous aragonite crystalline structure and preservation of calcification centers (cc). (C) X-radiograph of *G. hilli* exhibiting weakly defined thecal walls and annual growth bands. (D) Downward-looking view of corallite during microsampling at 39,700 µm. White lines denote positions of previous micromilling transects

Next, powdered subsamples were analyzed by X-ray diffraction (XRD) at Cornell College using a Scintag X-ray powder diffractometer with a DMS2000 diffraction management system and a copper target with a graphite monochromator. The scans were run at 40 kV and 30 mA from 25–35° 2Θ at a scan rate of 6°/minute, a technique capable of identifying calcite at abundances >1%. No calcite peak was detected in any of the scans from this *G. hilli* sample suggesting a coral composed of 99 + % aragonite. And finally, X-radiographs of 7 mm-thick coral slabs were taken at the University of Iowa College of Dentistry using an accelerating voltage of 60 kV and 25 mA in order to define growth banding. Although faint, both corallite walls and growth bands are visible in X-ray images (Fig. 3.3).

A slab of *G. hilli* was cut from its coral head parallel to the principal growth direction of the corallites using a water-cooled saw, and the slab was then ultrasonicated in distilled water to remove detritus, oven dried at 30°C, and vacuum-impregnated with UV-activated epoxy. No chemical pretreatment technique such as H_2O_2 or NaOH was used to remove organic contaminants (see below). As corallites in *G. hilli* grow curvilinearly, the slab was visually inspected and the individual corallite that remained consistently parallel to the surface of the slab was isolated using a thin-sectioning saw. The resulting ~5 cm-long section of corallite was then epoxied to a glass slide and polished to thickness of ~1 mm (Fig. 3.3).

3.3.2 *Stable Isotope Ratios*

Stable isotope analyses were performed at the University of Michigan Department of Geological Sciences using either a Finnigan MAT Kiel I preparation device coupled directly to the inlet of a Finnigan MAT 251 triple collector isotope ratio mass spectrometer, or in a Finnigan MAT Kiel IV preparation device coupled to the inlet of a MAT 253 mass spectrometer. Precision and accuracy of data were monitored through daily analysis of a variety of powdered carbonate standards. At least six standards were reacted and analyzed daily, bracketing the sample suite at the beginning, middle, and end of the day's run. Measured precision was maintained at better than 0.1‰ (1σ) for both carbon and oxygen isotope compositions with isotopic ratios reported relative to the Vienna Pee Dee Belemnite (VPDB) standard. Sample powders for stable isotope analysis were micromilled in parallel traverses across the entire corallite using a Merchantek micromill, with the first 120 samples incorporating 75 µm of growth and the subsequent 250 samples incorporating 150 µm of growth (Fig. 3.3). Powder was collected at the end of each pass and transferred to stainless steel vials. Given the age and burial conditions of Gurabo Formation corals, we deemed it necessary to ensure that organic contamination was minimized prior to isotopic analysis. Chemical pretreatment such as with H_2O_2 or NaOH may yield unpredictable shifts in coral aragonite $\delta^{18}O$ values (Grottoli et al., 2005). Roasting has also been tied to isotopic shifts in aragonite (Gaffey et al., 1991), as well. Experimental evidence suggests, however, that roasting *in vacuo* at 200°C for 1 hour does not cause the transformations of aragonite to calcite that lead to significant isotopic fractionation (Dauphin et al., 2006) but is effective in driving off volatile organic compounds. In order to assess the impact of roasting on our samples, the first 15 samples milled from the coral (distances 75–1,125 µm) were split, with one half undergoing roasting prior to isotopic analysis and the other half not. The results, displayed in Fig. 3.4, suggest a systematic shift in carbon isotopic values with all 15 carbon isotopic ratios becoming an average of 0.20‰ ± 0.15‰ higher after roasting. In contrast, only 10 of the oxygen isotopic ratios increased after roasting, while 5 decreased, with an average offset of +0.05‰ ± 0.10‰, less than the analytical uncertainty. As the shape and range in both the roasted and unroasted carbon and oxygen isotopic ratios remain largely consistent, and as seasonal ranges were the primary focus of this study, the remainder of the samples were roasted prior to analysis.

An additional obstacle to high-resolution geochemical analysis of corals is determining the sampling density (number of samples per year of growth) necessary to adequately capture the full seasonal temperature range. Previous studies of coral $\delta^{18}O$ seasonality have suggested minimum sampling densities of 50 (Leder et al., 1996), 40 (Watanabe et al., 2002), 14 (Leder et al., 1991), 8 (Klein et al., 1992; Ren

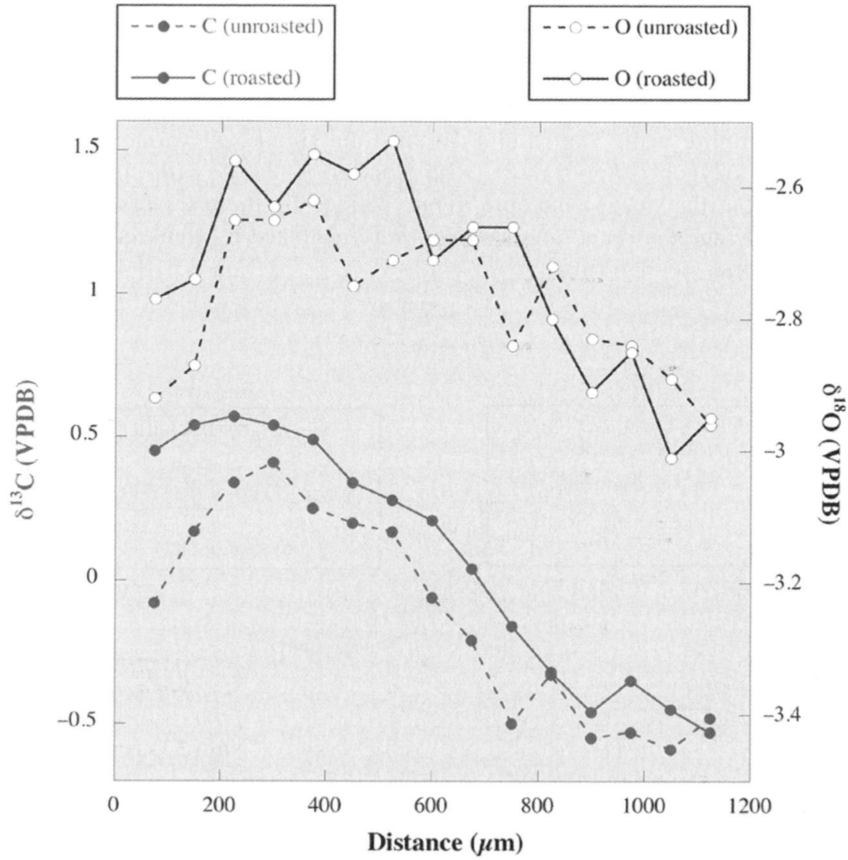

Fig. 3.4 Comparison of carbon and oxygen isotopic ratios in roasted and unroasted coral samples. Note that while $\delta^{13}C$ values are consistently elevated after roasting and the $\delta^{18}O$ values exhibit a more non-uniform response, the overall trends remain parallel

et al., 2002), and 6 samples per year of growth (Quinn et al., 1996). Measurement of growth banding and distances between peaks in *G. hilli* isotope profiles yields average annual growth rates of ~2mm/year. Thus, the 150 μm-deep traverses used here equates to a sampling density of >12 samples per year. Mathematically manipulating these *G. hilli* data demonstrated no signal attenuation at this sampling density relative to the 75 μm-deep traverses (24 samples/year) used for the first 120 samples.

Variations in the stable isotopic composition across the coral skeleton have been identified which can also lead to an artificially reduced seasonal signal. Leder et al. (1996) found reduced seasonal signals in $\delta^{18}O$ values across endothecal (dissepiments and columella) portions relative to thecal samples that they attributed to calcification of skeletal structures at different times throughout the year, and to time-averaging effects. Watanabe et al. (2002) found significant, but not consistent, differences between oxygen isotopic samples isolated solely from thecal walls as compared to the entire corallite. In order to test the importance of sampling position in *G. hilli*, we compared (1) traverses milled across the entire corallite at depths of 200 µm and full depth (~1,000 µm) across the entire corallite and (2) points drilled to full depth (~1,000 µm) but restricted to the (poorly-defined) corallite wall. The results demonstrate significant variability in both carbon and oxygen isotopic ratios with position in the *G. hilli* coral skeleton (Fig. 3.5), but the oxygen isotopic *trends* defined by these traverses remained similar, and thus the remaining samples were analyzed by transects across the entire corallite.

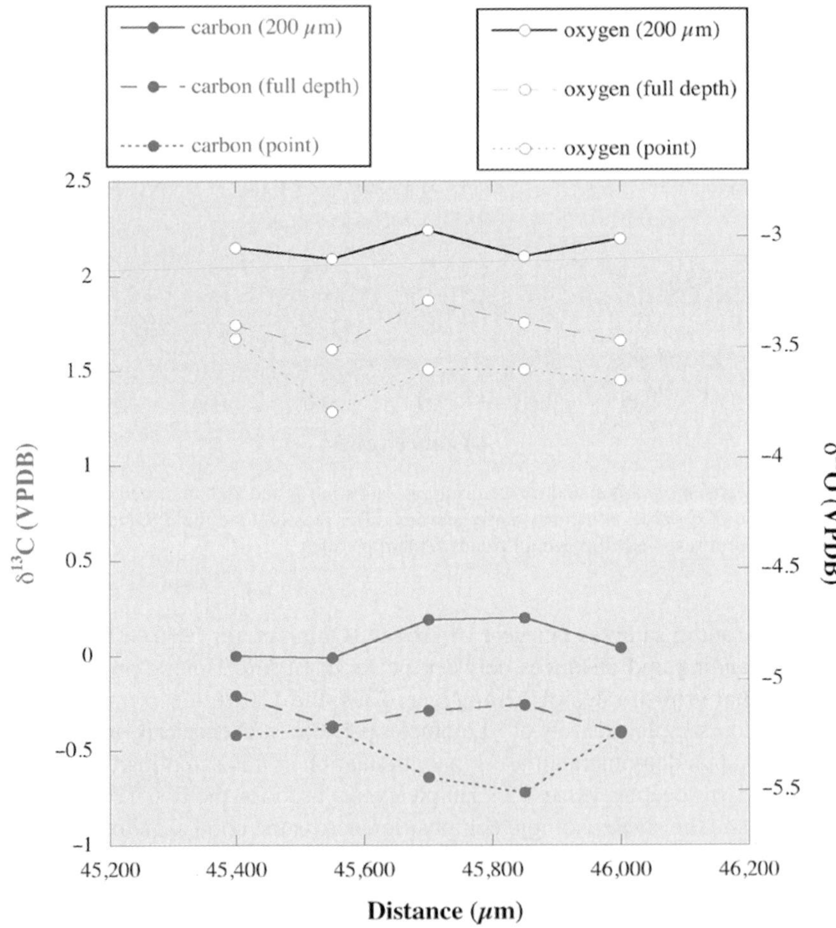

Fig. 3.5 Comparison of carbon and oxygen isotopic ratios with position in the corallite. Note the similarity in oxygen isotopic trends, despite the >0.5‰ offset between samples of different depth

3.3.3 *Strontium/Calcium Ratios*

Splits from alternate stable isotope samples were analyzed at the Keck Elemental Geochemistry Laboratory in the Department of Geological Sciences University of Michigan using a Finnigan MAT Element inductively coupled plasma-high resolution mass spectrometer (ICP-MS) and the method of Rosenthal et al. (1999); analytical precision averages 7‰ (1σ). Although the first three growth years were analyzed, problems with instrument calibration resulted in the offset of measured Sr/Ca ratios from those of adjacent samples, and thus we chose not to include them in this data set. In addition, insufficient powder was available for Sr/Ca analyses of the last ~2 growth years, resulting in a Sr/Ca record that is truncated relative to the stable isotope record.

3.4 Constraining Shallow Marine Conditions

When plotted versus distance along the corallite growth axis, carbon and oxygen isotopic ratios define clear and quasi-regular sinusoids (Fig. 3.6). The δ^{13}C and δ^{18}O values of coral aragonite are easily altered by replacement of the skeleton by

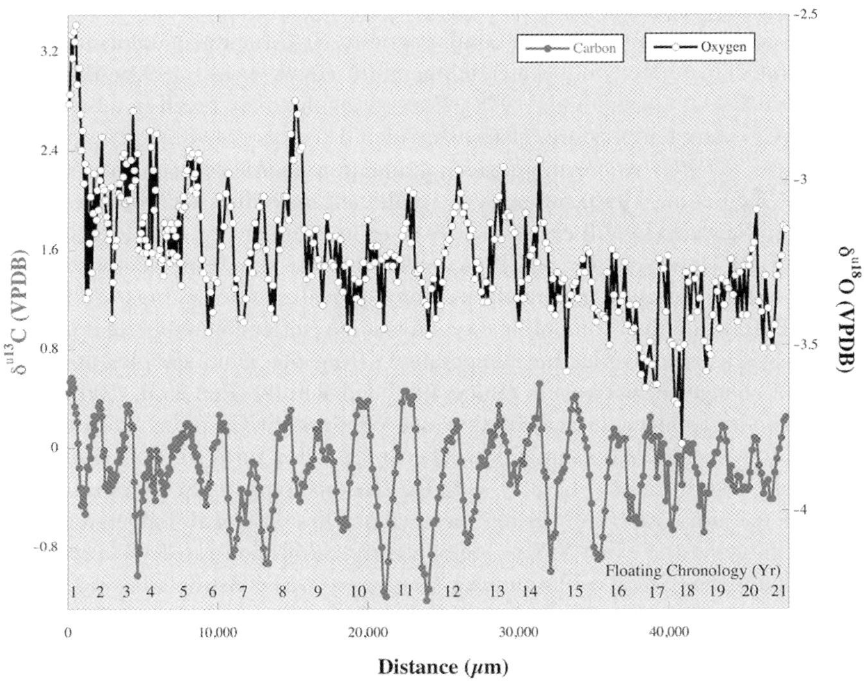

Fig. 3.6 Carbon and oxygen isotopic profiles from *G. hilli*. Floating chronology refers to individual growth years as defined by sinusoids in stable isotopic ratios

secondary calcite or infilling of pore space by marine or meteoric cements. Oxygen isotopic ratios are more likely to be diverted from their original values by diagenetic alteration than carbon isotopic ratios (Key et al., 2005), and this may explain the more regular sinusoids defined by the latter. But carbon isotopic compositions can also be shifted significantly by interaction with ^{12}C-enriched groundwaters reflecting a terrestrial vegetation fingerprint (Hurley and Lohmann, 1989). The well-behaved sinusoids in $\delta^{13}C$ values, coupled with limited visual evidence for diagenetic alteration, suggest that this sample of *G. hilli* preserves its primary isotopic signals, if not pristinely, then at least with high fidelity.

Oxygen isotopic ratios of coral aragonite reflect several parameters, but the primary controls are the temperature and oxygen isotopic composition of the ambient seawater (Emiliani et al., 1978; Fairbanks and Dodge, 1979; McConnaughey, 1989). Corrége (2006) calculated an average relationship of 0.18‰–0.22‰/°C from published oxygen isotopic studies of corals. Seasonal ranges in $\delta^{18}O$ values of the Gurabo *G. hilli* average 0.4‰ ± 0.1‰ (1σ) (Fig. 3.6), a value that if attributed solely to water temperature, corresponds to an average seasonal range of ~2°C, nearly identical to modern values from Haiti (Slutz et al., 1985). However, seasonal or long-term changes in ocean water $\delta^{18}O$ values and salinity, due to evaporation, meteoric precipitation (direct or via riverine discharge), or upwelling, can mask the temperature signals in coral $\delta^{18}O$ values (Swart et al., 2001; Watanabe et al., 2002).

While both the temperature and the oxygen isotopic composition of ambient seawater control the $\delta^{18}O$ value of coral aragonite, Sr/Ca ratios in corals have been demonstrated to reflect only water temperature (Beck et al., 1992; Alibert and McCulloch, 1997; Gagan et al., 1998). Exceptions that may result in a breakdown of the Sr/Ca-water temperature relationship include corals growing in exceptionally cool waters (<18°C) where inorganic fractionation and/or coral metabolisms are reduced (Shen et al., 1996), in areas of significant upwelling with deep waters of unusual Sr/Ca ratio (DeVilliers et al., 1994), or in symbiont-bearing forms (Cohen et al., 2002). However, the Sr/Ca paleothermometer has been demonstrated to record water temperature accurately in many typical marine settings (Schrag and Linsley, 2002), and thus combining oxygen isotope (reflecting temperature + salinity) and Sr/Ca ratios (reflecting temperature) from the same samples allows the parsing of changes in both water temperature and salinity (Ren et al., 2002).

In order to perform this calculation, we obtained Sr/Ca ratios from splits of alternate (odd-numbered) sample powders. Sr/Ca ratios form sinusoids coincident with stable isotope ratios (Fig. 3.7), with the seasonal range in Sr/Ca ratios averaging 0.3 ± 0.1 mmol/mol. Converting these values to a seasonal temperature range requires knowing the exact Sr/Ca – temperature calibration for *G. hilli*. However the only published Sr/Ca calibration for *Goniopora* was constructed using samples from the South China Sea (Yu et al., 2004) (*Goniopora* are extinct in the Caribbean). The relationship defined by Yu et al. (2004) is:

$$\mathrm{SST}(^\circ\mathrm{C}) = -32.8\ (\pm 3.5) \times \mathrm{Sr/Ca}_{(\mathrm{mmol/mol})} + 315(\pm 31)\ (r=-0.9999;\ n=48) \quad (1)$$

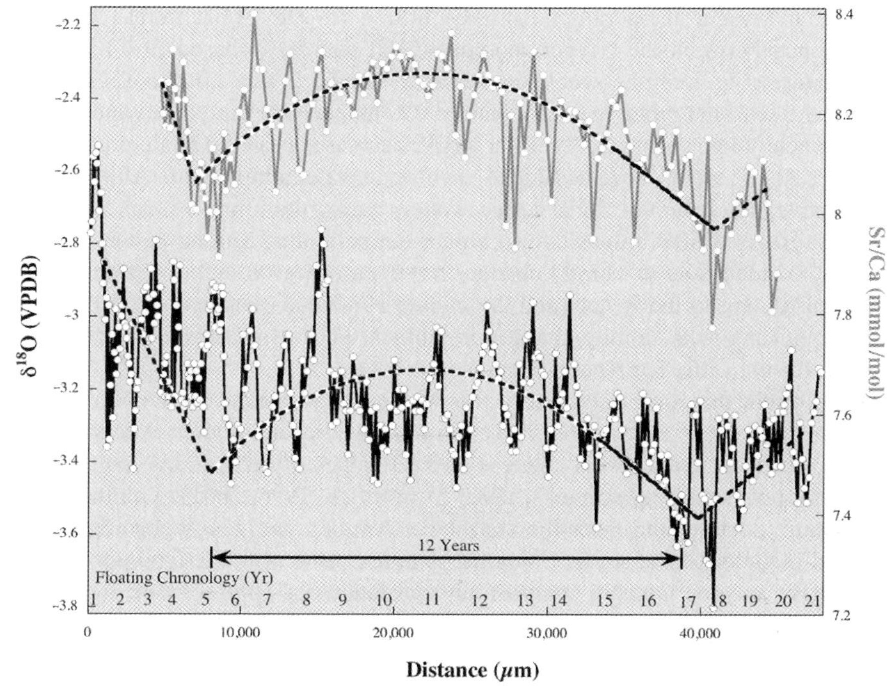

Fig. 3.7 Comparison of oxygen isotopic and Sr/Ca ratios. Sr/Ca analyses were conducted on splits of alternate powders, thus the relatively high sample density for oxygen isotopic analyses. Sr/Ca ratios for growth years 1–2 and 20–21 are not reported owing to analytical errors and insufficient sample powder, respectively. Note the suggested ~12-year "cycle" in both records (dashed line)

Using this relationship, the 0.3 mmol/mol average seasonal range in Sr/Ca ratios in the *G. hilli* coral translates to a ~9°C range in water temperature and an average annual water temperature of 114°C. But Sr/Ca – temperature calibrations have been shown to vary significantly based on factors including species. The Sr/Ca – temperature relationships for *Porites* and various other genera (*Pocillopora, Montipora, Pavona, Diploria, Montastraea, Diploastrea,* and *Goniopora*) range from –0.03 to –0.09 mmol/mol/°C and –0.03 to –0.08 mmol/mol/°C, respectively. Applying the maximum Sr/Ca – temperature relationship from *Porites* (–0.09 mmol/mol/°C) to *G. hilli*, on the other hand, corresponds to only ~3°C seasonal temperature range. Complicating matters further is the fact that marine Sr/Ca ratios have changed through the Cenozoic, which would affect Sr/Ca – temperature calibrations (Tripati et al., 2001; Steuber, 2002). Thus, applying Recent *Goniopora* Sr/Ca – temperature calibrations to a Miocene *G. hilli* introduces significant uncertainty, and temperature and salinity reconstructions must be viewed with caution.

However, keeping in mind these caveats, and assuming that the slope of the Sr/Ca – temperature calibration is correct, we can attempt to reconstruct late Miocene salinity and water temperature from G. hilli $\delta^{18}O$ and Sr/Ca ratios. The 7°C discrepancy between the oxygen isotope- (2°C) and Sr/Ca-based (9°C) seasonal temperature ranges can be accounted for by a ~1.4‰ (7°C × −0.2‰/°C) suppression of the seasonal range in ocean water $\delta^{18}O$ values. For example, a winter influx of isotopically light, low salinity water would decrease ocean $\delta^{18}O$ values and offset the increase in coral $\delta^{18}O$ values due to the drop in water temperature. Alternatively, evaporative enrichment of ^{18}O in surface waters during the summer would offset the decrease in coral $\delta^{18}O$ values due to higher temperatures. Muting of the range in coral $\delta^{18}O$ values due to salinity changes has been observed in the western Pacific for corals dating to the Recent and the middle Holocene (Gagan et al., 1998), and significant long-term salinity changes in ambient ocean surface water were documented for the Little Ice Age by Hendy et al. (2002).

How might this have occurred in the Miocene Caribbean? Studies linking the covariance of ocean water $\delta^{18}O$ values and salinity in the modern Atlantic along eastern North America reveal changes in $\delta^{18}O$ values of 0.2‰–0.6‰ per part per thousand (ppt) salinity (Fairbanks, 1982; Swart et al., 1996, 2001), with the lowermost value documented from the Caribbean Antilles, the geographically closest location included in the study. Using the Antilles value of 0.2‰ $\delta^{18}O$/ppt salinity, this shift in oxygen isotopic composition translates to a 7 ppt decrease in the seasonal range of ocean water salinity. Discharge from the Orinoco and Amazon rivers routinely forms plumes of low salinity waters (>2 ppt below the regional average) that travel more than 2,000 km from their sources and toward the central and eastern Caribbean (Hu et al., 2004), but discharge from these rivers would have to be increased significantly to sufficiently suppress ocean salinities in the central Caribbean. Alternatively, upwelling of isotopically distinct bottom waters could have played a role in shifting surface ocean $\delta^{18}O$ values, but the amount and seasonality of late Miocene upwelling and surface flow in the region currently occupied by the Dominican Republic is poorly constrained (Collins, 1996). Barium, with its distinct depth profile in most ocean basins, is a commonly used proxy for upwelling, and future coupling of Ba, Sr/Ca and $\delta^{18}O$ analyses might help to constrain the magnitude and seasonality of upwelling in the central Caribbean during the late Miocene.

Longer-term trends in temperature/salinity are also suggested by multi-annual $\delta^{18}O$ and Sr/Ca trends. The *G. hilli* record is insufficient to allow a statistical analysis, but the minima in both oxygen isotope and Sr/Ca ratios suggest a ~12-year cycle (Fig. 3.7). Greer and Swart (2006) report 18–20-year cycles in modern precipitation near the Dominican Republic (Haiti) and in cycles of similar scale preserved by oxygen isotopic anomalies in middle Holocene corals from the Dominican Republic. Both cycles are tied to shifts in the position of the ITCZ (Greer and Swart, 2006). In addition, these authors also suggest a 12–13-year temperature cycle based on Haitian temperature data, but the punctuated nature of their record precludes statistical analysis. The construction of a considerably longer *G. hilli* isotopic record, currently underway, will provide a clearer understanding of the nature of decadal-scale climate cycles.

3.5 Conclusions

Stable isotopic ratios from a well-preserved *Goniopora hilli* suggest that the seasonal range in shallow marine temperature during one 21-year span in the late Miocene was approximately 2°C. Considerably larger seasonal temperature ranges suggested by Sr/Ca ratios appear to reflect a misfit between a Miocene *G. hilli* and Recent *Goniopora sp.* The well-defined sinusoids in both Sr/Ca and carbon and oxygen isotopic ratios argue against diagenetic alteration having significantly overprinted the original isotopic signals, and thus the remaining and as yet unanalyzed portions of this coral may help to more clearly define late Miocene shallow marine seasonality in the central Caribbean.

Acknowledgments Stable isotope and Sr/Ca analyses performed at the University of Michigan Department of Geosciences by Lora Wingate and Ted Huston, respectively. X-radiography conducted at the University of Iowa School of Dentistry by Rosemary Stanley under the direction of Axel Ruprecht. Scanning electron microscopy performed by Mike Spilde at the University of New Mexico and by Troy Fadiga at the University of Iowa. Fieldwork completed with the assistance of James Klaus, Don McNeill, and Ross Nehm. Initial sample preparation conducted by Alexa Clements under the direction of Scott Carpenter in the Department of Geosciences at the University of Iowa. Acknowledgment is made to Cornell College and the Donors of The Petroleum Research Fund, administered by the American Chemical Society, for support of this research. We thank James Klaus for his constructive and helpful review of this manuscript.

References

Alibert, C. and McCulloch, M.T., 1997, Strontium/calcium ratios in modern Porites corals from the Great Barrier Reef as a proxy for sea surface temperature: calibration of the thermometer and monitoring of ENSO, *Paleoceanography*, **12**:345–363.

Ateweberhan, M., Bruggemann, J.H., and Breeman, A.M., 2006, Effects of extreme seasonality on community structure and functional group dynamics of coral reef algae in the southern Red Sea, *Coral Reefs*, **25**:391–406.

Beck, J.W., Edwards, R.L., Ito, E., Recy, F.W., Rougerie, F., Joannot, P., and Henin, C., 1992, Sea surface temperature from coral skeletal strontium/calcium ratios, *Science*, **257**:644–647.

Billups, K., Ravelo, A.C., Zachos, J.C., and Norris, R.D., 1999, Link between oceanic heat transport, thermohaline circulation, and the Intertropical Convergence Zone in the early Pliocene Atlantic, *Geology*, **27**:319–322.

Cheetham, A.H., 1986, Tempo of evolution in a Neogene Bryzoan: rates of morphologic change within and across species boundaries, *Paleobiology*, **12**:190–202.

Cheetham, A.H., 1987, Tempo of evolution in a Neogene Bryzoan: are trends in single morphologic characters misleading? *Paleobiology*, **13**:286–296.

Coates, A.G., Aubry, M.P., Berggren, W.A., Collins, L.S., and Kunk, M.J., 2003, Early Neogene history of the Central American arc from Bocas del Toro, western Panama, *Geol. Soc. Am. Bull.*, **115**:271–287.

Coates, A.G., Jackson, J.B.C., Collins, L.S., Cronin, T.M., Dowsett, H.J., Bybell, L.M., Jung, P., and Obando, J.A., 1992, Closure of the Isthmus of Panamana: the near-shore marine record of Costa Rica and western Panama, *Geol. Soc. Am. Bull.*, **104**:814–828.

Cohen, A.L., Owens, K.E., Layne, G.D., and Shimizu, N., 2002, The effect of algal symbionts on the accuracy of Sr/Ca paleotemperatures from coral, *Science*, **296**:1247–1249.

Collins, L.S., 1996, The Geological Evolution of the Central American Isthmus, in: *Evolution and Environment in Tropical America* (J.B.C. Jackson, A.F. Budd, and A.G. Coates, eds.), University of Chicago Press, Chicago, IL, pp. 130–167.

Collins, L.S. and Coates, A.G., 1999, A Paleobiotic Survey of Caribbean Faunas from Neogene of Isthmus of Panama, *Bull. Am. Paleontol.*, **356–358**:351.

Corrége, T., 2006, Sea surface temperature and salinity reconstruction from coral geochemical tracers, *Palaeogeogr. Palaeoclim. Palaeoecol.*, **232**:408–428.

Dauphin, Y., Cuif, J.P., and Massard, P., 2006, Persistent organic components in heated biogenic aragonites – implications for palaeoenvironmental reconstructions, *Geoph. Res. Abst.*, **8**:09364.

DeVilliers, S., Shen, G.T., and Nelson, B.K., 1994, The Sr/Ca-temperature relationship in coralline aragonite: influence of variability in $(Sr/Ca)_{seawater}$ and skeletal growth parameters, *Geochim. Cosmo. Acta*, **58**:197–208.

Emiliani, C., Hudson, J.H., Lidz, B., Shin, E.A., and George, R.Y., 1978, Oxygen and carbon isotopic growth record in a reef coral from Florida Keys and a deep-sea coral from Blake Plateau, *Science*, **202**:627–629.

Estevez, J.A., Vonhof, H.B., van Hinte, J.E., Troelstra, S.R., and Kroon, D., 2003, Late Cenozoic seasonal circulation patterns in the Caribbean Sea. *Abst. EGS-AGU-EUG Joint Assembly*, Nice, France, abstract 5526.

Evans, C.C., 1986, A field guide to the mixed reefs and the siliciclastic of the Neogene Yaque Group, Cibao Valley, Dominican Republic. Coral Gables, Fla.: comparative Sedimentology Laboratory, University of Miami.

Fairbanks, R.G., 1982, The origin of continental shelf and slope water in the New York Bight and Gulf of Maine: evidence from $H_2{}^{18}O/H_2{}^{16}O$ ratio measurements, *J. Geophys. Res.*, **102**:929–945.

Fairbanks, R.G. and Dodge, R.E., 1979, Annual periodicity of the 18O/16O and 13C/12C ratios in the coral Montastrea annularis, *Geochim. Cosmo. Acta*, **43**:1009–1020.

Gaffey, S.J., Kolak, J.J., and Bronnimann, C.E., 1991, Effects of drying, heating, annealing, and roasting on carbonate skeletal material, with geochemical and diagenetic implications, *Geochim. Cosmo. Acta*, **55**:1627–1640.

Gagan, M.K., Ayliffe, L.K., Hopley, D., Cali, J.A., Mortimer, G.E., Chappell, J., McCulloch, M.T., and Jead, M.J., 1998, Temperature and surface ocean water balance of the mid-Holocene tropical western Pacific, *Science*, **279**:1014–1018.

Goreau, T.F. and Wells, J.W., 1967, The shallow-water Scleractinia of Jamaica: revised list of species and their vertical distribution range, *Bull. Mar. Sci.*, **17**:442–453.

Graus, R.R. and Macintyre, I.G., 1989, The zonation pattern of Caribbean coral reefs as controlled by wave and light energy input, bathymetric setting and reef morphology: computer simulation experiments, *Coral Reefs*, **8**:9–18.

Greer, L. and Swart, P.K., 2006, Decadal cyclicity of regional mid-Holocene precipitation: evidence from Dominican coral proxies: *Paleoceanography*, **21**:PA2020, doi:10.1029/2005 PA001166.

Grottoli, A.G., Rodrigues, L.J., Matthews, K.A., Palardy, J.E., and Gibb, O.T., 2005, Pretreatment effects on coral skeletal ^{13}C and ^{18}O. *Chemical Geology*, **221**:225–242.

Gussone, N., Eisenhauer, A., Tiedemann, R., Haug, G.H., Heuser, A., Bock, B., Nagler, Th. F., and Muller, A., 2004, Reconstruction of Caribbean Sea surface temperature and salinity fluctuations in response to the Pliocene closure of the Central American Gateway and radiative forcing, using $\delta^{44/40}Ca$, $\delta^{18}O$, and Mg/Ca ratios, *Earth Planet. Sci. Lett.*, **227**: 201–214.

Haug, G.H. and Tiedemann R., 1998, Effect of th formation of the Isthmus of Panama on Atlantic Ocean thermohaline circulation. *Nature*, **393**:673–676.

Hendy, E.J., Gagan, M.K., Alibert, C.A., McCulloch, M.T., Lough, J.M., and Isdale, P.J., 2002, Abrupt decrease in tropical Pacific sea surface salinity at the end of the Little Ice Age, *Science*, **295**:1511–1514.

Hellweger, F.L. and Gordon, A.L., 2002, Tracing Amazon River water into the Caribbean Sea, *J. Mar. Res.*, **60**:537–549.

Hu, A., Meehl, G.A., and Han, W., 2004, Detecting thermohaline circulation changes from ocean properties in a coupled model. *Geophysical Research Letters*, **31**:L13204, doi:10.1029/2004 GLO20218.

Hurley, N.F. and Lohmann, K.C., 1989, Diagenesis of Devonian reefal carbonates in the Oscar Range, Canning Basin, Western Australia, *J. Sed. Petr.*, **59**:127–146.

Jackson, J.B.C., Budd, A.F., and Coates, A.G. (1996) *Evolution and Environment in Tropical America*. University of Chicago Press, Chicago, IL, 425 pp.

Johnson, K.G. and Perez, M.E., 2006, Skeletal extension rates of Cenozoic Caribbean reef corals, *Palaios*, **21**:262–271.

Keigwin, L.D., 1978, Pliocene closing of the Isthmus of Panama, based on biostratigraphic evidence from nearby Pacific Ocean and Caribbean Sea cores, *Geology*, **6**:630–634.

Key, M.M., Jackson, P.N.W., Patterson, W.P., and Moore, M.D., 2005, Stable isotope evidence for diagenesis of the Ordovician Courtown and Tramore limestones, south-eastern Ireland, *Irish J. Earth Sci.*, **23**:25–38.

Klein, R., Putzold, J., Wefer, G., and Loya, Y., 1992, Seasonal variations in the stable isotopic composition and the skeletal density pattern of the coral *Porites lobata* (Gulf of Eilat, Red Sea), *Mar. Biol.*, **112**:259–263.

Lear, C.H., Rosenthal, Y., and Wright, J.D., 2003, The closing of a seaway: ocean water masses and global climate change, *Earth Planet. Sci. Lett.*, **210**:425–436.

Leder, J.J., Swart, P.K., Szmant, A., and Dodge, R.E., 1996, The origin of variations in the isotopic record of scleractinian corals: I. Oxygen, *Geochim. Cosmochim. Acta*, **60**:2857–2870.

Leder, J.L., Szmant, A.M., and Swart, P.K., 1991, The effect of prolonged "bleaching" on skeletal banding and stable isotopic composition in *Montastrea Annularis*, *Coral Reefs*, **10**:19–27.

McClanahan, T.R., Muthiga, N.A., and Mangi, S., 2001, Coral and algal response to the 1998 bleaching and mortality: interaction with management and herbivores on Kenyan reefs, *Coral Reefs*, **19**:380–391.

McConnaughey, T.A., 1989, C and O isotopic disequilibrium in biological carbonates: I. Patterns, *Geochim. Cosmochim. Acta*, **53**:151–162.

Nehm, R.H. and Geary, D., 1994, A gradual morphologic transition during a rapid speciation event in marginellid gastropods (Neogene; Dominican Republic), *J. Paleontol.*, **68**:787–795.

Nehm, R.H., 2001, Calibrating spatial and temporal species richness patterns in tropical American Marginellid gastropods, *J. Paleontol.*, **75**:680–696.

Quinn, T.M. and Taylor, F.W., 2006, SST artifacts in coral proxy records produced by early marine diagenesis in a modern coral from Rabaul, Papua New Guinea, *Geophys. Res. Lett.*, **33**: L04601, doi:10.1029/2005GL024972.

Quinn, T.M., Taylor, F.W., Crowley, T.J., and Link, S.M., 1996, Evaluation of sampling resolution in coral stable isotope records: a case study using records from New Caledonia and Tarawa, *Paleoceanography*, **12**:529–542.

Ren, L., Linsley, B.K., Wellington, G.M., Schrag, D.P., and Hoegh-Guldberg, O., 2002, Deconvolving the $\delta^{18}O$ seawater component from subseasonal coral $\delta^{18}O$ and Sr/Ca at Rarotonga in the southwestern subtropical Pacific for the period 1726–1997, *Geochim. Cosmochim. Acta*, **67**:1609–1621.

Reuter, M., Brachert, T.C., and Kroeger, K.F., 2005, Diagenesis of growth bands in fossil scleractinian corals: identification and modes of preservation, *Facies*, **51**:146–159.

Romine, K., 1982, Late Quaternary history of atmospheric and oceanic circulation in the eastern equatorial Pacific, *Mar. Micropaleo.*, **7**:1163–1187.

Rosenthal, Y., Field, M.P., and Sherrell, R.M., 1999, Precise determination of element/calcium ratios in calcareous samples using sector field inductively coupled plasma mass spectrometry, *Anal. Chem.*, **71**:3248–3253.

Roulier, L.M. and Quinn, T.M., 1995, Seasonal- to decadal-scale climatic variability in Southwest Florida during the middle Pliocene; inferences from a coralline stable isotope record, *Paleoceanography*, **10**:429–443.

Saunders, J.B., Jung, P., and Biju-Duval, B., 1986, Neogene paleontology in the Northern Dominican Republic. Part 1, Field surveys, lithology, environment, and age, *Bull. Am. Paleontol.*, **89**:1–79.

Schrag, D.P. and Linsley, B.K., 2002, Corals, chemistry, and climate, *Science*, **296**:277–278.

Shen, C.C., Lee, T., Chen, C.Y., Wang, C.H., Dai, C.F., and Li, L.A., 1996, The calibration of D[Sr/Ca] versus sea surface temperature relationship for Porites corals, *Geochim. Cosmochim. Acta*, **60**:3849–3858.

Slutz, R.J., Lubker, S.J., Hiscox, S.D., Woodruff, R.L., Jenne, R.L., Joseph, D.H., Steuer, P.M., and Elms, J.D., 1985, Comprehensive ocean-atmosphere data set: Release 1, *Rep. NTIS PB86–105723*, NOAA Environ. Res. Lab., Boulder, CO, 268 pp.

Steuber, T., 2002, Plate tectonic control on the evolution of Cretaceous platform-carbonate production, *Geology*, **30**:259–262.

Swart, P.K., Dodge, R.E., and Hudson, H.J., 1996, A 240-year stable oxygen and carbon isotopic record in a coral from South Florida: implications for the prediction of precipitation in southern Florida, *Palaios*, **11**:362–375.

Swart, P.K., Price, R.M., and Greer, L., 2001, The relationship between stable isotopic variations (O, H, and C) and salinity in waters and corals from environments in South Florida: implications for reading the paleoenvironmental record, *Bull. Am. Paleontol.*, **361**:17–30.

Teranes, J.L, Geary, D.H., and Bemis, B.E., 1996, The oxygen isotopic record of seasonality in Neogene bivalves from the Central American Isthmus, in: *Evolution and Environment in Tropical America* (J.B.C. Jackson, A.F. Budd, and A.G. Coates, eds.), University of Chicago Press, Chicago, IL, pp. 168–204.

Tripati, A.K., Zachos, J.C., Allmon, W.D., and Schellenberg,, S.A., 2001, Evolution of seawater Sr: Ca ratios and seasonality during the Paleogene, *PaleoBios*, **21**:126–127.

Vermeij, G.J. and Petuch, E.J., 1986, Differential extinction in tropical American mollusks: endism, architecture, and the Panama land bridge, *Malacologia*, **27**:29–41.

Watanabe, T., Winter, A., Oba, T., Anzai, R., and Ishioroshi, H., 2002, Evaluation of the fidelity of isotope records as an environmental proxy in the coral *Montastrea*, *Coral Reefs*, **21**:169–178.

Yu, K.F., Zhao, J.X., Liu, T.S., Wei, G.J., Wang, P.X., and Collerson, K.D., 2004, High-frequency winter cooling and reef mortality during the Holocene climatic optimum, *Earth Planet. Sci. Lett.*, **224**:143–155.

Chapter 4
Assessing the Effects of Taphonomic Processes on Palaeobiological Patterns using Turbinid Gastropod Shells and Opercula

Ross H. Nehm[1] **and Carole S. Hickman**[2]

Contents

4.1 Introduction

For nearly 30 years the Dominican Republic Neogene has served as a productive research system for exploring a broad array of palaeobiological topics, including speciation (e.g., Cheetham, 1986, 1987; Nehm and Geary, 1994; Nehm, 2005), intraspecific morphological variation (e.g., Anderson, 1994, 1996; Foster, 1986; Nehm, 2001), palaeoecological reconstruction (e.g., Vokes, 1989; Costa et al., 2001), and faunal

[1]The Ohio State University, Columbus, OH, USA. Email: nehm.1@osu.edu

[2]Department of Integrative Biology and Museum of Palaeontology, University of California, Berkeley, CA, USA. Email: caroleh@berkeley.edu

R.H. Nehm, A.F. Budd (eds.) *Evolutionary Stasis and Change in the Dominican Republic Neogene,*

turnover (e.g., Budd et al., 1996). Despite a long-standing recognition that taphonomic processes may significantly influence palaeobiological patterns (e.g., Donovan and Paul, 1998; Martin, 1999) our study is the first to explore the role of taphonomy in this geological research system. We do so using a unique morphological model that allows us to compare palaeobiological patterns derived from skeletal hard-parts with different preservation potentials from the same animal. Specifically, we compare estimates of relative abundance, species richness, stratigraphic distribution, size distribution, and morphological change in shells and opercula from two different species of *Turbo* Linnaeus, 1758 (Gastropoda: Turbinidae). Both species were sampled using the same techniques as other invertebrates in the Río Cana and Río Gurabo stratigraphic sections of the Dominican Republic Neogene (see Saunders et al., 1986). We use species of *Turbo* to test two hypotheses: (1) That shells and opercula from the *same* species record the same palaeobiological signals (e.g., abundances, stratigraphic distributions, population structures, and morphological patterns), and (2) that shells and opercula from two morphologically similar and stratigraphically co-occurring species display similar palaeobiological patterns. We use our results to explore how the lack of preservation uniformity in morphologically similar clades may limit our ability to make taphonomic generalizations in the Dominican Republic Neogene, and how extensive sampling may not alleviate taphonomic bias.

4.2 Research System

4.2.1 Introduction

Turbinid gastropods (Gastropoda: Turbinidae) of the genus *Turbo* provide an ingenious morphological system for studying taphonomic patterns and processes in the fossil record because each animal contains two separate skeletal hardparts (shell and operculum) that are species-diagnostic but of different sizes, shapes, densities, masses, microstructures, and durabilities (Hickman, 1992, 2003). During life these two hard parts are connected by soft tissue and after death they dissociate. If the animal is not buried permanently prior to hard part dissociation, biostratinomic processes may affect the shell and operculum differently as a result of each structure's unique morphological properties (Hickman, 2003). Specifically, the *Turbo* operculum is an ovate, small, solid and plano-convex object generally two to three times thicker than the body whorl of an associated shell (Fig. 4.1). In contrast, the associated shell is hollow and significantly larger and thinner than the operculum (Fig. 4.1). Detailed comparisons of palaeobiological signals in *Turbo* shells and opercula have potential for suggesting the extent to which taphonomic processes may have influenced other invertebrate skeletons in the Dominican Republic Neogene and elsewhere.

At least five species of *Turbo* are represented in samples from the Dominican Republic. We focus on the two most abundant and stratigraphically long-ranging species: *Turbo crenulatoides* Maury, 1917a,b, and *T. dominicensis* Gabb, 1873. On the basis of shell and operculum morphology they can be assigned to the subgenera

Fig. 4.1 Neogene *Turbo* species from the Dominican Republic. Shells and corresponding opercula were found in two species (*T. dominicensis* and *T. crenulatoides*). A-C *T. dominicensis* shell (30.7 mm, TU 1354) and opercula; D-F. *T. crenulatoides* shell (27.9 mm, NMB 16857) and opercula

Marmarostoma Swainson, 1828, and *Taeniaturbo* Woodring, 1928, respectively. Correct association of opercular morphology and shell morphology in the two fossil species is confirmed by a small number of specimens in which the operculum was preserved in place within the aperture.

4.2.2 Turbo Crenulatoides Morphology

T. crenulatoides (Figs. 4.1D–F) is morphologically similar to the extant species *T. castanea* Gmelin, 1791. The operculum of *T. crenulatoides* is oval in shape and contains a deep, expanding spiral groove on its inner surface. The inner surface is smooth but contains fine spiral growth lines. The outer surface of the operculum is generally smooth, lacks a nucleus and concentric grooves, and unworn specimens display peripheral pustulosity. A shallow ridge is also present on the outer edge of the operculum. The shell of *T. crenulatoides* coils dextrally, contains four to five whorls, and is turbiform in shape (i.e., strongly shouldered, with a medium spire containing rectangular whorls). The aperture is large but proportionally smaller than in *T. domincensis*. The inner surface of the outer lip is generally smooth but contains a distinctive ridge that is most prominent on the inner outer lip. The parietal wall (inner aperture lip) is small, smooth, flat, and lacks ridges or beads. A deep groove occurs adjacent to the inner lip and the aperture rim contains grooves and ridges in some specimens. Fine growth lines are visible on the shell. Eleven to twelve spiral cords of varying widths are present on the body whorl. Within and among specimens, these cords contain varying degrees of beads, nodules, and spines. The spines are triangular, hollow, tube-like expansions or flanges of the larger spiral cords. These spines are projected at approximately 45 degree angles from the cord surface. The largest spiral cord near the body whorl shelf usually contains the best-developed spines. Spinosity increases through ontogeny: the apex usually lacks spines and displays beaded or smooth spiral cords, whereas the body whorl contains spines and lacks smooth spiral cords. On the body whorl near the aperture, margin spines are absent but rectangular beads are present. The shell lacks an umbilicus.

4.2.3 *Turbo Dominicensis Morphology*

T. dominicensis (Figs. 4.1A–C) is very similar to the extant species *T. canaliculatus* (Hermann, 1781). The operculum of *T. dominicensis* is oval but more circular in shape than in *T. crenulatoides*. The inner surface of the operculum is smooth and contains fine spiral growth lines and a deep spiral groove. The outer surface of the operculum contains a deep nucleus with a pustulose surface that is nearly circumscribed by five deep concentric grooves. On one quarter of the operculum the grooves are lacking; this region is solid, not as thick as the groove interspaces, and has a smooth or lightly pustulose surface. Six groove interspaces are present, and they decrease in width from the nucleus. The groove closest to the nucleus is the deepest. The groove interspaces are generally rounded, have a slightly pustulose surface, and contain different topographic surfaces. The innermost and second interspaces are rounded. The third interspace contains a very shallow groove medially. The fourth interspace is more prominent medially (towards the nucleus). The shell is coiled dextrally and contains four to five whorls. The aperture is large and contains a smooth outer lip and inner aperture surface. Some specimens contain ridges in the aperture that are most prominent on the inner outer lip. The parietal wall (inner aperture lip) is large, smooth, flat, and lacks ridges or beads. Nine smooth spiral cords of varying widths are present on the body whorl. Five large and thick cords and four small and thin cords are present on the body whorl. No beads are present on the body whorl spiral cords, but rectangular beads are present on the spire near the sutures. Like *T. crenulatoides*, the shell lacks an umbilicus.

4.2.4 Palaeoecology of Neogene Dominican Turbo

All of the specimens used in this study were collected from the Cibao Valley as part of the Dominican Republic Project (for details see Saunders et al., 1986; Nehm and Budd, this volume). Turbinid gastropod shells and opercula are most abundant in the Río Cana and Río Gurabo sections. The Río Cana section contains brackish, shallow marine, and deep marine sediments deposited in progressively offshore and more open marine conditions upsection. Brackish-water palaeoenvironments were identified by brackish-restricted ostracod species (see Bold, 1988) and the *Larkinia* (=*Anadara* [*Grandiarca*]), *Mytilus*, and *Melongena* mollusk assemblage (Saunders et al., 1986). Shallow marine palaeoenvironments (<30 m palaeodepth) were characterized by a lack of planktonic foraminifera and the presence of the shallow marine and intertidal mollusks *Anadara*, *Tellina*, *Strombina*, and *Pachycrommium* (Saunders et al., 1986; Nehm and Geary, 1994), benthic foraminifera (e.g., soritids, miliolids, and *Amphistegina*), and ostracods (e.g., *Cytherella*, *Randimella*, *Caudites*, *Proteoconcha*, *Loxoconcha*, and *Paracytheridea*). Moderately deep marine palaeoenvironments (30–100 m palaeodepths) were characterized by common marine mollusks (e.g., *Oliva*, *Prunum*, *Lyria*, and *Polystira*) and corals (Saunders et al., 1982). Deep marine palaeoenvironments (exceeding palaeodepths of 100 m) were characterized by rich assemblages of planktonic foraminifera and the deep-water ostracod *Krithe* (Bold, 1988; Saunders et al., 1986). Brackish-water ostracods occur below 150 m in the section and between 200 m and 230 m in the middle Cercado Formation (Bold, 1988). Sediments of shallow marine origin (<30 m depth) occur from 150 m to 200 m in the section, whereas sediments of deeper marine origin (>30 m depth) occur from 230 to 450 m in the section (Bold, 1988; Saunders et al., 1986; Anderson, 1996).

As in the Río Cana section, the Río Gurabo section contains brackish, very shallow marine, marine, and deep marine sediments deposited in progressively offshore and more open marine conditions upsection. The Cercado Formation contains brackish and very shallow marine deposits, whereas the Gurabo Formation contains shallow and deep marine deposits. The brackish water *Larkinia-Mytilus-Melongena* mollusk assemblage, which occurs in the Cercado Formation of the Río Cana, also occurs in the lower Cercado Formation of the Río Gurabo from approximately 20–50 m. Brackish water ostracod species also occur near 50 m in the middle Cercado Formation (Bold, 1988; Saunders et al., 1986). Very shallow marine conditions (<30 m palaeodepth) occur from 60–150 m in the section, whereas deeper marine conditions (30–100 m palaeodepth) occur from 150–380 m in the section.

4.2.5 Ecology of Living Turbo Species

Living species of the genus *Turbo* Linnaeus, 1758 (Turbinidae, Turbininae) occur worldwide, primarily at tropical to subtropical latitudes and in intertidal and shallow subtidal habitats on hard substrates (Hickman and McLean, 1990; Hickman, 1998).

They are opportunistic herbivores, feeding primarily on filamentous and encrusting macroalgae. At temperate latitudes they are often abundant on rock platforms, and at tropical latitudes they occur both in reef settings and in seagrass beds. In reef habitats they are most commonly found attached to coral slabs and large pieces of coral rubble. Some species have adapted for life of the flexible moving substrates of seagrass blades (Hickman, 2005), where they graze on epiphytic algae. Larvae tend to settle high in the intertidal zone, often on turfs of encrusting coralline red algae, migrating downward as they grow to adult size.

In Neogene environments of the Dominican Republic, species of *Turbo* would not have been expected in brackish settings or on unconsolidated sediment in deeper, offshore settings. They would have been expected in intertidal and subtidal settings with rock or coral rubble, and may have reached peak densities in seagrass beds. *Turbo castanea*, a living analog of the fossil *T. crenulatoides*, is often abundant in *Thalassia testudinum* meadows in the Caribbean and northwestern Atlantic (Engstrom, 1982). It is the most abundant grazer in *Thalassia* beds in Florida Bay, where mean densities range from 6.7 to 27.5 individuals per square meter (Frankovich and Zieman, 2005). In a study of subfossil remains at St. Croix (US Virgin Islands), *T. castanea* was one of a cluster of five species that all occurred live on seagrass blades (Miller, 1988). This cluster included the seagrass neritid gastropod *Smaragdia viridis* (Gmelin, 1791), a species that is locally abundant in Dominican Republic Neogene sections (Costa et al., 2001).

We predict that the two species of *Turbo* in this study would have lived at depths of <6 m in an embayment or lagoonal setting with a mix of seagrass, bare sand and patch reef, and perhaps bare limestone. In both the Río Cana and Río Gurabo sections, *Turbo* species occur in palaeoecological contexts that do not always match these predictions. Preservation of shells and opercula in these deeper settings appears to have resulted from down slope post-mortem transport.

4.3 Methods

We compared five palaeobiological patterns between *Turbo* shells and opercula: (1) Species richness, (2) abundance, (3) stratigraphic range, (4) population (age and size) structure, and (5) morphological (evolutionary and/or ecophenotypic) pattern.

1. Species Richness. Prior to documenting detailed patterns within and between *Turbo crenulatoides* and *T. dominicensis*, we first explored whether shells and opercula recorded similar patterns of turbinid species richness. Dominican turbinid shells and opercula were identified to the species level using (a) living analogs from the western Atlantic and eastern Pacific, (b) a review of the Neogene palaeontological literature, and (c) museum research as part of an ongoing study of the systematics and evolution of the western Atlantic Turbinidae. Species richness was measured using opercula only, shells only, and both shells and opercula.

2. Abundance. The second pattern that we explored was operculum and shell abundances to determine whether they differed within and between *T. crenulatoides* and *T. dominicensis* in the same stratigraphic sections. After species identification of all shells and opercula from the Río Cana and Gurabo sections, the total abundances of opercula and shells were tabulated for each species. Chi-squared goodness-of-fit tests were used to determine if shell and operculum abundances differed significantly within species, between species, or between stratigraphic sections.
3. Stratigraphic ranges. The third question that we explored was whether the stratigraphic ranges of opercula and shells from the same species were concordant or discordant within sections. After the identification of specimens to species, and tabulation of the number of shells and opercula in each sample, the first occurrences, last occurrences, and relative abundances of shells and opercula were plotted and compared in the Río Cana and Río Gurabo sections. The stratigraphic positions of samples within these sections were established using data from Saunders et al. (1986). A Kolmogorov-Smirnov test was used to determine whether the stratigraphic occurrences and abundances of opercula and shells for *T. crenulatoides* and *T. dominicensis* differed significantly within and between sections. This non-parametric test was used because it is sensitive to differences in dispersion (kurtosis and skewness; Sokal and Rohlf, 1995).
4. Population structure. The fourth question that we investigated was whether shells and opercula produced similar estimates of size distribution in the same stratigraphic intervals. In order to compare the size (and presumably age) distributions of shells and opercula, the major axis of the operculum and the major axis of the shell aperture were measured on approximately 1,200 specimens. Images of opercula were captured using a digital camera, imported to a computer, and measured using Scion Image. Prior to comparing size distributions within stratigraphic sections it was first necessary to explore whether the major axis of the shell operculum was in fact an equivalent measure to the major axis of the shell aperture. Aperture widths and the major axis of the operculum were compared in each species using F statistics and t-tests. These tests demonstrated that the major axis of the aperture and the major axis of the operculum of *Turbo dominicensis* ($F = .714$, $df = 548$, $p > 0.27$; $t = 1.99$, $df = 74$, $p > 0.05$) and *T. crenulatoides* ($F = .941$, $df = 283$, $p > 0.36$; $t = 1.45$, $df = 334$, $p > 0.145$) were not significantly different. Therefore, comparing the size distributions of aperture width and the major axis of the operculum within each stratigraphic section was deemed appropriate.

Because aperture width and opercular size were not normally distributed within NMB samples or stratigraphic intervals, the non-parametric Kolmogorov-Smirnov test was performed in order to determine if the distributions of aperture sizes and operculum sizes were significantly different within species and within stratigraphic intervals. This test is useful for comparing population distributions between opercula and shells because it is sensitive to differences in dispersion (Sokal and Rohlf, 1995).

5. Evolutionary patterns. The final question that we examined was whether opercula and shells of the same species displayed similar size changes through time. Kruskal-Wallis and Mann-Whitney U tests were performed on size distributions in order to answer this question. These methods were used to test the hypothesis that aperture width and operculum size have similar "locations" or "central tendencies". Kruskal-Wallis tests are the nonparametric version of an analysis of variation (ANOVA) and were used to determine if shell and operculum samples differed significantly among all stratigraphic intervals, whereas the Mann-Whitney U tests were used to compare operculum and shell "central tendencies" within the same stratigraphic intervals (that is, whether shell and opercula sizes of the same species differed in the same stratigraphic interval).

4.4 Results

4.4.1 Species Richness

Overall, shells underestimated turbinid gastropod species richness in the Neogene of the Dominican Republic by 60%. Opercula belonging to five species of *Turbo* (*Turbo crenulatoides* Maury, *T. dominicensis* Gabb, *T. rhectogrammicus* Dall, *T. castanea* Gmelin, and *T. species C*) occurred in the sections whereas shells belonging to only two species of *Turbo* were found (*Turbo crenulatoides* and *T. dominicensis*). Among stratigraphic sections, shells underestimated species richness by 0–50%. In the Río Cana and Río Mao sections, species richness estimates were the same using shells and opercula. In contrast, in the Río Gurabo section opercula belonging to four species were collected, whereas shells belonging to only two species were collected. Likewise in the Río Yaque del Norte section, opercula belonging to three species were collected whereas shells belonging to only two species were collected. Thus, species richness values differ greatly depending on whether shells or opercula from the same animal are used for analysis.

4.4.2 Abundance

In addition to representing greater species richness values, opercula are significantly more abundant than shells in the Dominican Republic Neogene (Chi squared $=502$, $df = 1$, $p < 0.01$). Overall, less than 26% of all sampled *Turbo* specimens were shells (309 shells *vs* 1,171 opercula). For all five *Turbo* species, opercula were more abundant than shells. Statistical tests confirm what seems apparent: for species that were preserved as both shells and corresponding opercula, there are significantly greater numbers of opercula (*T. dominicensis*, Chi squared $= 306$, $df = 1$, $p < 0.01$; *T. crenulatoides*, Chi squared $= 175.7$, $df = 1$, $p < 0.01$). Overall, opercula appear to have greater preservation potential than shells (Fig. 4.2).

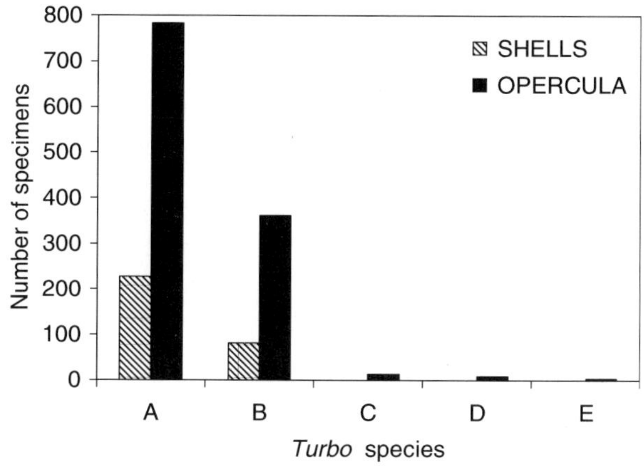

Fig. 4.2 Relative abundances of shells and opercula in the Neogene of the Dominican Republic for each *Turbo* species. A. = *T. dominicensis* Gabb, B. = *T. crenulatoides* Maury, C.= *T.* species C, D = *T.* species D., E = *T.* species E

4.4.3 Stratigraphic Ranges

Within and between species, the stratigraphic ranges of opercula are greater than the stratigraphic ranges of shells. As discussed above, only two of the five Dominican *Turbo* species are preserved as both shells and opercula. The stratigraphic ranges and relative abundances of shells and corresponding opercula of *T. dominicensis* and *T. crenulatoides* in the Río Cana and Río Gurabo sections are illustrated in Figs. 4.3–4.5. Comparisons of the first occurrences, last occurrences, and relative abundances of shells and opercula in the Río Cana and Río Gurabo sections indicate that there are no consistent relationships between the stratigraphic distributions of shells and opercula for the two *Turbo* species. In the Río Cana section, the opercula of *T. dominicensis* first occur at approximately 140 m in the section, whereas the shells first occur at approximately 230 m in the section. The last occurrences of shells and opercula of *T. dominicensis* are also discordant within the Río Cana section: opercula last occur at approximately 590 m in the section, whereas shells last occur at approximately 362 m in the section. Overall, shells significantly under-represent the stratigraphic range of *T. dominicensis* (132 m stratigraphic range for shells, 450 m stratigraphic range for opercula). In contrast to *T. dominicensis*, the shells and opercula of *T. crenulatoides* do not differ in their first occurrences or last occurrences (237 m stratigraphic range for both opercula and shells).

The Kolmogorov-Smirnov test was used to compare several components of stratigraphic distribution: the range (first and last occurrences), abundance (number of specimens at each horizon), and distribution (kurtosis and skewness) of shells and opercula. Unsurprisingly, the Kolmogorov-Smirnov test produced different results for the two *Turbo* species in the Río Cana section. The shape of the stratigraphic distributions and abundances of the shells and opercula of

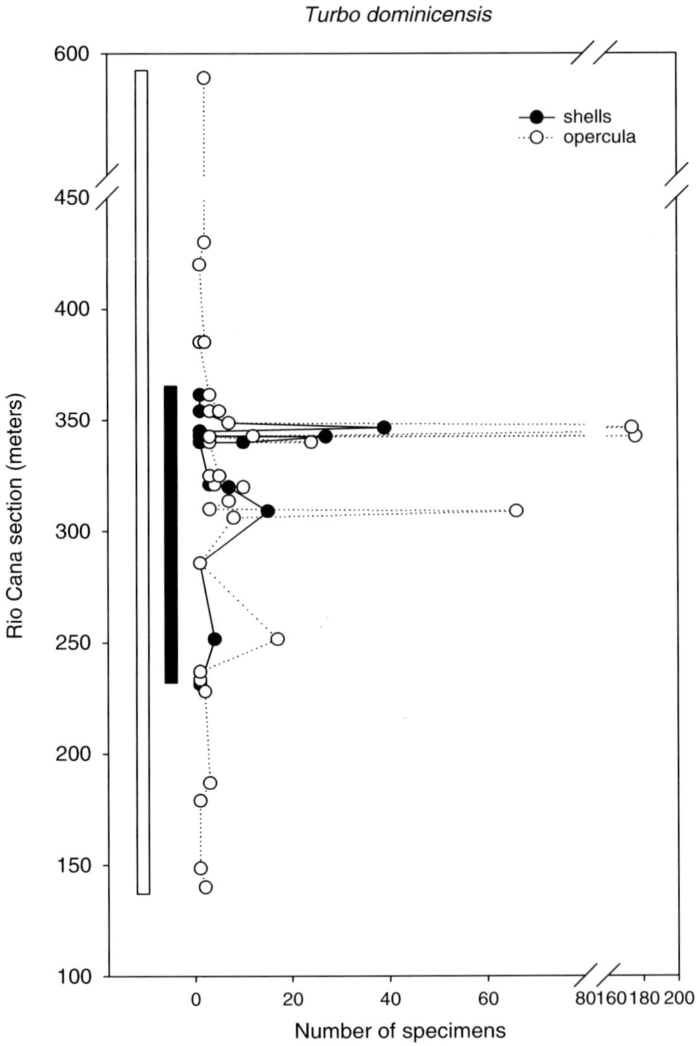

Fig. 4.3 Stratigraphic ranges and abundances of opercula and shells of *Turbo dominicensis* in the Río Cana section

T. dominicensis in the Río Cana section were significantly different (Group difference = 0.2418; $p < 0.0001$) whereas in *T. crenulatoides* the distributions of shells and opercula in the same section were not significantly different (Group difference = 0.0194; $p > 0.01$).

Different stratigraphic patterns occur in the Río Gurabo section than in the Río Cana section for *Turbo dominicensis* (Fig. 4.3). In the Río Gurabo section, the opercula from *T. dominicensis* first occur at approximately 177 m in the

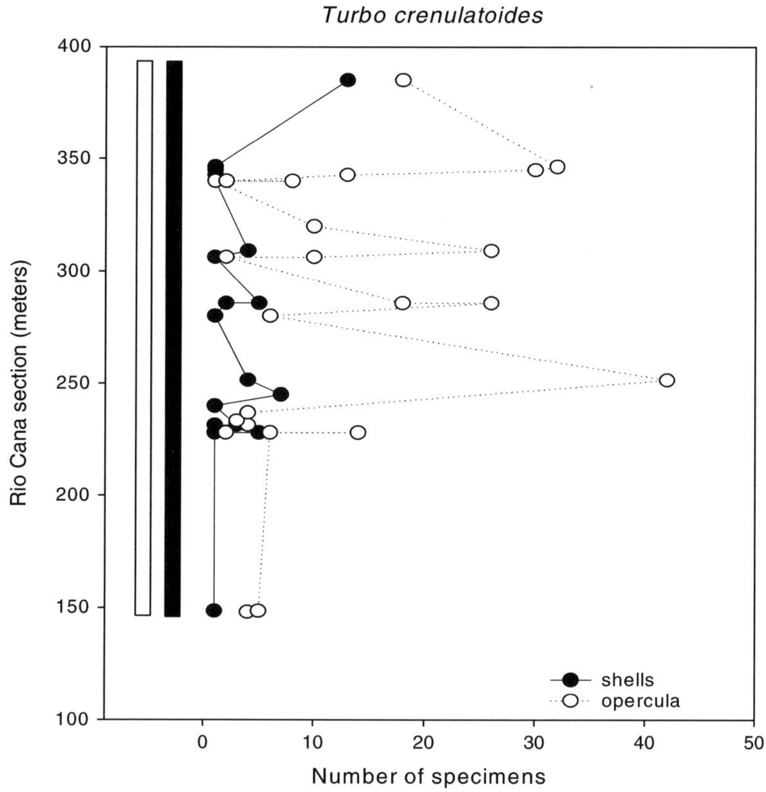

Fig. 4.4 Stratigraphic ranges and abundances of opercula and shells of *Turbo crenulatoides* in the Río Cana section

section, whereas the shells first occur at approximately 121 m in the section. The last occurrences of shells and opercula of *T. dominicensis* are also discordant within the Río Gurabo section: opercula last occur at approximately 208 m in the section whereas shells last occur at approximately 195 m in the section. In contrast to the stratigraphic patterns observed in the Río Cana section, shells provide a greater estimate of stratigraphic range than opercula for *T. dominicensis* in the Río Gurabo section.

Unlike *Turbo dominicensis*, the first and last occurrences of shells and opercula are generally concordant within and among sections for *T. crenulatoides* (Figs. 4.4 and 4.5). The opercula of *T. crenulatoides* first occur at approximately 123 m in the Río Gurabo section, whereas the shells first occur at approximately 180 m in the section. The last occurrences of shells and opercula of *T. crenulatoides* are concordant within the Río Gurabo section and last occur at approximately 673 m. In the Río Gurabo section, shells under- or overestimate the stratigraphic range of *T. crenulatoides* (494 m stratigraphic range for shells, 551 m stratigraphic range for opercula).

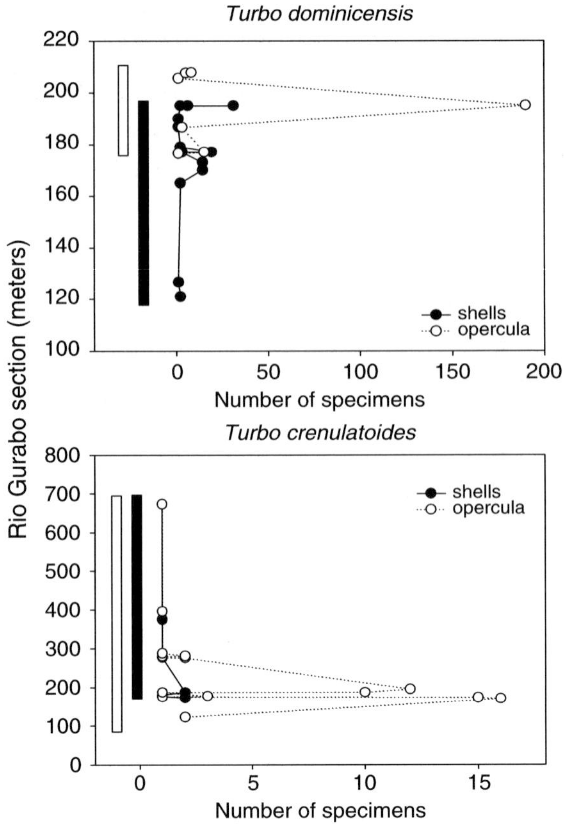

Fig. 4.5 Stratigraphic ranges and abundances of opercula and shells of *Turbo dominicensis* (top) and *T. crenulatoides* (bottom) in the Río Gurabo section. Note the different scales on the two plots

4.4.4 Population Structure

Pair-wise comparisons of the size (and presumably age) distributions of turbinid shells and opercula from a series of stratigraphic intervals in the Río Cana section using Kolmogorov-Smirnov tests indicate that the shapes of the distributions of aperture sizes and opercular sizes are generally not significantly different (Table 4.1). Pair-wise comparisons of the size distributions of the major axis of the operculum and the major axis of the aperture of *T. dominicensis* in five stratigraphic intervals of the Río Cana section produced only two significant differences (intervals 4 and 5: Table 4.1). Similarly, pair-wise comparisons of the distributions of the opercular major axis and the major axis of the aperture of *T. crenulatoides* in six stratigraphic intervals of the Río Cana section produced only one significant difference (interval 3: Table 4.1). Kolmogorov-Smirnov tests comparing the

Table 4.1 Kolmogorov-Smirnov and Mann-Whitney *U* tests comparing the distributions and central tendencies, respectively, of opercula and shells from each stratigraphic interval (indicated in meters). The greatest number of possible stratigraphic intervals was constructed depending sample size. Significant results ($p<0.01$) are indicated in bold

Interval	m	Kolmogorov-Smirnov	Mann-Whitney U Test
1	140–300	0.2664	0.3937
2	301–324	0.2889	0.1544
3	325–340	0.7314	0.7180
4	341–342	**0.0039**	**0.0050**
5	343–350	**0.0486**	0.0918
6	351–590	–	–
1	145–232	0.3748	0.6677
3	233–260	**0.0001**	**0.0001**
5	261–300	0.7494	0.6255
6	301–315	0.4817	0.2571
7–9	316–347	0.0925	0.4798
10	385	0.1602	0.0753

size distributions of all shells and opercula for each species indicate that the two distributions for both species are not significantly different (*T. dominicensis* distribution test statistic 0.1924; $p = 0.03$; *T. crenulatoides* distribution test statistic 0.1485; $p = 0.32$).

4.4.5 Evolutionary Patterns

In general, opercula and shells in the Río Cana section display similar size patterns (Figs. 4.6 and 4.7). Pair-wise Mann-Whitney U tests indicated that shells and opercula generally record similar morphological signals. Pair-wise comparisons of the central tendencies of the size of the opercula and aperture of *T. dominicensis* in five stratigraphic intervals of the Río Cana section produced only one significant difference (interval 4: 341–342 m in the section; Table 4.1). Similarly, Mann-Whitney U tests of opercular and aperture sizes of *T. crenulatoides* in six stratigraphic intervals of the Río Cana section produced only one significant difference (interval 3: 233–260 m in the section; Table 4.1). These tests indicate that the hypothesis that aperture width and opercular size have similar "locations" or "central tendencies" through time in the Río Cana section cannot be rejected. Kruskal-Wallis tests among all opercula and shell samples indicate that the shells of both *T. dominicensis* and *T. crenulatoides* do not show significant differences among samples, whereas the opercula of both *T. dominicensis* and *T. crenulatoides* do exhibit some significant differences among samples in the Río Cana section (Table 4.1). In general, no significant size changes occur through time in *Turbo* species.

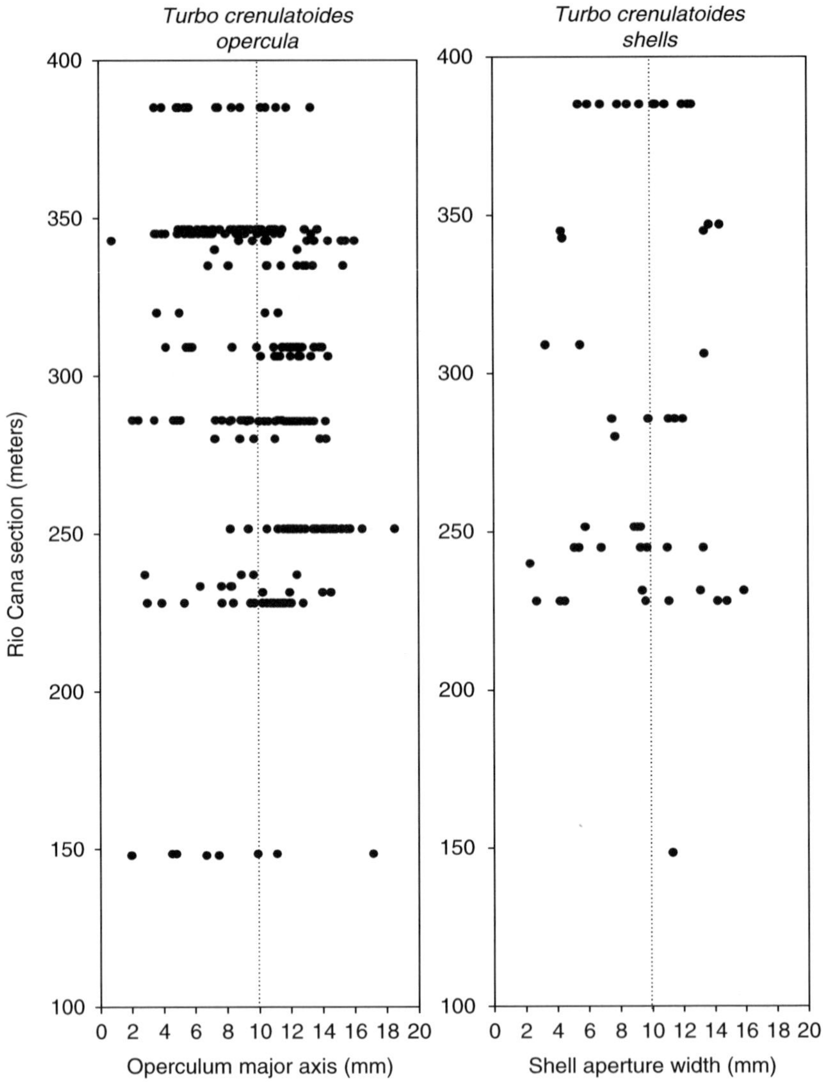

Fig. 4.6 Size distributions of the operculum major axis (in mm) and shell aperture width (in mm) of *Turbo crenulatoides* from the Río Cana section

4.5 Discussion

The Neogene stratigraphic sections of the northern Dominican Republic are being used to address an increasingly wide range of evolutionary, biostratigraphic, and palaeoecological questions. This study adds a taphonomic dimension to this research by using turbinid gastropods of the genus *Turbo* to better understand how

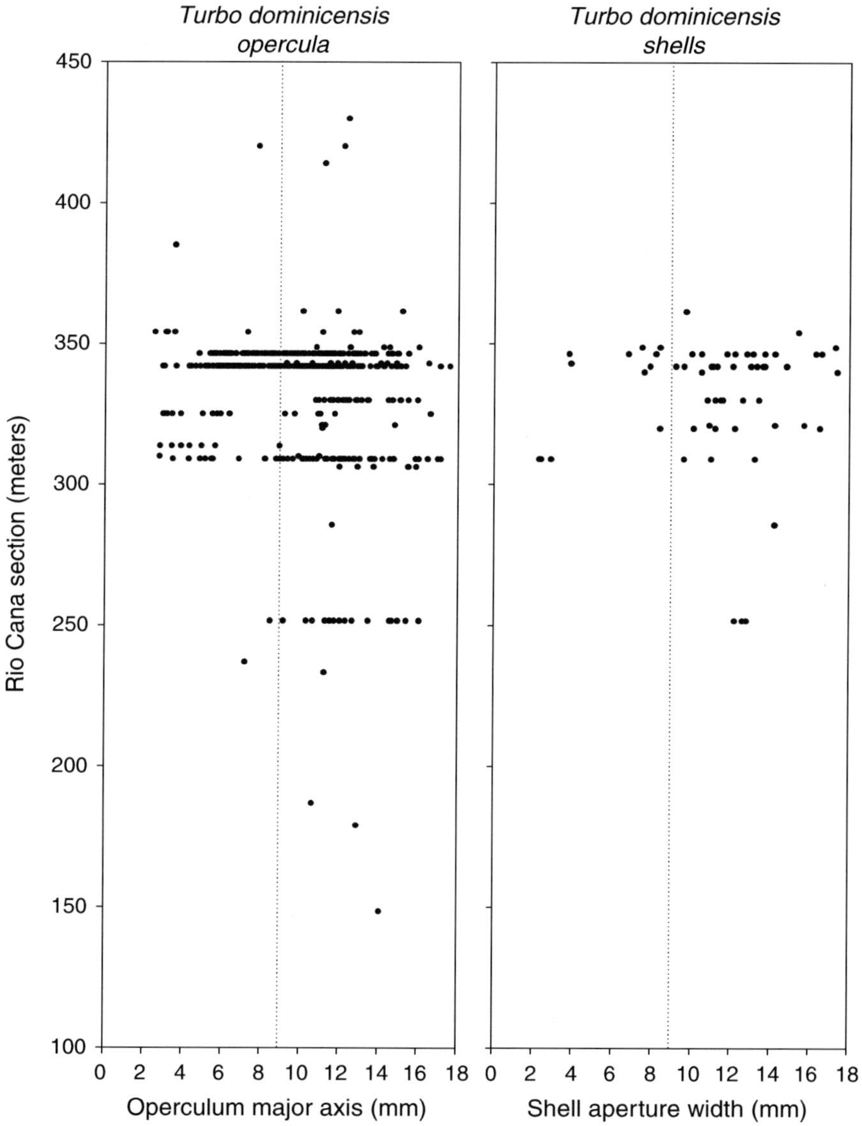

Fig. 4.7 Size distributions of the operculum major axis and aperture width (in mm) of *Turbo dominicensis* from the Río Cana section

taphonomic processes may be influencing palaeobiological patterns. Specifically, we analyzed taphonomic patterns in two *Turbo* species (*T. dominicensis* and *T. crenulatoides*) displaying similar sizes, shapes, and masses and containing two different hardparts (shell and operculum). Both species were sampled using the same techniques as other invertebrates in the Río Cana and Río Gurabo stratigraphic sections (see Saunders et al., 1986). We used *Turbo* species to test two hypotheses: (1) That shells and opercula from the same species record the same palaeobiological signals (e.g., abundances, stratigraphic distributions, population structures, and morphological patterns), and (2) that shells and opercula from morphologically similar and stratigraphically co-occurring species display similar patterns.

4.5.1 Differences and Similarities in Palaeobiological Patterns

Our analyses of approximately 1,400 specimens from 200 samples show that shells and opercula do not give similar estimates of species richness. Overall, while we identified opercula from five *Turbo* species, we found shells from only two corresponding species. In addition, in *T. dominicensis* and *T. crenulatoides*, for which both shells and corresponding opercula were identified, only about one shell was preserved and/or sampled for every four opercula. It is therefore not surprising that the three *Turbo* species with very few opercula (n < 10) have no corresponding shells. Indeed, differences in species richness values using shells and opercula between the Río Cana, Río Gurabo, Río Mao, and Río Yaque del Norte sections are primarily a result of the restriction of rare *Turbo* species to the Río Gurabo and Río Yaque del Norte sections. In the Río Cana and Río Mao sections, species richness values were the same using shells and opercula because the two most abundant species (*T. dominicensis* and *T. crenulatoides*) occurred in these sections.

Despite the significant under-representation of *Turbo* shells relative to opercula in the Dominican Neogene fossil record, quantitative studies of shell aperture size and operculum size patterns in *T. dominicensis* and *T. crenulatoides* indicated that, for the most part, the distributions of the two measures did not differ significantly within stratigraphic intervals or through time. These results suggest (1) the absence of shell preservation size-bias and (2) a lack of demonstrable evolutionary and/or ecophenotypic size change in the Río Cana section (recall that these species were not abundant enough to conduct tests of temporal patterns in the Río Gurabo section). These species richness and morphometric patterns indicate that despite the significant under-representation of shells in the fossil record, those shells that were preserved or sampled in most cases have similar aperture size distributions to the preserved opercula distributions.

Sampling protocols are one possible explanation for the different abundance (and resulting species richness) patterns that we document. In a study comparing mollusk species richness and abundance patterns (captured using both bulk and float sampling) in the same stratigraphic horizon, Jarrett et al. (2004) documented significant differences in species composition and abundance. The differences that they documented were primarily due to a greater number of large specimens present in the float sample and the near absence of large specimens in the bulk samples. Even though six bulk bags (~3 kg each) were collected randomly from the same horizon (NMB 15805) and completely processed and studied, some large-shelled species were not found in any of the bulk samples, not even as fragments (all specimens >5 mm were studied). Because *Turbo* shells are significantly larger than opercula, it is possible that the lack of shells in some samples containing opercula was a product of how the material was collected. This appears unlikely, however, because the vast majority of the NMB macrofossil samples from which our *Turbo* shells and opercula were derived appear to have been collected as float (Saunders et al., 1986: Appendix 3). The sampling details provided by Saunders et al. (1986) are insufficient for rejecting this idea outright, however. The apparent lack of shell size-distribution bias relative to opercula that we document above lends support to the idea that sampling bias may not be causing these patterns. Indeed, if shells are differentially sampled as a result of their large size then one would not expect to find a general morphometric concordance between the major axes of apertures and opercula from the same horizons.

It appears more likely that taphonomic processes are responsible for the differences in palaeobiological patterns between shells and opercula. Taphonomic studies have established that skeletons of different sizes, shapes, masses, and compositions may be affected differently by the same biostratinomic and diagenetic processes (Kidwell and Bosence, 1991). Therefore, it is likely that preservation potential or quality may vary significantly with any of these variables. As bioclasts, shells and opercula are of different shape, size, mass and composition; therefore they are likely to behave differently as sedimentary particles. The preservation patterns for all five *Turbo* species identified in the Neogene of the Dominican Republic support the idea that opercula are more durable bioclasts than shells. Indeed, the patterns that we document suggest that shells are destroyed by biostratinomic and/or diagenetic processes at a rate four times greater than opercula. This result has significant palaeobiological consequences. If turbinid gastropods had only been preserved as shells, then abundance would have been underestimated by 75% and turbinid diversity would have been underestimated by 60%. These results suggest diversity and abundance in other invertebrates from the Neogene of the Dominican Republic—even those with durable skeletons—are significantly underestimated.

4.5.2 Preservation Patterns in T. dominicensis and T. crenulatoides

There are several *a priori* reasons that taphonomic patterns should be similar in *T. dominicensis* and *T. crenulatoides:* (1) The size, shape, and mass of the shells and opercula of the two species are closely comparable; (2) Both species have generally concurrent stratigraphic ranges within stratigraphic sections; and (3) The two species were sampled using the same methods. Nevertheless, several observations suggest that *T. dominicensis* and *T. crenulatoides* behaved differently as bioclasts: (1) The first and last occurrences of opercula and shells of *T. crenulatoides* in the Río Cana and Gurabo sections are nearly concordant, whereas the first and last occurrences of opercula and shells of *T. dominicensis* are strikingly discordant. (2) The shapes of the stratigraphic distributions (kurtosis and skewness) of opercula and shells within the Río Cana section are significantly different in *T. dominicensis* but are not significantly different in *T. crenulatoides*. And (3) Shells and opercula of *T. dominicensis* are more abundant than *T. crenulatoides* but they exhibit fewer similarities in first and last appearances in the Río Cana and Río Gurabo sections.

We offer four possible explanations for differences in patterns between two morphologically similar turbinid species: (1) differences in strength of the attachment of the soft anatomy to both the shell and the operculum, (2) differences in causes of mortality that affect post-mortem dissociation; (3) differences resulting from post-mortem hermit crab occupation of shells, and (4) differences in microhabitats occupied in life. It is possible that combinations of these factors, or stochastic processes alone, could have generated the discordant patterns that we document.

Construction of the gastropod operculum is understood in a general sense (Fretter and Graham, 1994:81–85; Checa and Jiménez-Jiménez, 1998). Although it has been suggested that the shell and operculum are homologous (Adanson, 1757; Fleischmann, 1932), it is well-established that the operculum is secreted by a separate epithelium in a groove on the dorsal surface of the foot. Opercula may be calcified in several different ways, and in turbinids the calcareous portion is added to the exterior surface of the corneous layer by an extension of the foot epithelium located immediately anterior to the opercular groove (Kessel, 1941). The degree of envelopment of the operculum by the metapodium is highly variable in turbinids (Hickman and McLean, 1990), and it is possible that there are differences in strength of attachment of the operculum to the foot or the strength of the attachment of the animal (and its operculum) to the shell that could affect rates of postmortem dissociation of the two calcified elements. Experimental studies of extant analogues are required to test for such differences between taxa.

Hermit crabs are known to occupy western Atlantic *Turbo* shells (Nehm, personal observations). If one of the two *Turbo* species was a more preferable host, this could explain the transport of the shell away from the operculum, the more rapid destruction of shells as a result of wear and tear by crab inhabitants, and subsequent differences in shell and operculum preservation. This hypothesis appears unlikely, however, because of the great similarity between the shells of the two Dominican species. Nevertheless, investigations of hermit crab occupation and detailed morphological analyses of the surviving shells might shed some light on this possibility (e.g., Walker, 1989).

If predation was a significant cause of mortality, differences in predators and their modes of predation on *T. crenulatoides* and *T. dominicensis* could explain the patterns that we have documented. A variety of predators are known to attack turbinid gastropods (Vermeij, 1978), and different modes of postmortem separation of shells and opercula have been observed in avian, fish, octopus, gastropod, and human predation on turbinids (Hickman, personal observations). It is essential to distinguish whether soft parts (and operculum) are both removed from the shell, whether the soft parts are consumed at a distance from the shell, and whether the operculum is consumed along with the soft parts and subsequently eliminated by the predator. Although we found no evidence of drilling predation on the vast majority of the shells of the two species, other predatory causes of mortality may be strongly masked by taphonomic alteration of shells.

There are no clear indications of habitat differences between the two *Turbo* species, but it is possible that they may have occupied different microhabitats that we could not detect in our palaeontological data. Microhabitat properties, coupled with different patterns of habitat use and behaviour, could have promoted biostratinomic separation of opercula and shells in *T. dominicensis* but not in *T. crenulatoides*. Furthermore, differences in microhabitat could be connected to different predators and modes of predation (see above). Further research on living and fossil turbinid gastropods could be designed to search for signals of microhabitat and predators.

Overall, the size, shape, and mass similarities in the shells and opercula of *T. dominicensis* and *T. crenulatoides* fail to support an exclusive or primary role for physical biostratinomic processes as generators of the differences in palaeobiological patterns that we document. Biological and ecological explanations, such as soft-tissue operculum attachment differences and/or ecological differences in microhabitats and mortality, are more consistent with the observed differences in abundance and stratigraphic distribution in Dominican *Turbo*. This possibility underscores the need for integrating biological and ecological data into taphonomic research because experimental investigations of the physical behaviour of bioclasts may not illuminate the primary forces producing preservation differences within and among species.

4.5.3 The Fidelity of the Dominican Neogene Fossil Record

Previous taphonomic research suggests that shallow marine depositional environments, such as those preserved in the sediments of the Río Cana and Río Gurabo sections of the Dominican Republic, may accurately record species richness, relative abundance, and age and size frequency distributions (Kidwell and Bosence, 1991; Hartshorne et al., 1987; Fürsich and Flessa, 1987; Powell et al., 1982; Staff and Powell, 1990). For example, Hartshorne, Gillespie, and Flessa (1987) found that postmortem transportation and destruction of gastropod shells do not act in a selective manner: currents and tides were able to move both large and small shells, and abrasion, bioerosion and dissolution affected large and small shells equally. Likewise, Cummings et al. (1986) (see also Kidwell and Bosence, 1991) report on strong qualitative (and in many cases quantitative) concordance between life and death assemblage size frequency distributions. Nevertheless, qualitative and quantitative changes in size frequency distributions as a result of size sorting, abrasive reduction, diagenetic filtering, hermit crab and bird transport, and many other processes have also been extensively documented in molluscs (Kidwell and Bosence, 1991; Tanabe and Arimura, 1987; Shimoyama, 1985; Cadee, 1982). Thus, while it is possible to generalize about the fidelity of particular depositional environments, notable exceptions have also been documented.

We end with the question of which skeletal hard part provides a better estimate of palaeobiological patterns. Hickman's (1992) study of Dominican *Turbo* reported that opercula provide better estimates of population size structure than shells and that estimates of population size structure using opercula are skewed toward smaller sizes. In contrast to Hickman (1992), this study indicates that comparisons of size patterns between shells and opercula of *T. dominicensis* and *T. crenulatoides* are not significantly different. This suggests that, like Hartshorne, Gillespie, and Flessa's (1987) study of gastropod taphonomy, post-mortem transportation and destruction is not size-correlated and, as a result, the fidelity of size-frequency distributions appears to be high. Estimates of species richness and abundance appear to be most appropriate using opercula, whereas size patterns appear to be similar using either hard part. Overall, then, the lack of preservation consistency between these two morphologically similar *Turbo* species highlights the need for caution against taphonomic extrapolation even from one congener to another.

4.6 Conclusions

The Neogene stratigraphic sections of the northern Dominican Republic have been and will continue to serve as a productive research system for exploring macroevolutionary, palaeoecological, and biostratigraphic questions. Few palaeobiological studies of macroinvertebrates have involved comparable sampling intensity or collection magnitude: more than 200 samples containing more than three tons of material were collected. Despite such an unusually well-studied and sampled stratigraphic system, the hypothesis that shells and opercula from the same species produce similar estimates of diversity, abundance, and stratigraphic distribution was rejected. If only turbinid shells had been studied, abundance would have been underestimated by 75% and species richness would have been underestimated by 60%. These results suggest that estimates of diversity and abundance in other Dominican invertebrates—even those with durable skeletons—may be significantly underrepresented. Although significantly fewer shells were preserved and/or sampled than opercula, studies of turbinid population structure and morphological patterns in shells and opercula display similar patterns. This result is encouraging for studies of stasis and change within these well-studied sections. Overall, our study provides a lesson in the limits of taphonomic extrapolation from one related and morphologically similar species to another. Unique biological and ecological factors may influence palaeobiological signals to an equal or greater extent than physical biostratinomic processes.

Acknowledgments The material for this study was provided to by Peter Jung and Rene Pauchaud of the Naturhistorisches Museum, Basel, Switzerland; Jack and Winifred Gibson-Smith of Surrey, England; Emily Vokes of Tulane University; Roger Portell and Fred Thomson of the Florida Museum of Natural History at Gainesville; Robert Van Syk and Peter Roopnarine of the California Academy of Sciences and Gary Rosenberg of the Academy of Natural Sciences, Philadelphia. We are grateful for the access to these collections and the generous hospitality and assistance provided by these individuals during visits to the collections. Field assistance in the Dominican Republic by Brian Beck is also appreciated. We thank Laurie Anderson for reviews of the manuscript. Financial support by the National Science Foundation (DEB 9520457 and EAR Career) is gratefully acknowledged. This is a contribution of The University of California Museum of Palaeontology.

References

Adanson, M., 1757, *Histoire naturelle du Sénegal (Coquillages), Avec la relation abrégé d'un voyage fait en ce pays pendant les années* 1749–53, Paris, Bauche.

Anderson, L.C., 1994, Palaeoenvironmental control of species distributions and intraspecific variability in Neogene Corbulidae (Bivalvia: Myacea) of the Dominican Republic, *J. Palaeontol.*, **68**:460–473.

Anderson, L.C., 1996, Neogene Palaeontology in the northern Dominican Republic, 16, The Family Corbulidae (Mollusca: Bivalvia), *Bull. Am. Palaeontol.*, **110**:1–34.

Bold, W. A. van den, 1988, Neogene palaeontology of the northern Dominican Republic, 7, The subclass Ostracoda (Arthropoda: Crustacea), *Bull. Am. Palaeontol.*, **94**:1–105.

Budd, A.F., Johnson, K.G., and Stemann, T.A., 1996, Plio-Pleistocene turnover and extinctions in the Caribbean reef coral fauna, in: *Evolution and Environment in Tropical America* (Jackson, J.B.C., A.F. Budd, and A.G. Coates, eds.), University of Chicago Press, Chicago, IL, pp. 168–204.

Cadee, G.C., 1982, Low juvenile mortality in fossil brachiopods, some comments. *Interne Versalagen Ned. Inst. Onderzoek Zee,* **3**:1–29.

Checa, A.G. and Jiménez-Jiménez, A.P., 1998, Constructional morphology, origin, and evolution of the gastropod operculum, *Palaeobiology*, **24**:109–132.

Cheetham, A.H., 1986, Tempo of evolution in a Neogene Bryozoan: rates of morphologic change within and across species boundaries, *Palaeobiology*, **12**:190–202.

Cheetham, A.H., 1987, Tempo of evolution in a Neogene Bryozoan: are trends in single morphologic characters misleading?, *Palaeobiology*, **13**:286–296.

Costa, F., Nehm, R.H., and Hickman, C., 2001, Neogene Palaeontology in the northern Dominican Republic, 22, The Family Neritidae, *Bull. Am. Palaeontol.*, **359**:47–71.

Cummings, H., Powell E.N., Stanton, R.J., and Staff, G., 1986, The size frequency distribution in palaeoecology: effects of taphonomic processes during formation of molluscan death assemblages in Texas bays, *Palaeontology*, **29**:495–518.

Donovan, S.K. and Paul, C.R.C. (eds), 1998, *The Adequacy of the Fossil Record*, Wiley, New Tork.

Engstrom, N.A., 1982, Escape responses of *Turbo castanea* to the predatory gastropod *Fasciolaria tulipa*, *Veliger*, **25**:163–168.

Fleischmann, A., 1932, Vergleichende Betrachtungen über das Schalenwachstum der Weichtiere (Mollusca). II. Deckel (operculum) und Haus (Concha) der Schnecken (Gastropoden), *Zeitschrist für Morphologie und Ökologie der Tiere*, **25**:549–622.

Foster, A. B., 1986, Neogene palaeontology in the Northern Dominican Republic, 3, The Family Poritidae (Anthozoa: Scleractinia), *Bull. Am. Palaeontol.*, **90**: 47–123.

Frankovich, T.A. and Zieman, J.C., 2005, A temporal investigation of grazer dynamics, nutrients, seagrass leaf productivity, and epiphyte standing stock, *Estuaries*, **28**:41–52.

Fretter, V. and Graham, A., 1994, *British Prosobranch Molluscs*, 2nd ed., Ray Society, London.

Fürsich, F.T. and Flessa, K.W., 1987, Taphonomy of tidal flat molluscs in the northern Gulf of California: palaeoenvironmental analysis despite the perils of preservation, *Palaios*, **2**:543–559.

Hartshorne, P.M., Gillespie, W.B., and Flessa, K.W., 1987, Population structure of live and dead gastropods from Bahia la Choya, in: *Palaeoecology and Taphonomy of Recent to Pleistocene Intertidal Deposits, Gulf of California* (Flessa, K.W., ed.), Palaeontological Society Special Publication, **2**:139–149.

Hickman, C.S., 1992, Interpreting the separate taphonomic fates of turbinid gastropod shells and opercula in fossil mollusk assemblages, *Western Society of Malacologists Annual Report*, **24**:18–19.

Hickman, C.S., 1998, Subfamily Turbininae, in: *Mollusca: The Southern Synthesis, Fauna of Australia* (Beesley, P.L., Ross, G.J.B. and A. Wells, eds.), Vol. 5. Melbourne, CSIRO Publishing, pp. 675–676.

Hickman, C.S., 2003, Modes of formation of gastropod operculum concentrations, *Western Society of Malacologists Annual Report,* **34**:17.

Hickman, C.S., 2005, Evolution on flexible hard substrates: metazoan adaptations for life on seagrasses, *Geol. Soc. Am., Abstr. Prog.*, **37**, 7:181.

Hickman, C.S. and McLean, J.H., 1990, *Systematic revision and suprageneric classification of trochacean gastropods*, Natural History Museum of Los Angeles County, Science Series 35:1–169.

Jarrett, N., Nehm, R.H., Hidalgo, Y., and Cabrera, I., 2004, Quantifying the effects of sampling method and effort on estimates of species richness and relative abundance of Neogene benthic marine mollusks (Cibao Valley, Dominican Republic), *Geol. Soc. Am., Abstr. Prog.*, **36**(5):366.

Kessel, E., 1941, Über Bau und Bildung des Prosobranchier-Deckels, *Zeitschrift für Morphologie und ökologie der Tiere*, **38**:197–250.

Kidwell, S.M. and Bosence, D.W., 1991, Taphonomy and time-averaging of marine shelly faunas. In *Taphonomy: Releasing Data Locked in the Fossil Record* (Allison, P.A. and D.E.G. Briggs, eds.), Topics in Geobiology, Plenum, New York, pp. 115–209.

Maury, C.J., 1917a, Santo Domingo type sections and fossils, Part 1, *Bull. Am. Palaeontol.*, **5**, 29:1–251.

Maury, C.J., 1917b, Santo Domingo type sections and fossils, Part 2, *Bull. Am. Palaeontol.*, **5**, 30:1–43.

Martin, R.E., 1999, *Taphonomy: A Process Approach*, Cambridge University Press, Cambridge.

Miller, A.I., 1988, Spatial resolution in subfossil molluscan remains: implications for palaeobiological analysis, *Palaeobiology*, **14**:91–103.

Nehm, R.H., 2001, Neogene Palaeontology in the northern Dominican Republic, 21, The Genus *Prunum*, *Bull. Am. Palaeontol.*, **359**:1–46.

Nehm, R. H., 2005, Patterns and processes of evolutionary stasis and change in *Eratoidea* (Gastropoda: Marginellidae) from the Dominican Republic Neogene, *Carib. J. Sci.*, **41**:189–214.

Nehm, R.H., 2006, The Dominican Republic Project, Accessed online at: www.dominicanrepublic-project.org

Nehm, R.H. and Geary, D., 1994, A gradual morphologic transition during a rapid speciation event in marginellid gastropods (Neogene; Dominican Republic), *J. Palaeontol.*, **68**:787–795.

Powell, E.N., Stanton, R.J., Cummings, H., and Staff, G., 1982, Temporal fluctuations in bay environments—The death assemblage as a key to the past, *Proceedings Symposium Recent Benthological Investigations in Texas and adjacent states*, Academy of Science, Austin, TX.

Saunders, J.B., Jung, P., Geister, J., and Biju-Duval, B., 1982, The Neogene of the south lank of the Cibao Valley, Dominican Republic: a stratigraphic study, *Transactions of the 9th Caribbean Geological Conference, Santo Domingo*, 1980.

Saunders, J.B., Jung, P., and Biju-Duval, B., 1986, Neogene palaeontology of the northern Dominican Republic, 1, Field surveys, lithology, environment, and age, *Bull. Am. Palaeontol.*, **89**:1–79.

Shimoyama, S., 1985, Size-frequency distribution of living populations and dead shell assemblages in a marine intertidal sand snail, *Umbonium (Suchium) moniliferum* (Lamarck), and their palaeoecological significance, *Palaeogr. Palaeoclim. Palaeoecol.*, **49**:327–353.

Sokal, R.R. and Rohlf, F.J., 1995, *Biometry: The Principles and Practice of Statistics in Biological Research*, 3rd ed., W.H. Freeman, New York.

Staff, G.M. and Powell, E.N., 1990, Local variability of taphonomic attributes in a parautochthonous assemblage: can taphonomic signature distinguish a heterogeneous environment? *J. Palaeontol.*, **64**:648–658.

Tanabe, K. and Arimura, E., 1987, Ecology of four infaunal bivalve species in the Recent intertidal zone, Shikoku, Japan, *Palaeogr. Palaeoclim. Palaeoecol.*, **60**:219–230.

Vermeij, G.J., 1978, *Biogeography and Adaptation: Patterns of Marine Life*, Harvard University Press, Cambridge, MA.

Vokes, E.H., 1989, Neogene Palaeontology in the Northern Dominican Republic, 8, The Family Muricidae (Mollusca: Gastropoda), *Bull. Am. Palaeontol.*, **97**:1–94.

Walker, S.E., 1989, Hermit crabs as taphonomic agents, *Palaios*, **4**:439–452.

Chapter 5
Early Evolution of the *Montastraea "annularis"* Species Complex (Anthozoa: Scleractinia): Evidence from the Mio-Pliocene of the Dominican Republic

Ann F. Budd[1] and James S. Klaus[2]

Contents

5.1 Introduction

Our understanding of species boundaries in reef corals has changed considerably over the past decade due to new discoveries in the areas of molecular phylogenetics, population genetics, and reproductive biology (Knowlton and Budd, 2001; Willis et al., 2006). Several species, long thought to be highly variable, have been found to be complexes of multiple species, similar to syngameons in plants. Within these species

[1] Department of Geoscience, University of Iowa, Iowa City, IA 52242, USA.
Email: ann-budd@uiowa.edu

[2] Department of Geological Sciences, University of Miami, 43 Cox Science Building, Coral Gables, FL, 3133. Email: jsklaus@rsmas.miami.edu

R.H. Nehm, A.F. Budd (eds.) *Evolutionary Stasis and Change in the Dominican Republic Neogene,*

complexes, hybridization takes place most frequently in marginal habitats at the periphery of species ranges (Fukami et al., 2004a), and is believed to play an important role in range expansion, adaptation to changing environments, and evolutionary diversification (Willis et al., 2006). Nevertheless, due in part to disruptive selection (Wolstenholme et al., 2003; Willis et al., 2006), species are discrete and cohesive evolutionary entities with distinct ecological characteristics. Thus, despite occasional interspecific gene flow and reticulate evolution, species can be traced through geologic time (Budd and Klaus, 2001; Budd and Pandolfi, 2004). The patterns of speciation and extinction within any one complex over geologic time remain largely unexplored.

Here we examine evolutionary stasis within one reef coral species complex, the *Montastraea "annularis"* (Ellis and Solander, 1786) complex, which today consists of three species and dominates Caribbean coral reefs (Knowlton et al., 1992; Weil and Knowlton, 1994; Fukami et al., 2004a; Klaus et al., 2007). We consider stasis at both the level of individual species and the level of the complex as a whole. Previous work on the long-term evolution of the complex has shown that it was significantly more diverse (speciose) during the Plio-Pleistocene and that most species within the complex became extinct during the early to middle Pleistocene. The three modern species are survivors of this extinction episode. They do not represent a monophyletic group but instead each of the three species belongs to a separate subclade within the complex (Budd and Klaus, 2001). Study of late Pleistocene to Recent members of the complex reveals two or more other extinct species in addition to those living today, but diversity was never as high as during the Plio-Pleistocene (Pandolfi et al., 2002; Pandolfi, 2007). Hybridization has been inferred in the Bahamas during the late Pleistocene and has persisted there until today (Budd and Pandolfi, 2004; Fukami et al., 2004a).

Following a previous paper by Budd and Klaus (2001) on the long-term evolution of the complex during the Plio-Pleistocene, the present paper focuses on the early evolution of the complex during the Mio-Pliocene, a relatively environmentally stable time interval preceding closure of the Central American Isthmus (Coates et al., 1992; Collins et al., 1996; Allmon, 2001) and Plio-Pleistocene turnover of Caribbean reef communities (Budd and Johnson, 1999; Jackson and Johnson, 2000). We use samples from a densely collected, continuous, and richly fossiliferous sequence with high resolution age dates, to examine patterns of speciation and extinction within the complex, as well as long-term temporal changes within the complex in diversity, mean morphology, and morphologic disparity. We also examine variation within species through time. We compare our results with those observed in other organisms in the same sequence, and with those observed in the complex during the Plio-Pleistocene.

As in previous work (Budd and Klaus, 2001), our approach to recognizing species and tracing their distributions is based on quantitative analyses of corallite morphology in transverse thin sections. The raw data are two dimensional Cartesian coordinates of landmarks taken on individual corallites within colonies. We perform our analyses using two datasets derived from these data: (1) a geometric morphometric dataset consisting of Bookstein shape coordinates and centroid size, and (2) a dataset consisting of traditional linear measurements, ratios, and septal counts. The first dataset is used in initial analyses distinguishing species, and the second dataset is used to characterize them. As explained in Budd and Klaus (2001), the second dataset represents a first step in constructing a character matrix that will be analyzed in future phylogenetic work examining the evolutionary relationships among species. In the present paper, the results of the two datasets are combined to trace the distributions of species through time, and to examine patterns of variation within and among species through time. Following Cheetham (1986, 1987) and Cheetham et al. (2007), in order to reduce noise, our study of variation within species focuses on overall morphologic variables (i.e., canonical discriminant functions) that best distinguish species. A taxonomic monograph formally describing and naming the species recognized in this work will be prepared after the phylogenetic analyses are complete. Preliminary descriptions are available in the Neogene Marine Biota of Tropical America (NMITA) database, http://nmita.geology.uiowa.edu.

5.2 Materials

5.2.1 Taxa

The present study treats two previously described Neogene species: *Montastraea trinitatis* (Vaughan in Vaughan and Hoffmeister, 1926) and *Montastraea limbata* (Duncan, 1863). Both species have wide distributions across the central and southern Caribbean; *M. trinitatis* ranges from the early Miocene to the early Pliocene, and *M. limbata* ranges from early Miocene to late Pliocene (Budd, 1991). *M. trinitatis* was originally distinguished from modern *M. "annularis"* s.l. by having four septal cycles, paliform lobes, and greater variability in the size and shape of its calices (Vaughan and Hoffmeister, 1926; Budd, 1991). *M. limbata* was originally distinguished from modern *M. "annularis"* s.l. by having thick primary septa, paliform lobes, and more widely spaced calices (Vaughan, 1919; Budd, 1991; Budd et al., 1994). *M. trinitatis* formed hemispherical-shaped colonies; whereas *M. limbata* had a range of colony shapes similar to modern *M. "annularis"* s.l., including thin and thick columns, mounds, and plates (Budd, 1991).

Since this original work, modern *M. "annularis"* s.l. has been discovered to be a species complex composed of three species [*M. annularis* s.s., *M. faveolata* (Ellis and Solander, 1786), *M. franksi* (Gregory, 1895)], based on multiple lines of evidence integrating genetic, reproductive, ecologic, and morphologic data (Knowlton et al., 1992; Weil and Knowlton, 1994). Although statistically significant differences among species exist, traditional morphologic features overlap among species and do not provide enough resolution to recognize clusters of colonies that correspond with species within the complex. We have therefore been searching for new, more refined morphologic features that better match the genetic data. The most promising morphologic features that have been found to date are related to the corallite wall and the extension of costae beyond the wall, and are most effectively quantified using geometric morphometrics (Budd and Klaus, 2001; Knowlton and Budd, 2001).

Morphometric analyses of Plio-Pleistocene *M. limbata* from Costa Rica and Panama using these new features has shown that it too is a complex composed of numerous species. Moreover, preliminary phylogenetic analyses based on characters related to these new morphologic features indicate that the two complexes (*M. "annularis"* s.l. and *M. "limbata"* s.l.) do not form separate clades, but instead belong to the same monophyletic group (Budd and Klaus, 2001). In addition, similar morphometric analyses of late Miocene *M. limbata* from the Dominican Republic show that it also consisted of numerous species, and that these species are not the same as those in the Plio-Pleistocene of Costa Rica and Panama (Klaus and Budd, 2003). The present study deals with the gap in knowledge of the *M. "annularis"* (= "*limbata*") complex between the late Miocene of Dominican Republic and the Plio-Pleistocene of Costa Rica and Panama.

5.2.2 Geologic Setting

In the present study, samples from four different stratigraphic levels in the Cibao Basin of the northern Dominican Republic are analyzed: (1) Baitoa Formation (early to middle Miocene), (2) Cercado Formation (late Miocene, ~6.5 to 5.6 Ma), (3) Gurabo Formation (late Miocene to early Pliocene, 5.6 to ~4.5 Ma), and (4) Mao Formation (early to late Pliocene, ~4.5 to ~3.4 Ma). The latter three formations ("the DR sequence") comprise the Yaque Group, a thick, nearly continuous, well-preserved, and richly fossiliferous wedge of mixed carbonate and siliciclastic sediments deposited on the northern flank of the Cordillera Central during Mio-Pliocene time (McNeill et al., this volume). The present study focuses on exposures of the Yaque Group formations along the Río Gurabo and Río Cana. The ages of these exposures are in the process of being revised using up-to-date high-resolution techniques that integrate microfossil, paleomagnetic and strontium-isotopic data (McNeill et al., this volume). For the purposes of distinguishing species in the present analyses, the localities are grouped into four "levels" by formation. Stasis is then examined using both stratigraphic levels and 100 Kyr time bins.

As explained in Johnson et al. (this volume), the DR sequence is one of the few reef-bearing sequences in the Caribbean that covers the period of time between 6.5 and 3.4 Ma, and is critical for understanding the origination of the modern Caribbean reef coral fauna.

5.2.3 Sampling

Reef corals within the DR sequence have been intensively collected during more than six field expeditions between 1978–2000, and the samples analyzed in the present study include all of the well-preserved and identifiable specimens of the *M. "annularis"* – like corals that were collected before 2001. Klaus et al. (this volume)

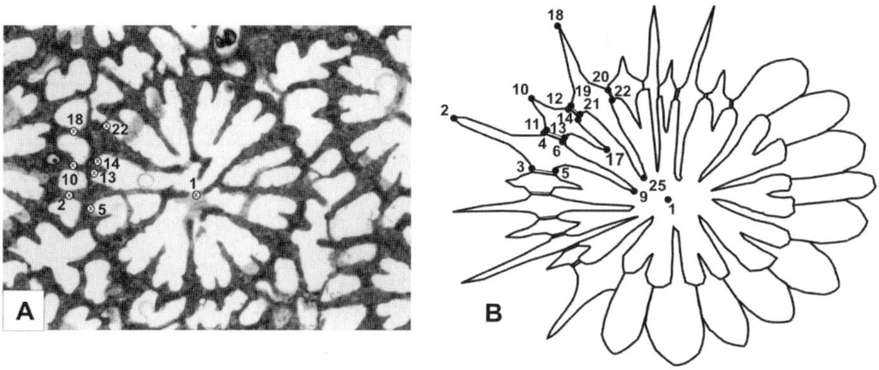

Fig. 5.1 Two-dimensional Cartesian coordinates collected for 27 landmarks on transverse thin-sections of corallites. Eight of the 27 landmarks are indicated on the thin-section on the left. Landmarks were selected to characterize the structure of the corallite wall and costal extensions beyond the wall. To facilitate morphologic interpretations, three different baselines (13–14, 5–6, 21–22) were used to calculate Bookstein shape coordinates

assess sampling adequacy for these collections, and determine that reef coral diversity within the sequence has been adequately sampled. In this assessment, rarefaction curves for each formation level off at 40–70 species (Fig. 5.1A of Klaus et al., this volume); cumulative number of species curves level off at ~100 species (Fig. 5.2D of Klaus et al., this volume). Estimates of completeness are in progress.

The samples in Level 1 (Baitoa Formation) consist of nine colonies that were collected at four localities (NMB localities 16943, 16945, 17283, 17284) within the Baitoa Formation of the López section along the Río Yaque del Norte (Table 5.1, Appendix 1). The stratigraphic locations of the four localities are given in text-figure 25 of Saunders et al. (1986). The colonies were part of the original collections made by the Saunders & Jung team in 1978–1980, and were originally identified as *Montastraea trinitatis* by Budd (1991).

The samples in Level 2 (Cercado Formation) consist of a total of 78 colonies. Four colonies were collected from the Cercado Formation along Río Cana (NMB locality 16853) and from the Arroyo López section on Río Yaque del Norte (NMB locality 17273) by the Saunders & Jung team in 1978–1980 (Table 5.1, Appendix 1; text-figures 15 and 25 of Saunders et al., 1986). In Budd (1991), the colonies from NMB locality 17273 were identified as *M. trinitatis*, and the colony from NMB locality 16853 as *Montastraea limbata*. The remaining 74 colonies were collected in the Cercado Formation along Arroyo Bellaco by Klaus and Budd (2003), who subdivided them into four morphospecies (AB1–4) using landmark-based morphometrics similar to those used in the present study.

The samples in Level 3 consist of a total of 64 colonies collected in the Gurabo Formation in the Río Gurabo (non-reefal environments) and in the Río Cana (reefal environments; Table 5.1, Appendix 1). The samples collected in non-reefal

Table 5.1 List of localities. "Lv", stratigraphic level. "NMB", localities registered by the Natural History Museum in Basel, Switzerland (see Saunders et al., 1986). "BK00", localities first collected by Budd and Klaus in 2000 (see Klaus and Budd, 2003)

Lv	Locality	No of cols	River	Formation	Section	Strat. Elevation	Age estimate
1	NMB16943	6	Yaque del Norte	Baitoa	Lopez		e–m. Mio
1	NMB16945	1	Yaque del Norte	Baitoa	Lopez		e–m. Mio
1	NMB17283	1	Yaque del Norte	Baitoa	Lopez		e–m. Mio
1	NMB17284	1	Yaque del Norte	Baitoa	Lopez		e–m. Mio
2	NMB17273	3	Yaque del Norte	Cercado	A. Lopez		l. Mio
2	BK00-CE(2)	2	Cana (A. Bellaco)	Cercado	Cana	125 m	6.23 Ma
2	BK00-CE(3)	1	Cana (A. Bellaco)	Cercado	Cana	128 m	6.22 Ma
2	BK00-CE(4)	1	Cana (A. Bellaco)	Cercado	Cana	130 m	6.22 Ma
2	BK00-CE(12)	50	Cana (A. Bellaco)	Cercado	Cana	135–140 m	6.21–6.22 Ma
2	BK00-CE(6)	20	Cana (A. Bellaco)	Cercado	Cana	135 m	6.21 Ma
2	NMB16853	1	Cana	Cercado	Cana	179 m	6.06 Ma
3	NMB16823	1	Cana (Zamba)	Gurabo	Cana	338 m	5.13 Ma
3	NMB16822	1	Cana (Zamba)	Gurabo	Cana	342 m	5.12 Ma
3	NMB16818	8	Cana (Zamba)	Gurabo	Cana	348 m	5.11 Ma
3	NMB16814	1	Cana (Zamba)	Gurabo	Cana	350 m	5.1 Ma
3	NMB16817	1	Cana (Zamba)	Gurabo	Cana	350 m	5.1 Ma
3	BK00–3	11	Cana (Zamba)	Gurabo	Cana	350 m	5.1 Ma
3	NMB15855	13	Gurabo	Gurabo	Gurabo	275 m	5.22 Ma
3	NMB15858	4	Gurabo	Gurabo	Gurabo	274 m	5.22 Ma
3	NMB16933	1	Gurabo	Gurabo	Gurabo	275 m	5.22 Ma
3	NMB16883	1	Gurabo	Gurabo	Gurabo	278 m	5.22 Ma
3	NMB15847	11	Gurabo	Gurabo	Gurabo	279 m	5.22 Ma
3	NMB16138	1	Gurabo	Gurabo	Gurabo	284 m	5.21 Ma
3	NMB16136	5	Gurabo	Gurabo	Gurabo	313 m	5.15 Ma
3	NMB15837	4	Gurabo	Gurabo	Gurabo	374 m	4.95 Ma
3	NMB15808	1	Gurabo	Gurabo	Gurabo	381 m	4.92 Ma
3	NMB16921	5	Gurabo	Gurabo	Gurabo	381 m	4.92 Ma
4	NMB16884	17	Cana (Cana Gorge)	Mao	Cana	988 m	3.48 Ma

environments (i.e., lacking reef framework) include 31 colonies collected in Río Gurabo at localities along the river bend 1500 m upstream from the Los Quemados bridge [NMB localities 15847, 15855, 15858, 16136, 16138, 16883, 16933], and 10 colonies collected in Río Gurabo along a coral-rich ledge <500 m downstream from the Los Quemados bridge [NMB localities 15808, 15837, 16921]. The samples collected in reefal environments include 23 colonies that were collected in Cañada de Zamba [NMB localities 16814, 16817, 16818, 16822, 16823; Locality BK00–3]. Fourteen of the 64 colonies were part of the original collections made by the Saunders and Jung team in 1978–1980 and identified as *Montastraea limbata* in Budd (1991); the remaining 50 colonies were collected by Budd and Klaus during a field trip in 2000.

The samples in Level 4 consist of a total of 15 colonies, all of which were collected in the Mao Formation at NMB locality 16884 in Cana Gorge on Río Cana (Table 5.1, Appendix 1). Three colonies were part of the original collections made by the Saunders and Jung team in 1978–1980 and identified as *Montastraea limbata* in Budd (1991); the remaining 12 colonies were collected by Budd and Klaus during a field trip in 2000.

For comparative purposes, data were also collected on 30 genetically-characterized modern colonies from a shallow protected fringing reef environment (~10 m in water depth) in the San Blas Islands of Panama, including 10 *M. annularis* s.s, 10 *M. faveolata*, and 10 *M. franksi* (Appendix 1). These collections are the same as those analyzed morphometrically in Budd and Klaus (2001), Pandolfi et al. (2002), Klaus and Budd (2003), Fukami et al. (2004a), Budd and Pandolfi (2004), and Klaus et al. (2007).

5.3 Methods

5.3.1 Data Collection

To distinguish species and examine changes within and among species through geologic time, the 2D Cartesian coordinates (x–y) of 27 landmarks were digitized on images of mature corallites in transverse thin-section (Fig. 5.1, Appendix 2). Three adjacent costosepta were digitized on six mature calices on the top of each colony. The landmarks consist of spatially homologous points selected to characterize the shape of the corallite wall and associated costosepta, and are the same as those used in Budd and Klaus (2001), Klaus and Budd (2003), and Budd and Pandolfi (2004). Of the 27 landmarks, only 19 were analyzed in the present study (Appendix 2); type 3 landmarks in the classification system of Bookstein (1991) and landmarks that are not located on the three adjacent costosepta were not analyzed.

5.3.2 Data Processing and Calculation of Two Morphologic Datasets

5.3.2.1 First Dataset

Centroid size and shape coordinates (Bookstein, 1991; Zelditch et al., 2004) were calculated for the landmark data using the computer program CoordGen6 in the IMP software series (Integrated Morphometrics Package, 2004, written by H. David Sheets, available at http://www2.canisius.edu/~sheets/morphsoft.html). Centroid size was calculated by summing the squared distances from each of the 27 landmarks to a common centroid. Shape coordinates were calculated for triplets of points using three baselines: (1) points 13–14, (2) points 5–6, and (3) points 21–22. To facilitate morphologic interpretation, 17 shape coordinates associated with the structure and development of the corallite wall and costosepta (Table 5.2) were selected for use in statistical analyses.

5.3.2.2 Second Dataset

For comparison with traditional measurements used in formal published species descriptions, 27 length measurements were calculated for the same landmark data using the computer program TMorphGen6 in the IMP software series (Integrated Morphometrics Package, 2004, written by H. David Sheets, available at http://www2.canisius.edu/~sheets/morphsoft.html). Seventeen traditional measurement variables (Table 5.3) were then calculated consisting of averages and ratios based

Table 5.2 Dataset 1. Variables (Bookstein shape coordinates and centroid size) used in statistical analyses distinguishing fossil clusters

Coordinate	Baseline	Morphologic feature
y1	13,14	Corallite diameter
y2	5,6	Primary costa extension and wall thickness
x4	13,14	Tertiary costa shape (incl outer wall dissepiment)
y4	13,14	Wall thickness
x6	13,14	Inner wall dissepiment thickness
y9	5,6	Primary septum length
y10	13,14	Tertiary costa extension and wall thickness
x11	13,14	Tertiary costa shape
y11	13,14	Wall thickness
x12	13,14	Tertiary costa shape
y12	13,14	Wall thickness
y17	13,14	Tertiary septum length
y18	21,22	Secondary costa extension and wall thickness
x19	13,14	Tertiary costa shape (incl outer wall dissepiment)
y19	13,14	Wall thickness
x21	13,14	Inner wall dissepiment thickness
y25	21,22	Secondary septum length
Csize		Centroid size (27 landmarks)

Table 5.3 Dataset 2. Traditional measurement variables and the results of Kruskal-Wallis (KW) and Tukey's HSD multiple comparisons tests comparing fossil clusters. In the measurement column, pairs of numbers refer to distances between landmarks. All of the Kruskal-Wallis chi-square values are significant at $p < 0.05$. Subsets of fossil clusters recognized by Tukey's tests are given in footnotes, in which numbers refer to fossil clusters

Morphologic feature	Measurement	KW chi-square	Tukey's HSD multiple comparisons test ($p < 0.05$)
Corallite diameter	cd=((1,14)+(1,22) + (1,13)+(1,5))/4	75.7	3 subsets[a]
Relative columella size	clwrat= (((1,9) + (1,25))/2)/cd	54.3	3 subsets[b]
Number of septa	ns= count of septa	56.7	2 subsets[c]
Prim. vs. sec. costa length	p_sclrat=pcl/scl	43.2	n.s.
Prim. vs. sec. costa width	p_scwrat=(5,6)/(21,22)	48.1	3 subsets[d]
Prim. vs. tert. costa length	p_tclrat=pcl/tcl	57.5	2 subsets[e]
Prim. vs. tert. costa width	p_tcwrat=(5,6)/(13,14)	25.0	2 subsets[f]
Wall dissepiment thickness	para=((4,11)+(12,19) + (6,13)+ (14,21))/4	130.0	5 subsets[g]
Primary costa length	pcl=((3,2)+(4,2))/2	62.5	3 subsets[h]
primary costa shape	pcwrat= (3,4)/(5,6)	56.4	5 subsets[i]
Secondary costa length	scl= ((19,18)+(20,18))/2	76.2	4 subsets[j]
Secondary costa shape	scwrat= (19,20)/(21,22)	38.0	3 subsets[k]
Tertiary costa length	tcl=((11,10)+(12,10))/2	75.0	4 subsets[l]
Tertiary costa width	tcw= (13,14)	82.9	4 subsets[m]
Tertiary costa shape	tcwrat= (11,12)/(13,14)	98.1	4 subsets[n]
Relative tert. septum length	tslrat=(1,17)/cd	45.3	n.s.
Wall thickness	wt= ((11,13)+(3,5)+(12,14) + (20,22))/4	114.4	4 subsets[o]

[a] [35=42=43=21=32=31=33=24=34=41=11]<[32=31=33=24=34=41=11=23] < [11=23=22]
[b] [11=31=24=43=34=23=35=22=21=32]<[31=24=43=34=23=35=22=21=32=41]<[43=34=23=35=22=21= 32=41=42=33]
[c] [24=42=34=31=33=32=23=21=43=11=22=35]<[41]
[d] [33=31=32=21=24=22=23=43]<[31=32=21=24=22=23=43=34=35=42=41]< [43=34=35=42=41=11]
[e] [31=22=11=35=21=43=34=42=41=32=23=24]<[43=34=42=41=32=23=24=33]
[f] [34=41=23=24=42=32=11=22=43=31=33=21]<[41=23=24=42=32=11=22= 43=31=33=21=35]
[g] [21=22=31=41=35=11=43]<[22=31=41=35=11=43=34=32]< [31=41=35=11=43=34=32=23]< [34=32 =23=42]<[33=24]
[h] [35=21=11=31=41=32=42=43=33=22=34]<[33=22=34=23]<[22=34=23=24]
[i] [24=11=23=42=33=32=34=21=22]<[23=42=33=32= 34=21=22=31] <[42 =33= 32=34=21=22= 31=41]<[32 =34=21=22=31=41=35]<[21=22=31=41=35=43]
[j] [11=21=35=42=41=31=32=43=22]<[35=42=41=31=32=43=22=33]<[42=41 =31=32=43=22=33=23=34]<[43=22=33=23=34=24]
[k] [24=22=42=33=11=21=32=23=31=34]<[32=23=31=34=41=35]<[23=31=34 =41=35 = 43]
[l] [33=41=32=21=42=35=11=43=31]<[21=42=35=11=43=31=34]<[35=11=43 =31=34=23=24]<[31=34=23=24=22]
[m] [33=35=41=42=43=32=21=31]<[35=41=42=43=32=21=31=24]<[24=11=23] <[11=23=34=22]
[n] [24=33=23=42=34=32=11]<[42=34=32=11=21]<[11=21=31=41=31=22=43 =53]<[41=31=22=43=53=35]
[o] [24=42=23=33=32=34=11=21]<[34=11=21=31]<[31=22=43=35]<[22=43=35=41]

on the 27 length measurements and septal counts. Many of the variables in the first dataset involve some aspect of wall thickness; whereas, in contrast, the variables in the second dataset each correspond with a unique skeletal feature.

5.3.3 *Statistical Analyses*

5.3.3.1 Distinguishing Fossil Clusters (CDA Set 1)

To distinguish species in the DR sequence, following Budd and Coates (1992) and Budd and Klaus (2001), analyses were first performed separately on each stratigraphic level in order to reduce noise caused by environmental differences among levels and evolutionary convergence. The first dataset (Table 5.2) was used in these analyses, because it has been found to have higher discriminatory power in previous work (Knowlton and Budd, 2001). In addition to using $p<0.05$ for F-values corresponding to Mahalanobis distances to determine significant differences among groups, clusters composed of genetically characterized colonies of the three modern species were added to the third step of the analysis to serve as a baseline for recognizing species. Using procedures modified from Cheetham (1986), the analysis for each level consisted of three steps: (1) size and shape coordinates for each corallite were used to calculate Mahalanobis distances between all of the colonies within each level; (2) the distance matrix was input into an average linkage cluster analysis, and clusters containing two or more colonies were recognized on the resulting dendrogram, in the present case using a rescaled distance of ~7–8 as the cutpoint; (3) a series of iterative canonical discriminant analyses (Fig. 5.2, CDA set 1) was performed using colony means beginning with the clusters in step 2 and the three modern species as *a priori* groups. In the iterative procedure, pairs of fossil clusters (labeled "FC") were combined if their Mahalanobis distance was statistically insignificant or if they overlapped more than the overlap observed among modern species. In addition, any individual fossil colonies that were misclassified with respect to fossil cluster were reassigned to the nearest fossil cluster. The first step in the analysis takes advantage of the fact that individuals within colonial organisms are genetically identical, and provides the initial framework for the series of iterative analyses that follow. The procedures are therefore uniquely designed for colonial organisms. Cheetham et al. (2007) later simplified these procedures by eliminating the second step: a practice followed by Schultz and Budd (this volume, n=95 colonies) and Beck and Budd (this volume, n = 63 colonies), but not by the present study due to the larger sample size (n=166 colonies).

5.3.3.2 Characterizing Morphologic Differences among Fossil Clusters

To characterize each of the resulting fossil clusters and further evaluate their differences, multiple comparisons tests (Tukey's HSD) and nonparametric analyses of variance (Kruskal-Wallis tests) were performed on each of the traditional measurements in the second dataset (Table 5.3).

5.3.3.3 Tracing Individual Morphospecies Between Levels (CDA Set 2)

To determine if any of the fossil clusters within levels continued between stratigraphic levels, a global canonical discriminant analysis (Fig. 5.2, CDA set 2) was performed combining the data for the four stratigraphic levels and the modern colonies, and using the fossil clusters within each level and the three modern species as *a priori* groups ("16 groups"). Two separate analyses were performed, one for each of the two morphologic datasets described above. Differences among fossil clusters were evaluated using Mahalanobis distances; F-values ($p<.05$) corresponding to pairwise Mahalanobis distances among fossil clusters between levels were used to determine if fossil clusters in different levels belonged to the same species. These results were integrated with the revised age dates for each locality (Table 5.1; McNeill et al., this volume) to determine the stratigraphic range of each species.

5.3.3.4 Speciation, Extinction, Diversity, and Disparity Through Geologic Time (CDA set 2)

The fossil clusters in different stratigraphic levels that were found to be the same (see above) were combined into morphospecies, and speciation and extinction events within the complex were examined in levels 2–4 using first and last occurrences of morphospecies. Level 2 was not included in speciation estimates because of the long hiatus between levels 1 and 2; level 4 was not included in extinction estimates because of the lack of samples in the present study between it and the Recent. Species diversity was estimated by counting the number of morphospecies that occur within each stratigraphic level.

Temporal changes in the mean morphology and morphologic disparity of the whole complex were evaluated using scores on functions in the initial global canonical discriminant analysis (16 groups). Multiple comparisons tests (Tukey's HSD) were used to test for differences in mean morphology between levels. Disparity was evaluated by calculating pairwise Mahalanobis distances among fossil clusters within each level, and then comparing these distances among levels using nonparametric analyses of variance (Kruskal-Wallis and Mann-Whitney tests).

5.3.3.5 Morphologic Changes within Species Through Geologic Time (CDA Sets 2 and 3)

Various methods have been used to evaluate morphologic rates of evolution in the fossil record, including: (1) comparisons of ancestor – descendant pairs using canonical discriminant analysis (e.g., Cheetham, 1986; Geary, 1992; Nehm and

Geary, 1994; Cheetham et al., 2007;); (2) comparisons within species among stratigraphic levels or time bins using Mahalanobis distances (e.g., Stanley and Yang, 1987; Lieberman et al., 1995); (3) comparisons within species among stratigraphic levels or time bins using scores on canonical discriminant functions distinguishing species within clades (e.g., Foster, 1986; Budd, 1991); and (4) comparisons within species among stratigraphic levels or time bins using principal components or individual morphologic traits (e.g., Cheetham, 1987; Geary, 1990, 1992; Nehm and Geary, 1994; Nehm, 2005). In the present study, due to the lack of phylogenetic context and the large amount of noise observed in individual morphologic traits, we

CDA SET 1:
A series of iterative canonical discriminant analyses for each stratigraphic level, in which groups that are not significantly different are combined and the analysis rerun until all groups are significantly different. Data for individual corallites within each colony are used in the first analysis; colony means are used in all subsequent analyses.

A priori groups in initial analysis: colonies within each stratigraphic level
A priori groups in final analysis: a set of "fossil clusters" for each stratigraphic level

Note: In the present study, one fossil cluster (FC11) was found in level 1, 4 fossil clusters (FC21-24) in level 2, 5 fossil clusters in level 3 (FC31-35), and 3 fossil clusters in level 4 (FC41-43), making a grand total of 13 fossil clusters.

↓

CDA SET 2:
A series of "global" (= all-inclusive) canonical discriminant analysis, in which groups that are not significantly different are combined and the analysis rerun until all groups are significantly different

A priori groups in initial analysis: all of the "fossil clusters" from step 1
A priori groups in final analysis: "morphospecies"

Note: In the present study, the 13 fossil clusters from step 1 and three modern species formed the groups in the initial analysis ("16 groups" analysis); 9 fossil morphospecies and three modern species formed the groups in the final analysis ("12 groups" analysis).

↓

CDA SET 3:
A single within-species canonical discriminant analysis

A priori groups: (1) stratigraphic levels or (2) 100 kyr time bins

Note: In the present study, CDA was performed on the morphospecies FC 21=32=42, and the groups consisted of (1) three levels or (2) six time bins.

Fig. 5.2 Canonical discriminant analyses performed in the present study. CDA set 1 was performed using the first dataset only; CDA sets 2 and 3 were performed using both the first and second datasets have used methods two and three of those listed above, and have focused on those aspects of morphology that best distinguish species (see also Schultz and Budd, this volume, and Beck and Budd, this volume, which also both use method three). Using the two methods, two separate sets of analyses were performed, one for each of the two morphologic datasets described above.

For the second method listed above, a separate discriminant analysis was first performed within morphospecies using *a priori* groups defined by (1) stratigraphic levels and (2) "time bins" consisting of 100 Kyr time intervals (Fig. 5.2, CDA set 3). Mahalanobis distances between pairwise combinations of levels and between pairwise combinations of time bins were evaluated using $p < .05$ for F-values corresponding to each Mahalanobis distance.

For the third method listed above, a final global canonical discriminant analysis (12 groups) was performed using *a priori* groups defined by morphospecies (Fig. 5.2, CDA set 2). Patterns of variation within species in the resulting discriminant functions were examined through geologic time using 100 Kyr time bins. Nonparametric analyses of variance (Kruskal-Wallis and Mann-Whitney tests) were performed comparing scores for colonies in different time bins.

The canonical discriminant analyses, analyses of variance, and nonparametric tests were performed using SPSS 14.0 for Windows. Mahalanobis distances were calculated using SAS 9.1 for Windows.

5.4 Results

5.4.1 Distinguishing Fossil Clusters (CDA Set 1)

5.4.1.1 Modern

To provide a baseline for recognizing species in the DR sequence, a canonical discriminant analysis was performed on the 30 genetically-characterized modern colonies using the 17 shape coordinates and centroid size (dataset 1) as variables. The results (Fig. 5.3) show that the three modern species have statistically significant differences and that no overlap occurs among species. Correlations of original variables with all of the canonical variates in canonical discriminant analyses distinguishing both modern species (section 4.1.1) and fossil clusters within each level (sections 4.1.2–4.1.5) are given in Appendix 3.

5.4.1.2 Level 1

The initial dendrogram revealed only one cluster composed of two or more colonies, and two unclassified colonies (NMB D5730, D5732). In the subsequent series of iterative discriminant analyses (Fig. 5.2, CDA set 1), the two unclassified colonies

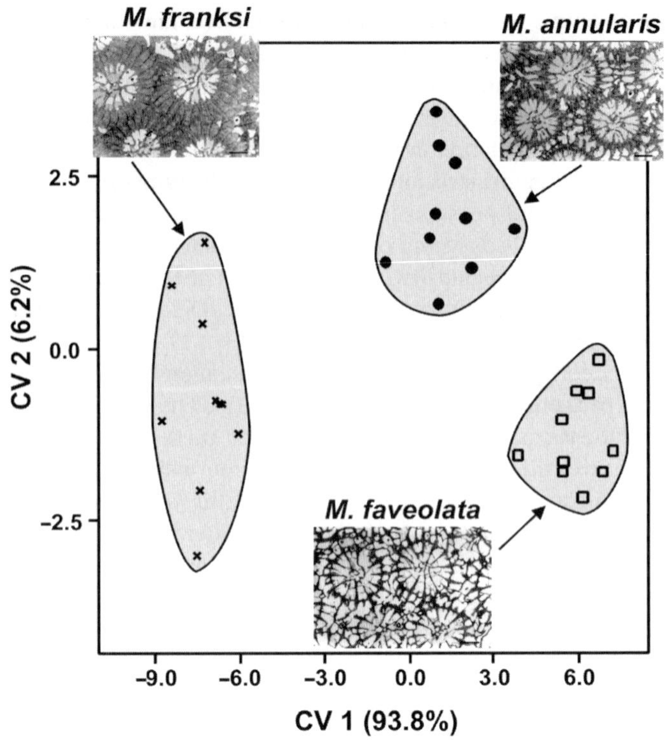

Fig. 5.3 Plots of scores on canonical discriminant functions comparing the three modern species. Each point on the plot represents one colony; polygons enclose each species. Multiple comparisons tests (Tukey's HSD) indicate that all three species differ from one another on CV1 ($p < <0.001$), and that *M. annularis* differs from the other two species on CV2 ($p < <0.001$ for *M. annularis* vs. *M. faveolata*, $p < <0.001$ for *M. annularis* vs. *M. franksi*, $p = .380$ for *M. faveolata* vs. *M. franksi*). The first canonical variate (CV1) is most strongly correlated with four original variables related to wall thickness (y4, y11, y12, y19); whereas the second canonical variate (CV2) is strongly correlated with the inverse of centroid size and, to a lesser extent, the shape of the tertiary costae (x11, x4; Appendix 3)

were assigned to the one fossil cluster (FC11) based on classification probabilities. In the final discriminant analysis (Fig. 5.4A), FC11 (9 colonies) and the three modern species have statistically significant differences ($p<.05$) and no overlap occurs among groups.

5.4.1.3 Level 2

The initial dendrogram revealed eight clusters composed of two or more colonies, and nine unclassified colonies. In the subsequent series of iterative discriminant analyses (Fig. 5.2, CDA set 1), four of the original clusters were combined into one, as were two other original clusters; unclassified colonies were assigned to the remaining four clusters based on classification probabilities. In the final discriminant analysis (Fig. 5.4B), the four fossil clusters (FC21, FC22, FC23, FC24) and the three modern species have statistically significant differences ($p < .05$). No overlap occurs among the four fossil clusters on the plot of the first two canonical variates (Fig. 5.4B), but overlap does occur between two modern species (*M. annularis* and *M. faveolata*) and between FC22 and *M. franksi*. CV4 separates *M. annularis* from *M. faveolata*, and FC22 from *M. franksi*.

Klaus and Budd (2003) also found four morphospecies in their analyses of the same 74 colonies; however, these species differ in definition from the four fossil clusters recognized herein. Species AB-2 and AB-3 of Klaus and Budd (2003), which differ primarily in tertiary septum length, were lumped into one fossil cluster (FC22) in the present analysis. Species AB-4 of Klaus and Budd (2003) was split into two fossil clusters (FC23, FC24), based on differences in CV1. In addition, the species assignments of three of the 74 colonies differ in the present work (Appendix 1). The differences between the present study and Klaus and Budd (2003) may be attributed to the differences in datasets used by the two studies. Although the methods in both studies were similar and the same landmark data were analyzed, different baselines were used in the two studies. Klaus and Budd (2003) used 13 Bookstein coordinates as variables, whereas we used 17 coordinates in the present study.

5.4.1.4 Level 3

The initial dendrogram revealed nine clusters composed of two or more colonies, and 8 unclassified colonies. In the subsequent series of iterative discriminant analyses (Fig. 5.2, CDA set 1), four of the original clusters were combined into one, as were two other original clusters; the unclassified colonies were assigned to the remaining five clusters based on classification probabilities. In the final discriminant analysis (Fig. 5.4C), the five fossil clusters (FC31, FC32, FC33, FC34, FC35) and the three modern species have statistically significant differences ($p < .05$). No overlap occurs among the five fossil clusters or the three modern species on the plot of the first two canonical variates (Fig. 5.4C), but overlap does occur between *M. annularis* and FC32, and between *M. annularis* and FC34.

5.4.1.5 Level 4

The initial dendrogram revealed three clusters composed of two or more colonies, and 2 unclassified colonies. In the subsequent series of iterative discriminant analyses, none of the three clusters could be combined; the unclassified colonies were assigned to these three clusters based on classification probabilities. In the final discriminant analysis (Fig. 5.4D), the three fossil clusters (FC41, FC42, FC43) have statistically significant differences ($p < .05$) and are separated from one another and

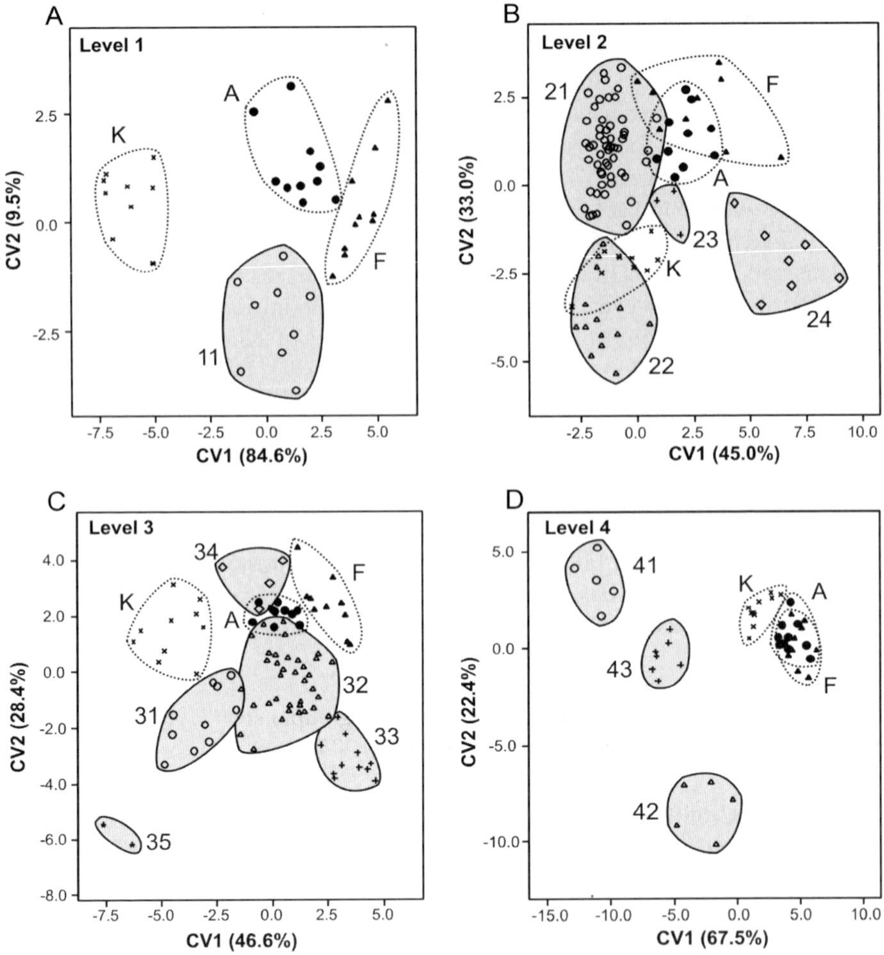

Fig. 5.4 Plots of scores on canonical discriminant functions comparing final fossil clusters ("FC") within each level and the three modern species using the first dataset. Each point on the plots represents one colony. Solid polygons enclose fossil clusters; dotted polygons enclose the three modern species. Numbers refer to fossil clusters; A= *M. annularis* s.s., K = *M. franksi*, F = *M. faveolata*.

A. Level 1. Multiple comparisons tests (Tukey's HSD) indicate that the one fossil cluster (FC11) differs from two modern species on CV1 ($p < <0.001$ for FC11 vs. *M. faveolata*, and for FC11 vs. *M. franksi*; $p = 0.088$ for FC11 vs. *M. annularis*) and from all of the modern species on CV2 ($p < <0.001$). CV1 distinguishes the three modern species; whereas CV2 distinguishes the modern species from FC11. CV1 is most strongly correlated with four original variables related to wall thickness (y4, y11, y12, y19); CV2 is most strongly correlated with amount of inner wall dissepiment (x21, –x6; Appendix 3)

B. Level 2. With the exception of *M. annularis* and *M. faveolata*, the seven *a priori* groups differ on the first two canonical variates. Multiple comparisons tests (Tukey's HSD) indicate three subsets of similar fossil clusters ("FC") and species for CV1 using $p < 0.05$: (1) FC21, FC22, *M. franksi*; (2) FC23, *M. annularis*, *M. faveolata*; and (3) FC24. There are also three subsets for CV2: (1) FC22; (2) *M. franksi*, FC24, FC23; (3) FC23, *M. annularis*, M. *faveolata*. CV1 is most strongly correlated with amount of inner wall dissepiment (x21, –x6); CV2 is strongly correlated with the costa extension (y2, y18, –csize; Appendix 3) the three modern species by gaps. *M. franksi* is distinguished from *M. annularis* and *M. faveolata* on CV1; *M. annularis* and *M. faveolata* are distinguished on CV4.

5.4.2 Characterizing Morphologic Differences among Fossil Clusters

Multiple comparisons tests (Tukey's HSD) revealed significant differences among the 13 fossil clusters in 15 of the 17 traditional measurements in dataset 2; Kruskal-Wallis tests indicated significant differences in all 17 measurements (Table 5.3). The patterns of variation of six measurements with high Kruskal-Wallis chi-square values are shown in Fig. 5.5; and representative corallites are shown in Fig. 5.6.

5.4.3 Tracing Individual Morphospecies Between Levels (CDA Set 2)

The results of the initial global canonical discriminant analysis in the first dataset for the four fossil levels and the modern colonies (16 groups) show that the fossil clusters within each level are statistically different ($p<.05$) from the other fossil clusters in the dataset. The one exception is FC23 in level 2 and FC34 in level 3, which are not statistically significantly different ($p=0.596$ for the Mahalanobis distance). Although statistically significant FC21 and FC32 have a slightly lower Mahalanobis distance (8.981) than the distance between FC23 and FC34 (9.328), and are therefore also considered to be the same in the present analysis. In the 16-group CDA for dataset 1, the first nine functions are significant and seven are needed to explain 95% of the variation. The plot of CV1 versus CV2 (Fig. 5.7A) shows that the three fossil clusters in level 4 are separated on CV1 from all of the other fossil clusters by a

Fig. 5.4 (continued) C. Level 3. With the exception of *M. annularis* and FC34, the eight *a priori* groups differ on the first two canonical variates. Multiple comparisons tests (Tukey's HSD) indicate four subsets of similar fossil clusters ("FC") and species for CV1 using $p<0.05$: (1) FC35, (2) FC31, *M. franksi*; (3) FC34, 32, *M. annularis*; (4) *M. faveolata*, FC33. There are three subsets for CV2: (1) FC35; (2) FC31–33; (3) FC34, *M. franksi*, *M. annularis*, M. *faveolata*. CV1 is strongly correlated with wall thickness (y4, y11, y12, y19, csize), and CV2 with many of the same variables as CV1 except csize (Appendix 3)
D. Level 4. With the exception of *M. annularis* and *M. faveolata*, the six *a priori* groups differ on the first two canonical variates. Multiple comparisons tests (Tukey's HSD) indicate five subsets of similar fossil clusters ("FC") and species for CV1 using $p <0.05$: one composed of *M. annularis* and *M. faveolata*, and four others composed of each of the three fossil clusters and *M. franksi*. There are three subsets for CV2: (1) FC42; (2) FC43, *M. franksi*, *M. annularis*, M. *faveolata*; (3) FC41. CV1 is strongly correlated with costa extension (x2, x18) and CV2 with inner wall dissepiment (x6, −x21; Appendix 3)

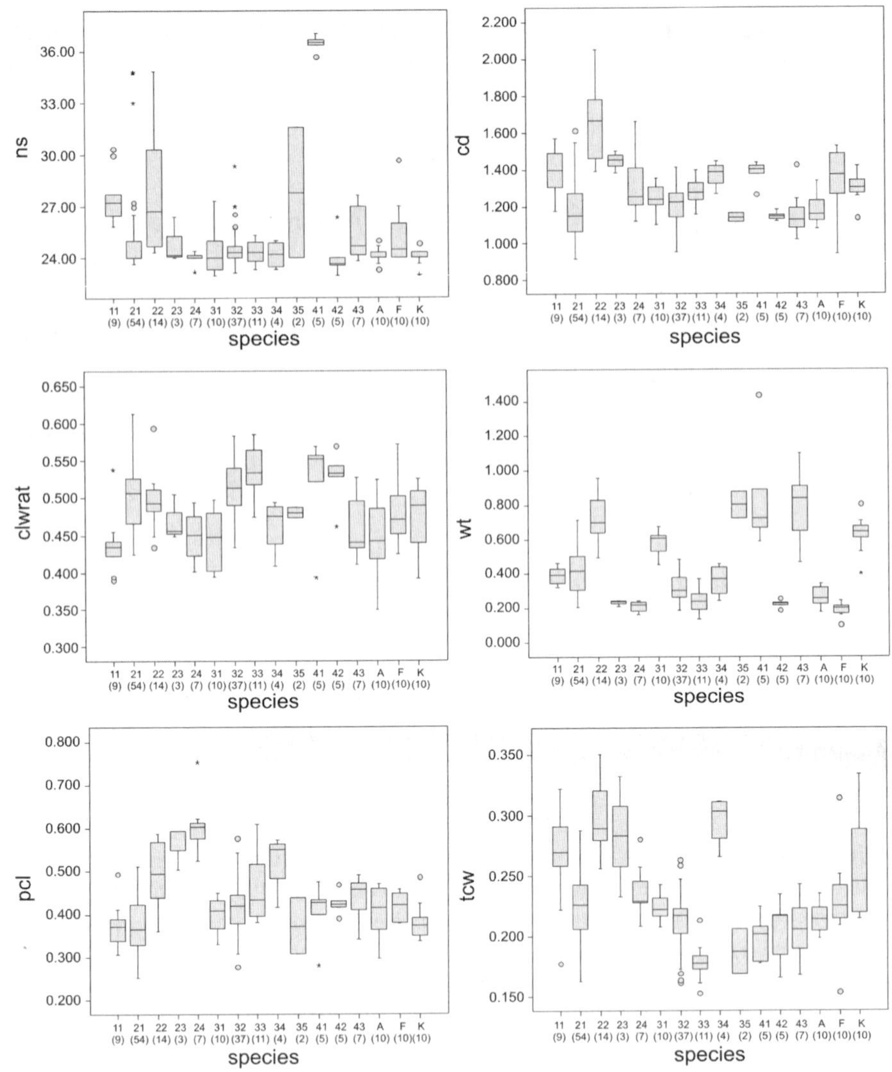

Fig. 5.5 Boxplots showing the median, quartiles, and extreme values for each fossil cluster and modern species. Each box represents the interquartile range containing 50% of the values; the line across each box represents the median. Whiskers represent the highest and lowest values, excluding outliers. Outliers are indicated by dots; extreme cases by asterisks. Numbers refer to fossil clusters; A = *M. annularis* s.s., K = *M. franksi*, F = *M. faveolata*. Sample sizes are given in parentheses. Multiple comparisons tests (Table 5.3) indicate that: (a) numbers of septa (ns) are high in FC41; (b) corallite diameters (cd) are high in FC22 and low in FC21, FC35, FC42, and FC43; (c) columellae (clwrat) are relatively large in FC21, FC32, FC33, FC41, and FC42 and small in FC11; (d) wall thickness (wt) is high in FC22, FC35, FC41, and FC43 and intermediate in FC31; (e) primary costae (pcl) are long in FC22, FC23, FC24, and FC34; (f) tertiary costae (tcw) are wide in FC11, FC22, FC23, and FC34

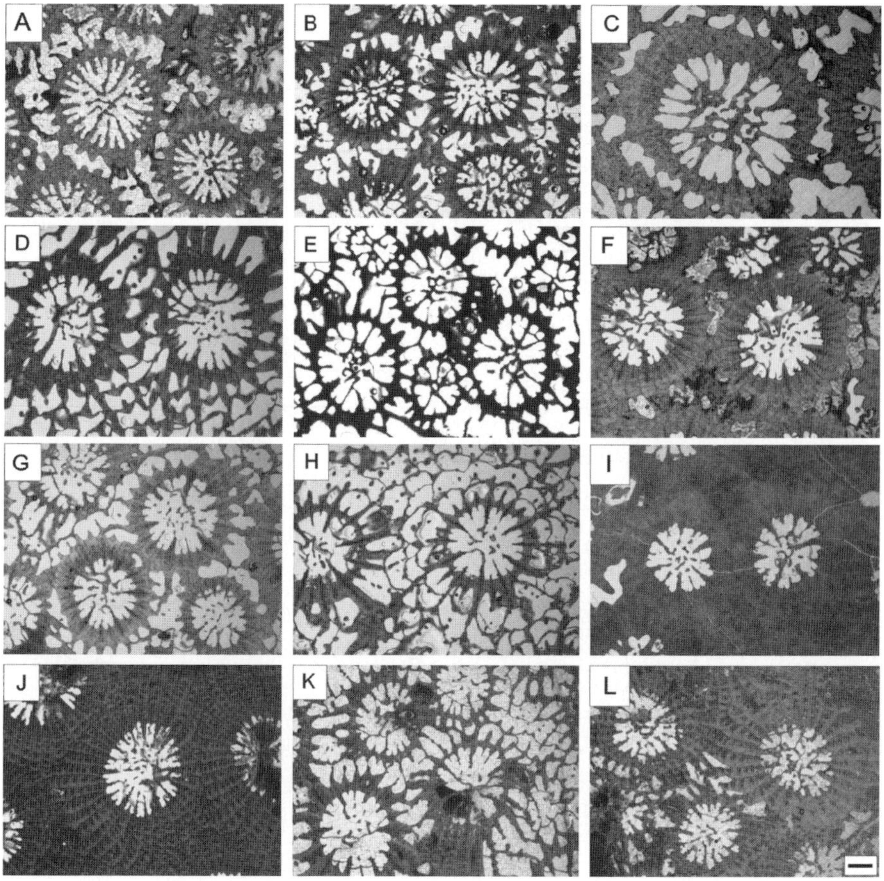

Fig. 5.6 Transverse thin-sections of representative corallites of the 13 fossil clusters. Scale bar = 1 mm. Study of traditional measurements indicates that FC11 is distinguished by a narrow columella and thick costae; FC21, FC32, and FC42 by small corallites and thick columellae; FC22 by large corallites and thick costae; FC23 and FC34 by long, thick costae; FC24 by a thin, dissepimental wall; FC31 by intermediate wall thickness; FC33 by a well-developed porous coenosteum; FC35 and FC43 by small corallites and thick walls; FC41 by ~36 septa per corallite and a thick wall. A. FC11. NMB D5758 (Colony 226). B. FC21. SUI 104528 (Colony 739). C. FC22. SUI 104540 (Colony 798). D. FC23= 34. SUI 104563 (Colony 938). E. FC24. SUI 104523 (Colony 1463). F. FC31. SUI 104593 (Colony 600). G. FC32. SUI 104565 (Colony 940). H. FC33. SUI 104572 (Colony 536). I. FC35. SUI 104570 (Colony 534). J. FC41. SUI 104610 (Colony 583). K. FC42. SUI 104614 (Colony 592). L. FC43. SUI 104604 (Colony 561)

distinctive gap. The fossil clusters in levels 2 and 3 form an intergrading continuum along CV2. The plot of CV2 versus CV3 (Fig. 5.7B) also shows the intergrading continuum of fossil clusters in levels 2 and 3 along CV2. CV3 distinguishes FC41 and FC42 from the rest of the fossil clusters.

The results of the initial global canonical discriminant analysis in the second dataset for the four fossil levels and the modern colonies (16 groups) reveal greater

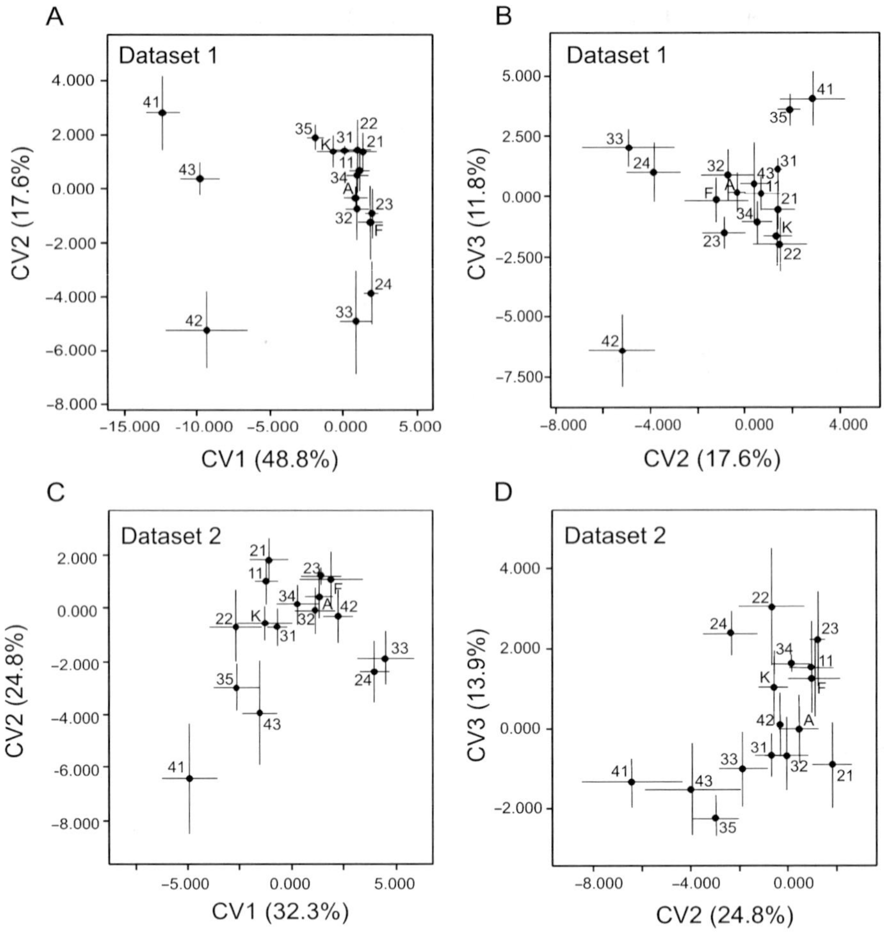

Fig. 5.7 Plots of scores on canonical discriminant functions (global analyses) comparing final fossil clusters in all four levels and the three modern species (16 groups). Means with error bars (one standard deviation) are shown for each fossil cluster and modern species. Numbers refer to fossil clusters; A = *M. annularis* s.s., K = *M. franksi*, F = *M. faveolata*

A. Dataset 1, CV1 vs. CV2. Multiple comparisons tests (Tukey's HSD) for CV1 show that FC41, FC42, and FC35 differ from the other fossil clusters and from each other. Tests for CV2 show that FC42, FC33, and FC24 are the same and differ from the other fossil clusters. CV1 accounts for 48.8% of the variation and is negatively correlated with costa extension (y18, y2, y11), and positively with wall thickness (y4, y11, y12, y19). CV2 accounts for 17.6% of the variation, and is correlated with inner wall dissepiment thickness (x6, x21)

B. Dataset 1, CV2 vs. CV3. Multiple comparisons tests (Tukey's HSD) for CV3 show that FC42 differs from all other clusters, and that FC35 (=FC41) intergrades with the remaining fossil clusters. CV3 accounts for 17.6% of the variation, and is negatively correlated with costa extension (y18, y2, y11), and positively with tertiary costa shape (x11, x12, x4, x19)

C. Dataset 2, CV1 vs. CV2. Multiple comparisons tests (Tukey's HSD) for CV1 show that FC41 and FC24(=FC33) differ from the other fossil clusters, which form two groups that intergrade: (1) FC22, FC35, FC43, FC11, FC21, FC31, and (2) FC34, FC32, FC23, FC42. Tests for CV2 show that FC41 differs from all other clusters, and that FC43, FC35, FC24, and FC33 are intermediate overlap among clusters. Not only are FC23 and FC34 not significantly different ($p = 0.5337$ for the Mahalanobis distance), but so are FC32 and FC42 ($p = 0.1449$ for the Mahalanobis distance), FC42 and *M. faveolata* ($p = 0.0535$ for the Mahalanobis distance), and FC35 and FC43 ($p = 0.3735$ for the Mahalanobis distance). In the 16-group CDA for dataset 2, the first eight functions are significant and seven are needed to explain 95% of the variation. The plot of CV1 versus CV2 (Fig. 5.7C) shows that the three clusters in level 4 are no longer separated from the other species by a gap; however FC41 remains distinct. The plot of CV2 versus CV3 (Fig. 5.7D) shows FC22 to differ from the other fossil clusters on CV3.

In the initial global canonical discriminant analyses (16 groups) for both datasets, the Mahalanobis distances among the three modern species are minimal. In the analysis for the second dataset, p-values for the distances between the three modern species are significantly different. The distance between *M. annularis* s.s. and *M. faveolata* is shortest with $p = 0.0208$. However, in the analysis for the first dataset, the distance between *M. annularis* s.s. and *M. faveolata* is less significant with $p = 0.0945$, indicating that the second dataset provides useful, additional information for distinguishing species.

5.4.4 Speciation, Extinction, Diversity, and Disparity Through Geologic Time

Based on the Mahalanobis distance results reported above, the following fossil clusters were linked into morphospecies: FC21 = FC32 = FC42, FC23 = FC34, FC35 = FC43. This linkage resulted in a total of nine fossil morphospecies. Study of the geologic ages of localities using McNeill et al. (this volume) indicates that eight species existed within the complex between 6.23 and 3.48 Ma. Three of the eight species (FC21 = FC32 = FC42, FC23 = FC34, FC35 = FC43) had durations within the sequence that were greater than one million years, and two species in level 3 (FC31, FC33) had durations less than 0.5 million years (Fig. 5.8). Diversity within the complex remained more or less constant between 6.23 and 3.48 Ma, with total numbers of co-occurring species ranging from 3–5. Speciation and extinction rates appear also to be constant through the interval. Three new species appeared in level 3 and one new species appeared in level 4; whereas two species became extinct in level 2 and two species in level 3.

Fig. 5.7 (continued) between FC41 and the rest of the clusters. CV1 accounts for 32.3% of the variation and is negatively correlated with wall thickness (wt), and positively with dissepiment thickness (para). CV2 accounts for 24.8% of the variation, and is negatively correlated with wall thickness (wt), dissepiment thickness (para), and number of septa (ns)
D. Dataset 2, CV2 vs. CV3. Multiple comparisons tests (Tukey's HSD) for CV3 show that the clusters completely intergrade with FC31, FC32, FC21, FC35, FC41, and FC43 on one end of the series, and FC22, FC24, FC23, FC34, and FC11 on the other. CV3 accounts for 13.9% of the variation, and is correlated with corallite diameter (cd) and costa width (tcw)

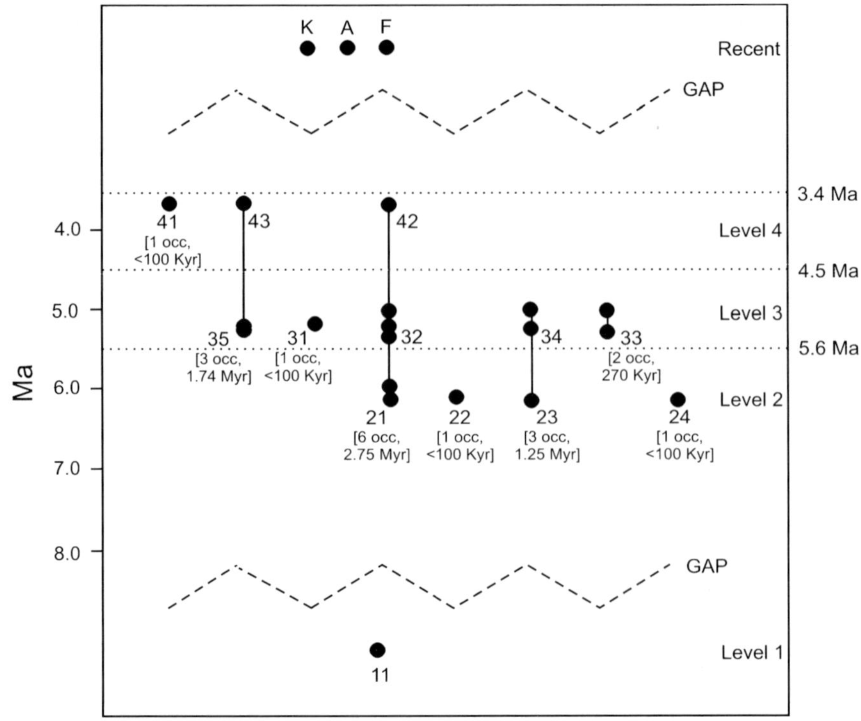

Fig. 5.8 Diagram showing the ranges of the eight resulting morphospecies in the DR sequence. Number of occurrences in 100 Kyr time bins and species durations are indicated in brackets for each species

Temporal changes in mean overall morphology within the complex were analyzed using scores for discriminant functions in the initial global canonical discriminant analysis (16 groups total). Multiple comparisons tests (Tukey's HSD) comparing levels using dataset 1 showed that level 4 differed from the other three levels and the Recent on CV1, level 3 differed from levels 1 and 2 on CV2 (with level 4 and the Recent being intermediate), levels 4 and 2 differed from level 3 on CV3, and level 1 and the Recent differed from the other three levels on CV5. No significant differences were detected on CV4 and CV6. However, examination of these differences through time (Fig. 5.9A, C) indicates that morphologic change was not directional. Multiple comparisons tests (Tukey's HSD) comparing levels using dataset 2 showed that levels 4 and 1 differed from level 3 and the Recent on CV1, level 4 differed from level 3 and level 3 differed from levels 1 and 2 and the Recent on CV2, levels 4 and 3 differed from level 1 and the Recent on CV3, level 1 differed from level 4 which differed from levels 3 and 2 on CV4, and the Recent differed from levels 4, 1, and 2 (but not 3) on CV5. No significant differences were detected on CV6. Examination of these differences through time (Fig. 5.9B, D) indicates that morphologic change was not directional.

Temporal changes in the morphologic disparity of the complex were analyzed by comparing within-level Mahalanobis distances between fossil clusters among

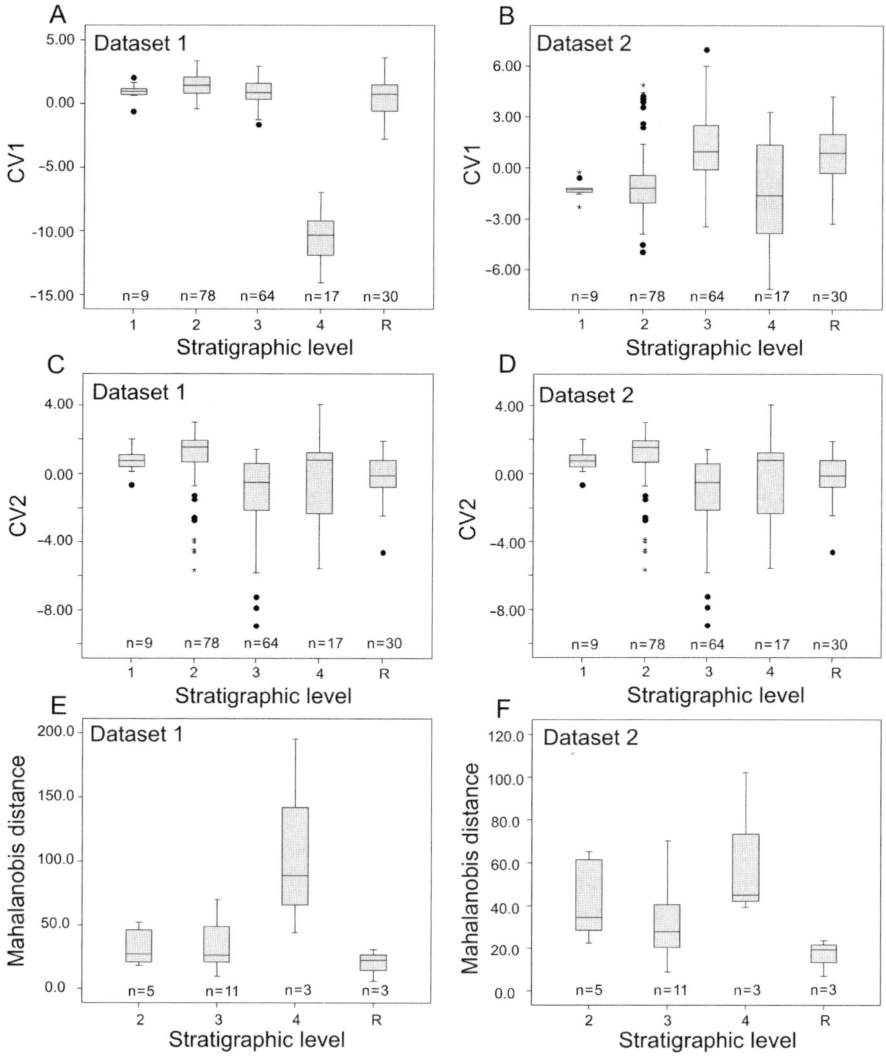

Fig. 5.9 Mean morphology (CV1, CV2) and disparity (Mahalanobis distance) within the complex as a whole. Boxplots showing the median, quartiles, and extreme values of CV scores and Mahalanobis distances among species for each sratigraphic level. The distances include between all pairwise combinations of species within each level. Each box represents the interquartile range containing 50% of the values; the line across each box represents the median. The whiskers represent the highest and lowest values. Outliers are indicated by dots; extreme cases by asterisks

levels using nonparametric statistics. Kruskal-Wallis tests indicate that differences among levels were not significant ($p = 0.066$ for dataset 1, $p=0.124$ for dataset 2). Pairwise Mann-Whitney tests (Fig. 5.9E, F) indicate that disparity is higher in level 4 than in the Recent ($p=05$ for both datasets). It is also slightly higher in level 4 than in level 3 ($p = 0.052$ for dataset 1, $p = 0.073$ for dataset 2). Disparity in levels

2 and 3 are the same (p = 0.955 for dataset 1, p = 0.336 for dataset 2). Disparity in the Recent is lower than in levels 2 and 3 using the second dataset (p = 0.053), but the same using the first dataset (p = 0.456).

5.4.5 *Changes Within Species Through Time*

Of the eight species in levels 2–4, only one occurred in more than three time bins: (FC21 = FC32 = FC42). In the canonical discriminant analysis comparing stratigraphic levels (3 total) within this species (Fig. 5.2, CDA set 3, method two in section 5.3.3.5), Mahalanobis distances were significantly different for all pairwise combinations using both datasets. Dataset 1 revealed a directional pattern among levels through the sequence, whereas dataset 2 revealed a zigzag pattern. However, only three levels are involved in these analyses, and more time intervals are needed to examine the patterns in greater detail. In the canonical discriminant analysis comparing time bins (6 total) within this species (Fig. 5.2, CDA set 3), Mahalanobis distances comparing the oldest and youngest time bins were significant for both datasets, but distances between adjacent time bins were not significantly different using $p<0.05$, the exceptions being time bins 3 versus 4 in both datasets and 5 versus 6 in the first dataset (Fig. 5.10A, B).

Using a different approach to assessing stasis (method three in section 5.3.3.5), a canonical discriminant analysis was performed on each of the two datasets using the nine fossil species and the three modern species as *a priori* groups (Fig. 5.2, CDA set 2). Patterns of variation among time bins were examined within species FC21 = FC32 = FC42. Kruskal-Wallis tests for both datasets revealed insignificant differences between time bins for CV1 and significant differences for CV2–4. Kruskal-Wallis tests for dataset 2 were also significant for CV5 and CV6, but not dataset 1. Mann-Whitney tests comparing the oldest and youngest time bins indicate significant differences ($p < 0.05$) on CV3 and CV6 for dataset 1, and on CV3, CV4 and CV6 for dataset 2.

5.5 Discussion

In comparison with other marine invertebrates (e.g., bryozoans, mollusks), tracing the evolution of reef coral species stratigraphically through the fossil record is complicated by a number of factors: (1) the general paucity of morphologic characters [e.g., 56 characters in the bryozoan *Metrarabdotos* (Cheetham et al., 2007)

Fig. 5.10 (continued) functions are significant and seven are needed to explain 95% of the variation. CV1 accounts for 32.3% of the variation and is negatively correlated with wall thickness (wt). CV2 accounts for 24.8% of the variation, and is negatively correlated with corallite diameter (cd), tertiary costa width (tcw), and tertiary costa length (tcl). CV3 accounts for 13.9% of the variation, and is correlated with dissepiment thickness (para) and primary and secondary costa length (pcl, scl)

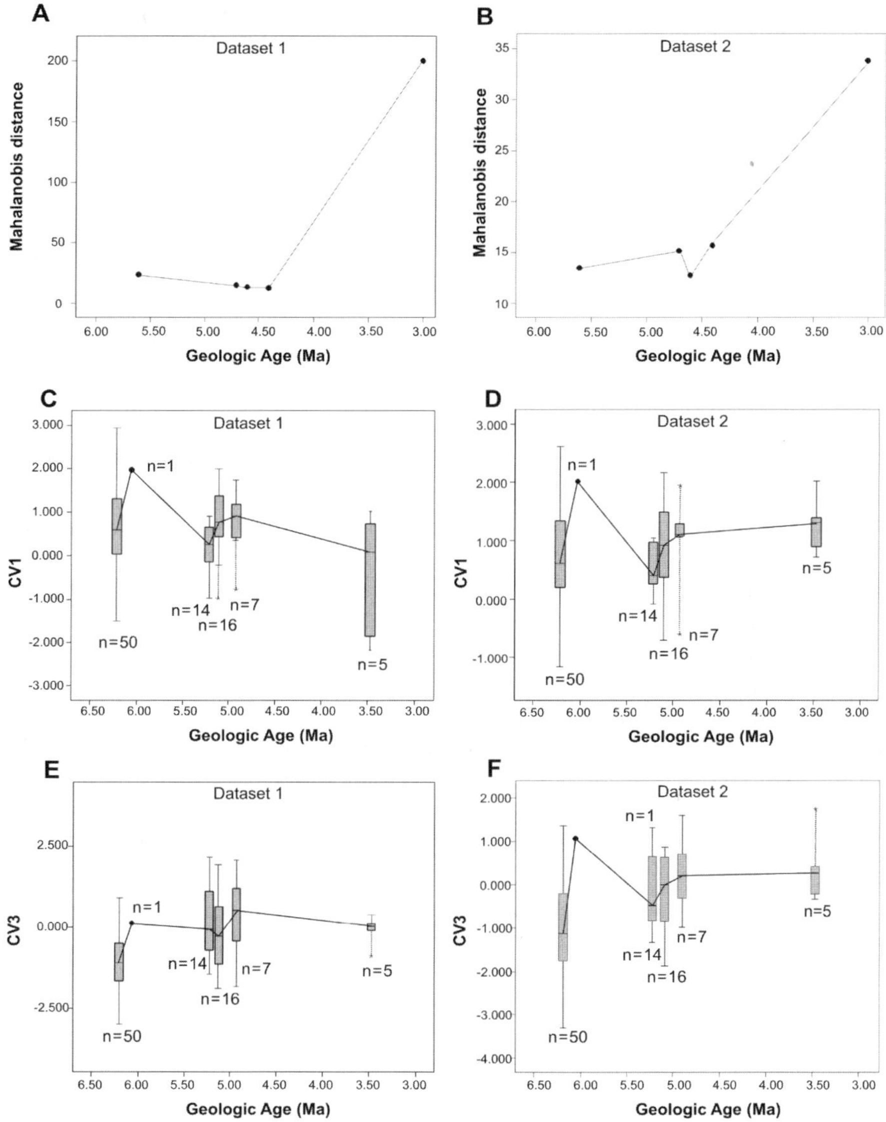

Fig. 5.10 Morphology within species. Scatterplots showing Mahalanobis distances from the oldest occurrences, and boxplots showing the median, quartiles, and extreme values of CV scores (12 groups) for time bins within one species (FC21 = 32= 42, n = 93). Each box represents the inter-quartile range containing 50% of the values; the line across each box represents the median. The whiskers represent the highest and lowest values, excluding outliers. Outliers are connected by dotted lines. In the canonical discriminant analysis using the first dataset, the first eight functions are significant and seven are needed to explain 95% of the variation. CV1 accounts for 35.4% of the variation and is correlated with wall thickness (y4, y11, y12, y19), costa extension (y2, y10, y18), and tertiary costa shape (x12, x19). CV2 accounts for 18.9% of the variation, and is correlated with csize. CV3 accounts for 15.1% of the variation, and is correlated with wall dissepiment thickness (x6, x21). In the canonical discriminant analysis using the second dataset, the first eight

versus 17 in each of the two datasets herein], (2) the continuous, quantitative nature of species diagnostic morphologic characters, (3) high ecophenotypic plasticity (Klaus et al., 2007; Foster, 1979), and (4) the patchy spatial distribution and restricted environments of coral reefs, and the correspondingly limited stratigraphic occurrences of reef coral species. In the present study, these four factors are further confounded by the fact that species belong to a potentially hybridizing species complex, and are therefore highly morphologically similar. Nevertheless, the present study shows that, even under these circumstances, morphologically distinct entities (=morphospecies) can be recognized in the fossil record, and the distributions of these morphospecies can be traced over approximately three million years of geologic time. Evolutionary independence despite "fuzzy" (=not absolute) species boundaries has also been detected in plants and freshwater fish since the Miocene and Pliocene (Smith, 1992; Arnold, 1997).

Comparisons between geometric morphometric and traditional measurement datasets indicate that geometric morphometrics is more effective in initially grouping colonies into morphospecies in sympatry (see also Knowlton and Budd, 2001). However, traditional measurements appear to be more effective at tracing their distributions through time and interpreting their evolutionary relationships. The eight species recognized in levels 2–4 in the present study were not recognized in previous analyses using traditional measurements (Budd, 1991), in part because of the use of traditional measurements in initial cluster analyses (which were also based on Mahalanobis distances), and in part because of the smaller sample size (21 colonies in Budd, 1991, as opposed to 157 in the present study). Nevertheless, in analyses comparing clusters from different stratigraphic levels, which have different environments (see McNeill et al., this volume), differences among groups may be accentuated using geometric morphometrics [e.g., FC41, FC42, FC43 on CV1 in Fig. 5.7A, which is correlated with costa length and tertiary costa shape; CV1 is related to stratigraphic level and thus to environment]. The reason for this accentuation may be because individual shape coordinates often contain information about several different skeletal features and different coordinates may contain information about the same features. As a result, certain skeletal features may be more heavily weighted in canonical discriminant analyses. Traditional measurements involve averaging linear distance measures and ratios made on many different features, which are more evenly represented in discriminant analysis. They therefore are less sensitive to ecophenotypic plasticity and environmental differences among levels. More experimental work (e.g., transplantation experiments) needs to be done on modern corals examining the effects of the environment on skeletal extension and thickening in order to better understand how environment affects skeletal architecture and hence these two datasets.

The low number of stratigraphic occurrences of each species (six or less horizons), despite the increased sampling effort, indicates that estimates of the timing of speciation and extinction events and species durations are subject to error. Marshall (1990) confidence intervals are roughly equivalent (1.06 times) to the observed range of the species for occurrences in six horizons (e.g., FC21 = FC32 = FC42), and they are 4.83 times the observed species range for occurrences in three horizons (e.g., FC35 = FC43; FC23 = FC34). Data from the present study suggest that the species complex may have initially diversified in the gap between level 1 and level 2 (between ~12.3 and 6.2 Ma), and that four species were present in the complex by 6.2 Ma. During the early to middle Miocene, only one *Montastraea "annularis"* – like species (*M. trinitatis*) is represented in the DR sequence, although *M. "limbata"* has been reported from the middle Miocene of Trinidad (Johnson, 2001; Vaughan and Hoffmeister, 1926). Little can be said about the pattern of the initial diversification of the complex, be it a burst of speciation or something more gradual, without additional morphometric study of *M. trinitatis* and *M. limbata* in the middle Miocene of Trinidad, Panama, and Curaçao (Johnson, 2001; Johnson and Kirby, 2006; Budd et al., 1998). However, occurrences of species in levels 2–4 suggest that speciation and extinction rates were both 1–2 species per million years between 6.23 and 3.48 Ma, and that diversity remained relatively constant (3–5 species) within the complex.

Comparisons with the Plio-Pleistocene of Costa Rica and Panama (Budd and Klaus, 2001) indicate that similar numbers of species co-occurred in these younger reef sequences. Three species have been recognized from the Quebrada Chocolate Formation of Costa Rica (3.5–2.9 Ma), two from the Moin Formation of Costa Rica (2.9–1.5 Ma), and four from Isla Bastimentos and Colon in Panama (2.2–0.8 Ma). For the most part, species differ between these three units, with only one species continuing from the Quebrada Chocolate Formation to the Moin Formation of Costa Rica. Phylogenetic analyses, however, suggest that as many as seven species may have existed at any one time during the Plio-Pleistocene. Speciation and extinction rates during the Plio-Pleistocene therefore appear to be similar to or higher than rates in the Mio-Pliocene of the Dominican Republic. Clearly the next step in tracing the evolution of the species complex involves combining data from the Mio-Pliocene of the Dominican Republic with the Plio-Pleistocene of Costa Rica and Panama, and performing phylogenetic analyses.

Even after additional analyses of older and younger material are performed, one problem that will continue to plague this research is the paucity of late Miocene and early Pliocene reefal deposits elsewhere in the Caribbean, which would permit assessment of geographic variation of the species recognized herein. The older and younger occurrences of *M. "limbata"* in the southern Caribbean suggest that the complex may not have been restricted to the northern Dominican Republic during the late Miocene and early Pliocene.

Morphometric analyses of bryozoans in the DR sequence (Cheetham, 1986; Cheetham and Jackson, 1998; Cheetham et al., 2001; Cheetham et al., 2007) have shown that 12 species of the genus *Metrarabdotos* occur in levels 2 and 3, nine species of which occur in seven or more stratigraphic bins (mean sampling density =12.8 bins per species). Using Saunders et al. (1986) age dates, first occurrences of the nine species reveal accelerated speciation between 6.5 and 8 Ma, with confidence intervals averaging 0.15–0.55 Myr. Median species ranges for the nine species are 3.5 Myr with a standard deviation of 1.04 (Table 5.12, Cheetham et al., 2007, p. 22). After adjusting for the new age estimates for the DR sequence (McNeill et al., this volume), which are 1–1.5 Myr younger for levels 2–3, seven of the nine species are estimated to have originated in level 2 and two in level 3. The *Montastraea "annularis"* complex in the present study, therefore, differs from *Metrarabdotos* by consisting of a third to a half as many species. Speciation rates and species durations are difficult to evaluate due to small sample sizes. Speciation rates appear to be roughly equivalent (5 of 12 species of *Metrarabdotos* arise in level 3, and 3 of 7 species of the *Montastraea "annularis"* complex arise in level 3) and maximum species durations may be slightly higher in *Metrarabdotos* with 5 of the 9 species with more occurrences having ranges longer than FC21 = FC32 = FC42.

Analyses of the mean morphology of the complex indicate that the complex as a whole experienced no directional change (i.e., morphologic stasis) between the time it originated and today. However, there was more morphologic disparity within the complex (i.e., greater differences among species) during the Mio-Pliocene than today, with the highest levels of disparity occurring during level 4. Fewer differences and more overlap among species have been found to result from hybridization in previous work (Fukami et al., 2004a; Budd and Pandolfi, 2004). Greater disparity during the Mio-Pliocene would therefore suggest that hybridization may have played a more important role later during the evolution of the complex, after closure of the Central American Isthmus and the onset of northern hemisphere glaciation.

Analyses of a single, common, long-ranging species indicate that it did, however, experience directional change (i.e., phyletic gradualism) in some morphologic features (e.g., CV3) through the 2.75 Myr over which it occurred, in addition to oscillatory stasis in other features (e.g., CV1). The first canonical variate distinguishing species was strongly correlated with wall thickness in both the first and second datasets, and did not experience directional change. However, higher canonical variates in both datasets did experience directional change, the magnitude of which was equivalent to differences among species. These variates were correlated with costae length, corallite diameter, and wall dissepiment thickness. This result appears to contrast with the oscillatory stasis observed in the DR sequence in bryozoans (Cheetham, 1986, 1987; Cheetham et al., 2007), gastropods (Nehm and Geary, 1994; Nehm, 2001, 2005), and bivalves (Anderson, 1994), as well as other reef corals (Beck and Budd, this volume; Schultz and Budd, this volume). However, it should be noted that different methods were used to test for stasis in these different studies, and that lack of a common methodology makes comparisons among taxonomic groups problematic.

Cheetham (1987) found directional trends in a few single morphologic features (46 total) within nine species of *Metrarabdotos,* but he found stasis when analyzing these features multivariately using canonical discriminant analysis. These directional trends, however, occur in features that are poorly related to the morphology that distinguishes between inferred ancestor and descendant species. Cheetham (1987) concludes that single-character changes should not be evaluated in isolation but as statistical samples of change in overall morphology. In the present study, ancestor and descendant species have not yet been inferred. CV1 (which exhibits stasis) explains 32–36% of the variation for the two datasets, and CV3 (which exhibits directional change) explains 13–16% of the variation, indicating that the characters that are experiencing directional changes are indeed important in distinguishing species.

Given the fact that directional change occurred over a time period when the Cibao Basin was experiencing environmental changes caused by regional changes in climate and circulation patterns (see McNeill et al., this volume), it cannot be determined whether morphological change was caused simply by ecophenotypic plasticity, or whether it had a genetic basis and was caused by selection (see Geary, 1995, for examples of the latter). Eldredge et al. (2005) suggest that differences in patterns of temporal variation among taxonomic groups may be explained by the genetic and geographic structure of species (metapopulations). As may be the case in the present example, local populations (e.g., FC21 = FC32 = FC42) are more likely to experience directional change, whereas more widely distributed species (e.g., the *Montastraea "annularis"* complex as a whole) experience stasis.

One of the shortcomings of the present study is the lack of phylogenetic context, which is still in progress at this stage in this research. In view of the patchy nature of the fossil record of the *Montastraea "annularis"* complex, phylogenetic analysis will need to play a more important role than stratigraphy in making interpretations regarding its origin and subsequent speciation patterns. As described above, comparisons of the two datasets in the present study (geometric morphometrics versus traditional measurements) suggest that traditional measurements should provide a starting point in recognizing appropriate characters and their states. However, additional characters will need to be added to the 15 variables with significant species differences (Table 5.3) in the present study in order to more completely characterize the morphology of these corals. Two aspects of morphology that are missing from the present study include colony shape and longitudinal elements of skeletal architecture (see Budd and Klaus, 2001; Holcomb et al., 2004). Because many of the specimens in the present study consist of fragments of colonies, the arrangement of corallites within colonies (their spacing and size variation) could serve as a proxy for colony shape. Longitudinal elements of corallite architecture include the spacing of dissepiments in both the endotheca and exotheca and the structure of the coenosteum. The addition of these two aspects of morphology could bring the total number of characters up to 25–30, a number still far lower than in other marine invertebrate groups but 3–4 times higher than the eight characters used in Budd and Klaus (2001).

Part of the challenge in performing phylogenetic analyses will involve determining appropriate ingroup and outgroup taxa. Recent molecular analyses (Fukami et al., 2004b) indicate that the genus *Montastraea* does not represent a monophyletic group and that *M. annularis* s.l. and *M. cavernosa* s.l. belong to separate families. The corallite architectures of modern *M. annularis* s.l. and *M. cavernosa* s.l. are distinct. Members of the *M. annularis* complex have three cycles of septa (24) per corallite, and corallite diameters <5 mm; whereas *M. cavernosa*-like corals have 4–5 cycles of septa (48–60) per corallite and corallite diameters of 5–10 mm. However, several species in the fossil record (e.g., *Montastraea brevis*) are intermediate between these two extremes. Additional characters (e.g., septal microarchitecture and microstructure) will need to be found to distinguish these two major lineages of Caribbean corals.

5.6 Summary

1. The *Montastraea "annularis"* complex consisted of a total of eight species in the northern Dominican Republic during the Mio-Pliocene, only one of which (*M. limbata*) is previously described. The species were recognized using a geometric morphometrics approach focusing on the structure of the corallite wall. They differ in a number of traditional measurements including: number of septa per corallite, corallite diameter, columella width, wall thickness, costa length and width.
2. Four species were present in the complex at ~6.23 Ma, and diversity remained more or less constant in the complex between 6.23–3.48 Ma. Speciation and extinction rates were approximately 1–2 species per million years over that time interval, and constant. Observed species ranges vary from < 100 Kyr to 2.73 Myr. These results are generally similar to those of bryozoans in the same sequence, which are more diverse and have more evenly distributed stratigraphic occurrences.
3. Although the complex as a whole exhibited morphologic stasis between 6.23 Ma and today, disparity was generally higher during the Mio-Pliocene than today (especially in level 4). Study of one long-ranging species within the complex indicates both stasis and directional morphologic change. The observed directional change could be related to environmental changes interpreted for the sequence, and may or may not have a genetic basis involving selection. The observed directional change differs from the stasis observed in other marine invertebrates studied in the same sequence.
4. In view of the extensive sampling efforts made in the present study, yet still low numbers of stratigraphic occurrences, phylogenetic analyses will need to play an important role in understanding the origin of the complex and speciation events within it. Additional morphologic characters are needed to perform these analyses, and distinguish between *M. annularis*-like and *M. cavernosa*-like corals.

Acknowledgments We thank Alan Cheetham and Ken Johnson for comments of the manuscript; S.C. Wallace for digitizing many of the colonies; B.K. Saville for thin sections; J. Golden, T. Adrain, R. Panchaud, and A. Ziems for curating and loaning specimens. This research was supported by a collaborative grant from the National Science Foundation (EAR 0445789 to A.F. Budd, and EAR 0446768 to D.F. McNeill).

Appendix 1

List of specimens analyzed. All fossil specimens were collected in the Dominican Republic; all recent specimens were collected in the San Blas Islands of Panama. NMB, Natural History Museum in Basel, Switzerland. SUI, University of Iowa Paleontology Repository

Museum	Cat no	CCD Specimen no	Col no	FC	Previous Indentification	Locality	River
Level 1 (n = 9)							
NMB	D5732	16943-19	197	11	*trinitatis*	NMB16943	Yaq. Norte
NMB	D5729	16943-1	199	11	*trinitatis*	NMB16943	Yaq. Norte
NMB	D5734	16943-21	201	11	*trinitatis*	NMB16943	Yaq. Norte
NMB	D5735	16943-22	203	11	*trinitatis*	NMB16943	Yaq. Norte
NMB	D5737	16943-24	205	11	*trinitatis*	NMB16943	Yaq. Norte
NMB	D5730	16943-2	207	11	*trinitatis*	NMB16943	Yaq. Norte
NMB	D5741	16945-1	209	11	*trinitatis*	NMB16945	Yaq. Norte
NMB	D5754	17243-4	224	11	*trinitatis*	NMB17243	Yaq. Norte
NMB	D5758	17244-7	226	11	*trinitatis*	NMB17244	Yaq. Norte
Level 2 (n = 78)							
NMB	D5624	16853-2	158	21	*limbata*	NMB16853	Cana
NMB	D5742	17273-1	212	21	*trinitatis*	NMB17273	Yaq. Norte
NMB	D5743	17273-5	214	21	*trinitatis*	NMB17273	Yaq. Norte
NMB	D5744	17273-6	216	21	*trinitatis*	NMB17273	Yaq. Norte
SUI	102445	dr40326	626	21	AB-1	BK00-CE(12)	Ar. Bellaco
SUI	102446	dr40129	629	22	AB-3	BK00-CE(12)	Ar. Bellaco
SUI	102447	dr40130	630	22	AB-3	BK00-CE(12)	Ar. Bellaco
SUI	104477	dr40131	631	22	AB-2	BK00-CE(12)	Ar. Bellaco
SUI	104478	dr40135	635	21	AB-1	BK00-CE(12)	Ar. Bellaco
SUI	104479	dr40136	636	21	AB-1	BK00-CE(12)	Ar. Bellaco
SUI	104480	dr40138	638	21	AB-1	BK00-CE(12)	Ar. Bellaco
SUI	104481	dr40139	639	21	AB-1	BK00-CE(12)	Ar. Bellaco
SUI	104482	dr40140	640	21	AB-1	BK00-CE(12)	Ar. Bellaco
SUI	104483	dr40143	643	22	AB-3	BK00-CE(12)	Ar. Bellaco
SUI	104484	dr40168	668	21	AB-1	BK00-CE(12)	Ar. Bellaco
SUI	104485	dr40179	679	21	AB-1	BK00-CE(12)	Ar. Bellaco
SUI	104486	dr40180	680	21	AB-1	BK00-CE(12)	Ar. Bellaco
SUI	104487	dr40181	681	21	AB-1	BK00-CE(12)	Ar. Bellaco
SUI	104488	dr40182	682	21	AB-1	BK00-CE(12)	Ar. Bellaco
SUI	104489	dr40183	683	21	AB-1	BK00-CE(12)	Ar. Bellaco
SUI	104490	dr40185	685	21	AB-1	BK00-CE(12)	Ar. Bellaco
SUI	104491	dr40187	687	21	AB-1	BK00-CE(12)	Ar. Bellaco
SUI	104492	dr40188	688	21	AB-1	BK00-CE(12)	Ar. Bellaco
SUI	104493	dr40195	695	21	AB-1	BK00-CE(12)	Ar. Bellaco
SUI	104494	dr40205	705	21	AB-1	BK00-CE(12)	Ar. Bellaco
SUI	104495	dr40206	706	22	AB-3	BK00-CE(12)	Ar. Bellaco
SUI	104496	dr40207	707	21	AB-1	BK00-CE(12)	Ar. Bellaco
SUI	104497	dr40208	708	21	AB-1	BK00-CE(12)	Ar. Bellaco
SUI	104498	dr40209	709	21	AB-1	BK00-CE(12)	Ar. Bellaco
SUI	104499	dr40210	710	21	AB-1	BK00-CE(12)	Ar. Bellaco
SUI	104500	dr40219	719	21	AB-1	BK00-CE(12)	Ar. Bellaco
SUI	104501	dr40223	723	21	AB-1	BK00-CE(12)	Ar. Bellaco
SUI	104502	dr40225	725	21	AB-1	BK00-CE(12)	Ar. Bellaco
SUI	104503	dr40227	727	21	AB-3	BK00-CE(12)	Ar. Bellaco
SUI	104504	dr40224	724	22	AB-3	BK00-CE(12)	Ar. Bellaco

(continued)

Appendix 1 (continued)

SUI	104505	dr40229	729	21	AB-1	BK00-CE(12)	Ar. Bellaco
SUI	104506	dr40230	730	21	AB-1	BK00-CE(12)	Ar. Bellaco
SUI	104507	dr40231	731	21	AB-1	BK00-CE(12)	Ar. Bellaco
SUI	104508	dr40232	732	22	AB-1	BK00-CE(12)	Ar. Bellaco
SUI	104509	dr40233	733	21	AB-1	BK00-CE(12)	Ar. Bellaco
SUI	104510	dr40234	734	22	AB-3	BK00-CE(12)	Ar. Bellaco
SUI	104511	dr40235	735	21	AB-1	BK00-CE(12)	Ar. Bellaco
SUI	104512	dr40236	736	22	AB-2	BK00-CE(12)	Ar. Bellaco
SUI	104513	dr40939	1,439	23	AB-4	BK00-CE(12)	Ar. Bellaco
SUI	104514	dr40940	1,440	21	AB-1	BK00-CE(12)	Ar. Bellaco
SUI	104515	dr40944	1,444	23	AB-4	BK00-CE(12)	Ar. Bellaco
SUI	104516	dr40945	1,445	22	AB-4	BK00-CE(12)	Ar. Bellaco
SUI	104517	dr40946	1,446	23	AB-4	BK00-CE(12)	Ar. Bellaco
SUI	104518	dr40947	1,447	24	AB-4	BK00-CE(12)	Ar. Bellaco
SUI	104519	dr40948	1,448	24	AB-4	BK00-CE(12)	Ar. Bellaco
SUI	104520	dr40951	1,451	24	AB-4	BK00-CE(12)	Ar. Bellaco
SUI	104521	dr40960	1,460	24	AB-4	BK00-CE(12)	Ar. Bellaco
SUI	104522	dr40962	1,462	24	AB-4	BK00-CE(12)	Ar. Bellaco
SUI	104523	dr40963	1,463	24	AB-4	BK00-CE(12)	Ar. Bellaco
SUI	104524	dr40943	1,443	21	AB-1	BK00-CE(2)	Ar. Bellaco
SUI	104525	dr40980	1,480	24	AB-4	BK00-CE(2)	Ar. Bellaco
SUI	104526	dr40937	1,437	22	AB-2	BK00-CE(3)	Ar. Bellaco
SUI	104527	dr40925	1,425	21	AB-1	BK00-CE(4)	Ar. Bellaco
SUI	104524	dr40239	739	21	AB-1	BK00-CE(6)	Ar. Bellaco
SUI	104529	dr40248	748	21	AB-1	BK00-CE(6)	Ar. Bellaco
SUI	104530	dr40250	750	21	AB-1	BK00-CE(6)	Ar. Bellaco
SUI	104531	dr40252	752	21	AB-1	BK00-CE(6)	Ar. Bellaco
SUI	104532	dr40254	754	22	AB-2	BK00-CE(6)	Ar. Bellaco
SUI	104533	dr40255	755	21	AB-1	BK00-CE(6)	Ar. Bellaco
SUI	104534	dr40264	764	21	AB-1	BK00-CE(6)	Ar. Bellaco
SUI	104535	dr40274	774	21	AB-1	BK00-CE(6)	Ar. Bellaco
SUI	104536	dr40275	775	21	AB-1	BK00-CE(6)	Ar. Bellaco
SUI	104537	dr40291	791	21	AB-1	BK00-CE(6)	Ar. Bellaco
SUI	104538	dr40295	795	21	AB-1	BK00-CE(6)	Ar. Bellaco
SUI	104539	dr40297	797	21	AB-1	BK00-CE(6)	Ar. Bellaco
SUI	104540	dr40298	798	22	AB-2	BK00-CE(6)	Ar. Bellaco
SUI	104541	dr40299	799	21	AB-1	BK00-CE(6)	Ar. Bellaco
SUI	104542	dr40301	801	21	AB-1	BK00-CE(6)	Ar. Bellaco
SUI	104543	dr40303	803	21	AB-1	BK00-CE(6)	Ar. Bellaco
SUI	104544	dr40307	807	22	AB-2	BK00-CE(6)	Ar. Bellaco
SUI	104545	dr40316	816	21	AB-1	BK00-CE(6)	Ar. Bellaco
SUI	104546	dr40318	818	21	AB-1	BK00-CE(6)	Ar. Bellaco
SUI	104547	dr40324	824	21	AB-1	BK00-CE(6)	Ar. Bellaco
Level 3 (n = 64)							
NMB	D5546	15808–1	140	32	*limbata*	NMB15808	Gurabo
NMB	D5556	15837–5	143	34	*limbata*	NMB15837	Gurabo
SUI	104548	dr40067	567	32		NMB15837	Gurabo
SUI	104549	dr40069	569	33		NMB15837	Gurabo
SUI	104550	dr40071	571	33		NMB15837	Gurabo
SUI	104551	dr40449	949	32		NMB15847	Gurabo
SUI	104552	dr40466	966	32		NMB15847	Gurabo
SUI	104553	dr40476	976	33		NMB15847	Gurabo

(continued)

Appendix 1 (continued)

SUI	104554	dr40477	977	33		NMB15847	Gurabo
SUI	104555	dr40478	978	31		NMB15847	Gurabo
SUI	104556	dr40480	980	32		NMB15847	Gurabo
SUI	104557	dr40481	981	32		NMB15847	Gurabo
SUI	104558	dr40484	984	32		NMB15847	Gurabo
SUI	104559	dr40485	985	31		NMB15847	Gurabo
SUI	104560	dr40486	986	32		NMB15847	Gurabo
SUI	104561	dr40490	990	33		NMB15847	Gurabo
NMB	D5592	15855–1	146	34	*limbata*	NMB15855	Gurabo
SUI	104562	dr40437	937	33		NMB15855	Gurabo
SUI	104563	dr40438	938	34		NMB15855	Gurabo
SUI	104564	dr40439	939	32		NMB15855	Gurabo
SUI	104565	dr40440	940	32		NMB15855	Gurabo
SUI	104566	dr40441	941	31		NMB15855	Gurabo
SUI	104567	dr40442	942	32		NMB15855	Gurabo
SUI	104568	dr40446	946	31		NMB15855	Gurabo
SUI	104569	dr40448	948	32		NMB15855	Gurabo
SUI	104570	dr40034	534	35		NMB15855	Gurabo
SUI	104571	dr40035	535	32		NMB15855	Gurabo
SUI	104572	dr40036	536	33		NMB15855	Gurabo
SUI	104573	dr40037	537	32		NMB15855	Gurabo
SUI	104574	dr40375	875	32		NMB16136	Gurabo
SUI	104575	dr40405	905	33		NMB16136	Gurabo
SUI	104576	dr40430	930	33		NMB16136	Gurabo
SUI	104577	dr40436	936	32		NMB16136	Gurabo
SUI	104578	dr40124	624	32		NMB16138	Gurabo
NMB	D5608	16814–1	149	32	*limbata*	NMB16814	Cana
NMB	D5617	16817–3	151	32	*limbata*	NMB16817	Cana
NMB	D5622	16818–5	153	32	*limbata*	NMB16818	Cana
SUI	104579	dr40327	827	31		NMB16818	Cana
SUI	104580	dr40324	824	32		NMB16818	Cana
SUI	104581	dr40329	829	31		NMB16818	Cana
SUI	104582	dr40333	833	32		NMB16818	Cana
SUI	104583	dr40334	834	32		NMB16818	Cana
SUI	104584	dr40335	835	31		NMB16818	Cana
SUI	104585	dr40121	621	33		NMB16822	Cana
NMB	D5626	16823–1	155	32	*limbata*	NMB16823	Cana
NMB	D5643	16883–3	161	32	*limbata*	NMB16883	Gurabo
NMB	D5714	16921–10	184	32	*limbata*	NMB16921	Gurabo
NMB	D5715	16921–11	186	32	*limbata*	NMB16921	Gurabo
NMB	D5717	16921–14	188	32	*limbata*	NMB16921	Gurabo
NMB	D5710	16921–2	191	32	*limbata*	NMB16921	Gurabo
NMB	D5712	16921–4	194	32	*limbata*	NMB16921	Gurabo
NMB	D5722	16933–9	196	34	*limbata*	NMB16933	Gurabo
SUI	104586	dr40068	568	35		NMB16136	Gurabo
SUI	104587	dr40094	594	32		BK-00–3	Cana
SUI	104588	dr40095	595	33		BK-00–3	Cana
SUI	104589	dr40096	596	32		BK-00–3	Cana
SUI	104590	dr40097	597	32		BK-00–3	Cana
SUI	104591	dr40098	598	32		BK-00–3	Cana
SUI	104592	dr40099	599	32		BK-00–3	Cana
SUI	104593	dr40100	600	31		BK-00–3	Cana
SUI	104594	dr40101	601	31		BK-00–3	Cana

(continued)

Appendix 1 (continued)

SUI	104595	dr40102	602	32		BK-00–3	Cana
SUI	104596	dr40103	603	31		BK-00–3	Cana
SUI	104597	dr40104	604	32		BK-00–3	Cana
Level 4 (n = 15)							
NMB	104598	16884–10	167	42	*limbata*	NMB16884	Cana
NMB	104599	16884–11	171	42	*limbata*	NMB16884	Cana
NMB	104600	16884–7	178	41	*limbata*	NMB16884	Cana
SUI	104601	dr40048	548	43		NMB16884	Cana
SUI	104602	dr40057	557	43		NMB16884	Cana
SUI	104603	dr40058	558	43		NMB16884	Cana
SUI	104604	dr40061	561	43		NMB16884	Cana
SUI	104605	dr40062	562	41		NMB16884	Cana
SUI	104606	dr40073	573	42		NMB16884	Cana
SUI	104607	dr40080	580	42		NMB16884	Cana
SUI	104608	dr40081	581	43		NMB16884	Cana
SUI	104609	dr40082	582	41		NMB16884	Cana
SUI	104610	dr40083	583	41		NMB16884	Cana
SUI	104611	dr40084	584	41		NMB16884	Cana
SUI	104612	dr40086	586	43		NMB16884	Cana
SUI	104613	dr40091	591	43		NMB16884	Cana
SUI	104614	dr40092	592	42		NMB16884	Cana
Recent (n = 30)							
SUI	95199	a455	7	1	*annularis*		San Blas Is.
SUI	95200	a457	8	1	*annularis*		San Blas Is.
SUI	95201	a464	9	1	*annularis*		San Blas Is.
SUI	95202	a465	10	1	*annularis*		San Blas Is.
SUI	95203	a95–10	3	1	*annularis*		San Blas Is.
SUI	95204	a95–14	4	1	*annularis*		San Blas Is.
SUI	95205	a95–18	5	1	*annularis*		San Blas Is.
SUI	95206	a95–39	6	1	*annularis*		San Blas Is.
SUI	95207	a95–07	2	1	*annularis*		San Blas Is.
SUI	95208	a95–01	1	1	*annularis*		San Blas Is.
SUI	95209	f409	17	2	*faveolata*		San Blas Is.
SUI	95210	f424	18	2	*faveolata*		San Blas Is.
SUI	95211	f438	19	2	*faveolata*		San Blas Is.
SUI	95212	f490	20	2	*faveolata*		San Blas Is.
SUI	95213	f95–37	11	2	*faveolata*		San Blas Is.
SUI	95214	f96–16	13	2	*faveolata*		San Blas Is.
SUI	95215	f96–07	12	2	*faveolata*		San Blas Is.
SUI	95216	f96–29	14	2	*faveolata*		San Blas Is.
SUI	95217	f96–47	15	2	*faveolata*		San Blas Is.
SUI	95218	f97–05	16	2	*faveolata*		San Blas Is.
SUI	95219	k312	27	3	*franksi*		San Blas Is.
SUI	95220	k408	24	3	*franksi*		San Blas Is.
SUI	95221	k417	29	3	*franksi*		San Blas Is.
SUI	95222	k427	30	3	*franksi*		San Blas Is.
SUI	95223	k95–13	23	3	*franksi*		San Blas Is.
SUI	95224	k95–15	24	3	*franksi*		San Blas Is.
SUI	95225	k95–04	22	3	*franksi*		San Blas Is.
SUI	95226	k95–25	25	3	*franksi*		San Blas Is.
SUI	95227	k95–03	21	3	*franksi*		San Blas Is.
SUI	95224	k97–01	26	3	*franksi*		San Blas Is.

Appendix 2

Landmarks on transverse thin sections of corallites of *Montastraea*. Types are: 1 = juxtaposition of structures; 2 = maxima of curvature; 3 = extremal points

No	Type	Analyzed	Description
1	Na	Y	Center of corallite
2	2	Y	Outermost point on secondary costa
3	1	y	Outer left junction of secondary costoseptum with wall dissepiment
4	1	Y	Outer right junction of secondary costoseptum with wall dissepiment
5	1	Y	Inner left junction of secondary costoseptum with wall dissepiment
6	1	Y	Inner right junction of secondary costoseptum with wall dissepiment
7	3	N	Left point of maximum curvature associated with secondary septal thinning
8	3	N	Right point of maximum curvature associated with secondary septal thinning
9	2	Y	Innermost point on secondary septum
10	2	Y	Outermost point on tertiary costa
11	1	Y	Outer left junction of tertiary costoseptum with wall dissepiment
12	1	Y	Outer right junction of tertiary costoseptum with wall dissepiment
13	1	Y	Inner left junction of tertiary costoseptum with wall dissepiment
14	1	Y	Inner right junction of tertiary costoseptum with wall dissepiment
15	3	n	Left point of maximum curvature associated with tertiary septal thinning
16	3	n	Right point of maximum curvature associated with tertiary septal thinning
17	2	y	Innermost point on tertiary septum
18	2	y	Outermost point on primary costa
19	1	Y	Outer left junction of primary costoseptum with wall dissepiment
20	1	Y	Outer right junction of primary costoseptum with wall dissepiment
21	1	Y	Inner left junction of primary costoseptum with wall dissepiment
22	1	Y	Inner right junction of primary costoseptum with wall dissepiment
23	3	N	Left point of maximum curvature associated with primary septal thinning
24	3	N	Right point of maximum curvature associated with Primary septal thinning
25	2	Y	Innermost point on primary septum
26	1	N	Outer left junction of tertiary costoseptum with wall dissepiment
27	1	N	Inner left junction of tertiary costoseptum with wall dissepiment

Appendix 3

Structure matrices for canonical discriminant analyses distinguishing fossil clusters within each level and the modern species (Figs. 5.3 and 5.4). * Most strongly correlated original variables

	Modern only		Level 1		Level 2			
	CV1	CV2	CV1	CV2	CV1	CV2	CV3	CV4
% variance explained	93.8	6.2	84.6	9.5	45.0	33.0	11.4	7.7
csize	−0.241	−0.392*	−0.314	−0.333	−0.193	−0.692*	0.091	0.323
x11	0.151	0.282*	0.215	0.062	0.142	0.273	−0.191	−0.004
x12	−0.118	0.083	−0.158	−0.005	−0.253	0.063	−0.117	−0.278*
x19	−0.059	0.091	−0.080	0.133	−0.072	0.064	−0.123	−0.277*
x21	0.093	0.039	0.120	0.471*	0.643*	0.025	0.262	0.036
x4	0.107	0.267*	0.153	−0.039	−0.046	0.278	−0.182	0.020
x6	−0.062	0.008	−0.077	−0.444*	−0.624*	0.039	−0.234	0.090
y1	0.071	−0.085	0.069	0.036	0.063	0.010	0.057	0.159
y10	0.281	−0.057	0.376	−0.145	0.237	0.255	−0.017	0.287*
y11	0.359*	0.124	0.480*	0.113	0.369	0.227	−0.027	0.302*
Y12	0.403*	0.163	0.528*	0.048	0.367	0.291	−0.056	0.275*
Y17	−0.099	0.052	−0.122	0.096	0.059	−0.252	0.265	−0.061
Y18	0.321	−0.027	0.397	−0.276	0.010	0.445*	−0.406	0.426*
Y19	0.418*	0.180	0.549*	0.052	0.386	0.292	−0.071	0.263*
Y2	0.256	0.040	0.330	−0.198	0.113	0.394*	0.131	0.309*
Y25	0.040	0.011	0.056	−0.033	0.330	−0.086	0.711*	0.174
Y4	0.354*	0.123	0.474*	0.127	0.381	0.223	−0.020	0.299*
Y9	0.095	0.108	0.132	−0.150	0.371	−0.087	0.151	0.184

	Level 3					Level 4			
	CV1	CV2	CV3	CV4	CV5	CV1	CV2	CV3	CV4
% variance explained	46.6	28.4	12.0	6.8	3.0	67.5	22.4	6.1	3.1
csize	−0.444*	−0.056	−0.325	0.016	0.525*	−0.168	0.240	−0.346*	−0.155
x11	0.159	−0.135	0.417*	−0.026	−0.046	−0.133	0.063	0.388*	−0.043
x12	−0.412	−0.228	0.344*	0.083	0.099	−0.400*	0.160	0.058	−0.512*
x19	−0.202	−0.401	0.191	0.174	−0.019	−0.371*	0.048	0.136	−0.473*
x21	0.293	−0.216	−0.305*	0.223	−0.178	0.050	−0.254*	0.162	0.140
x4	−0.051	0.068	0.601*	−0.064	0.092	−0.118	0.139	0.343*	−0.077
x6	−0.308	0.235	0.423*	−0.182	0.207	0.211	0.570*	−0.077	−0.061
y1	0.207	−0.326	−0.180	0.249	−0.109	−0.223	−0.108	0.284*	0.225
y10	0.556*	0.474*	0.031	−0.242	−0.055	0.278	−0.098	0.079	0.365*
y11	0.675*	0.457*	−0.021	−0.137	−0.096	0.300*	−0.162	0.100	0.345*
y12	0.666*	0.440*	0.079	−0.139	−0.047	0.253	−0.165	0.096	0.295
y17	−0.068	−0.004	−0.096	−0.123	0.036	−0.084	0.007	−0.164	0.098
y18	0.376	0.223	0.225	−0.007	0.309	0.379*	−0.259	0.142	0.192
y19	0.656*	0.415*	0.095	−0.145	−0.071	0.243	−0.168	0.118	0.276
y2	0.290	0.403	0.073	0.257	−0.119	0.373*	−0.207	0.070	0.264
y25	−0.001	0.135	−0.033	0.359*	−0.002	−0.055	0.119	0.230	0.288
y4	0.678*	0.447*	−0.048	−0.100	−0.095	0.304*	−0.170	0.096	0.352*
Y9	0.094	0.115	−0.003	0.209	0.107	−0.025	0.075	0.351*	0.125

References

Allmon, W.D., 2001, Nutrients, temperature, disturbance, and evolution: a model for the late Cenozoic marine record of the western Atlantic, *Palaeogeogr. Palaeoclim. Palaeoecol.*, **166**:9–26.

Anderson, L.C., 1994, Paleoenvironmental control of species distributions and intraspecific variability in Neogene Corbulidae (Bivalvia: Myacea) of the Dominican Republic, *J. Paleontol.*, **68**:460–473.

Arnold, M.L., 1997, *Natural Hybridization and Evolution*, Oxford University Press, New York, 211 pp.

Beck, B.R. and Budd A.F., This volume, Evolutionary patterns within the reef coral *Siderastrea* in the Mio-Pliocene of the Dominican Republic, in: *Evolutionary Stasis: Species and Communities Through Geologic Ttime* (R.H. Nehm and A.F. Budd, eds.), Kluwer/Plenum, New York.

Bookstein, F.L. 1991. *Morphometric tools for landmark data.* Cambridge University Press, Cambridge, 435 pp.

Budd, A.F., 1991, Neogene paleontology in the Northern Dominican Republic. 11. The Family Faviidae (Anthozoa: Scleractinia), *Bulls. Am. Paleontol.*, **101**:5–83.

Budd, A.F. and Coates A.G., 1992, Non-progressive evolution in a clade of Cretaceous *Montastraea* – like corals, *Paleobiology*, **18**:425–446.

Budd, A.F. and Klaus, J.S., 2001, The origin and early evolution of the *Montastraea "annularis"* species complex (Anthozoa: Scleractinia), *J. Paleontol.*, **75**:527–545.

Budd, A.F. and Johnson, K.G., 1999, Origination preceding extinction during Late Cenozoic turnover of Caribbean reefs, *Paleobiology*, **25**:188–200.

Budd, A.F. and Pandolfi, J.M., 2004, Overlapping species boundaries and hybridization within the *Montastraea "annularis"* reef coral complex in the Pleistocene of the Bahama Islands, *Paleobiology*, **30**:396–425.

Budd, A.F., Petersen, R.A., and McNeill, D.F., 1998, Stepwise faunal change during evolutionary turnover: a case study from the Neogene of Curaçao, The Netherlands Antilles, *Palaios*, **13**:167–185.

Budd, A.F., Stemann, T.A., and Johnson, K.G., 1994, Stratigraphic distributions of genera and species of Neogene to Recent Caribbean reef corals, *J. Paleontol.*, **68**:951–977.

Cheetham, A.H., 1986, Tempo of evolution in a Neogene bryozoan: rates of morphometric change within and across species boundaries, *Paleobiology*, **12**:190–202.

Cheetham, A.H., 1987, Tempo of evolution in a Neogene bryozoan: are trends in single morphologic characters misleading?, *Paleobiology*, **13**:286–296.

Cheetham, A. H., Jackson, J. B. C., and Sanner, J., 2001, Evolutionary significance of sexual and asexual modes of propagation in Neogene species of the bryozoan *Metrarabdotos* in tropical America, *J. Paleontol.*, **75**:564–577.

Cheetham, A.H. and Jackson, J.B.C., 1998, The fossil record of cheilostome bryozoa in the Neogene and Quaternary of tropical America: Adequacy for phylogenetic and evolutionary studies, in: *The Adequacy of the Fossil Record* (S. K. Donovan and C.R.C. Paul, eds.), Wiley, Chichester/England, pp. 227–242.

Cheetham, A.H., Sanner, J., and Jackson, J.B.C., 2007, *Metrarabdotos* and related genera (Bryozoa: Cheilostomata) in the late Paleogene and Neogene of Tropical America, *Paleontol. Soc. Mem.*, **67**:1–96.

Coates, A.G., Jackson, J.B.C., Collins, L.S., Cronin, T.M., Dowsett, H.J., Bybell, L.M., Jung, P., and Obando, J.A., 1992, Closure of the Isthmus of Panama: the near-shore marine record of Costa Rica and western Panama, *Geol. Soc. Am. Bull.*, **104**:814–828.

Collins, L.S., Coates, A.G., Berggren, W.A., Aubry, M.P., and Zhang, J., 1996, The late Miocene Panama isthmian strait, *Geology*, **24**:687–690.

Duncan, P.M., 1863, On the fossil corals of the West Indian Islands. Part 1, *Quart. J. Geol. Soc. Lond.*, **19**:406–458, pls. 13–16.

Eldredge, N., Thompson, J.N., Brakefield, P.M., Gavrilets, S., Jablonski, D., Jackson, J.B.C., Lenski, R.E., Lieberman, B.S., McPeek, M.A., and Miller W. III, 2005, The dynamics of evolutionary change, in: *Macroevolution: Diversity, Disparity, Contingency* (E.S. Vrba and N. Eldredge, eds.), *Paleobiology* (Suppl.) **31**:133–145.

Ellis, J. and Solander, D., 1786, *The Natural History of Many Curious and Common Zoophytes*, Benjamin White and Peter Elmsly, London, 208 pp., 63 pls.

Foster, A.B., 1986, Neogene palaeontology in the Northern Dominican Republic. 3. The Family Poritidae (Anthozoa: Scleractinia), *Bulls. Am. Palaeontol.*, **90**:47–123.

Foster, A.B., 1979, Phenotypic plasticity in the reef corals *Montastraea annularis* (Ellis and Solander) and *Siderastrea siderea* (Ellis and Solander), *J. Exp. Mar. Biol. Ecol.*, **39**:25–54.

Fukami, H., Budd, A.F., Levitan, D.R., Jara, J., Kersanach, R., and Knowlton, N., 2004a, Geographic differences in species boundaries among members of the *Montastraea annularis* complex based on molecular and morphological markers, *Evolution*, **58**:324–337.

Fukami, H., Budd, A.F., Paulay, G., Solé-Cava, A., Chen C.A., Iwao,K., and Knowlton, N., 2004b, Conventional Taxonomy Obscures Deep Divergence between Pacific and Atlantic Corals, *Nature*, **427**:832–835.

Geary, D.H., 1990, Patterns of evolutionary temp and mode in the radiation of *Melanopsis* (Gastropoda; Melanopsidae), *Paleobiology*, **16**:492–511.

Geary, D.H., 1992, An unusual pattern of divergence between two fossil gastropods: ecophenotypy, dimorphism, or hybridization?, *Paleobiology*, **18**:93–109.

Geary, D.H., 1995, The importance of gradual change in species-level transitions, in: *New Approaches to Speciation in the Fossil Record* (D.H. Erwin and R.L. Anstey, eds.), Columbia University Press, New York, pp. 67–86.

Gregory, J.W., 1895, Contributions to the palaeontology and physical geology of the West Indies, *Quart. J. Geol. Soc. Lond.*, **51**:255–312.

Holcomb, M., Pandolfi, J.M., Macintyre, I.G., and Budd, A.F., 2004, Use of X-radiographs to distinguish members of the *Montastraea "annularis"* reef coral species complex, in: D.G. Fautin, J.A. Westfall, P. Cartwright, M.Daly and C.R. Wyttenbach, eds. Coelenterate biology 2003: trends in research in Cnidaria and Ctenophora, *Hydrobiologia*, **530**:211–222.

Jackson, J.B.C. and Johnson, K.G., 2000, Life in the last few million years, in: *Deep Time: Paleobiology's Perspective* (D.H. Erwin and S.L. Wing eds.), *Paleobiology* (Suppl.) **26**:221–235.

Johnson, K.G., 2001, Middle Miocene recovery of Caribbean reef corals: new data from the Tamana Formation, Trinidad, *J. Paleontol.,* **75**:513–526.

Johnson, K.G., Budd, A.F., Klaus, J.S., and McNeill, D.F., This volume, The impact of fossil from the northern Dominican Republic on origination estimates for Miocene and Pliocene Caribbean reef corals, in: *Evolutionary Stasis and Change in the Dominican Republic Neogene.*

Johnson, K.G. and Kirby, M.X., 2006, The Emperador Limestone rediscovered: early Miocene corals from the Culebra Formation, Panama, *J. Paleontol.*, **80**:283–293.

Klaus, J.S. and Budd., A.F., 2003, Comparison of Caribbean coral reef communities before and after Plio-Pleistocene faunal turnover: Analyses of two Dominican Republic reef sequences, *Palaios*, **18**:3–21.

Klaus, J.S., Budd, A.F., Heikoop, J.M., and Fouke, B.W., 2007, Environmental controls on corallite morphology in the reef coral *Montastraea annularis*, *Bull. Mar. Sci.*, **80**:233–260.

Klaus, J.S., Budd, A.F., and McNeill, D.F., This volume, Assessing community change in Miocene to Pliocene Coral Assemblages of the northern Dominican Republic, in: *Evolutionary Stasis and Change in the Dominican Republic Neogene.*

Knowlton, N., Weil, E., Weigt, L.A., and Guzman, H.M., 1992, Sibling species in *Montastraea annularis*, coral bleaching, and the coral climate record, *Science*, **255**:330–333.

Knowlton, N. and Budd, A.F., 2001, Recognizing coral species past and present, in: *Evolutionary Patterns: Growth, Form, and Tempo in the Fossil Record* (J.B.C. Jackson, S. Lidgard, and F.K. McKinney eds.), University of Chicago Press, Chicago, IL, pp. 97–119.

Lieberman, B.S., Brett, C.E., and Elredge, N., 1995, A study of stasis and change in two species lineages from the Middle Devonian of New York state, *Paleobiology*, **21**:15–27.

Marshall, C.R., 1990, Confidence intervals on stratigraphic ranges, *Paleobiology*, **16**:1–10.
McNeill, D.F., Klaus, J.S., Evans, C.C., Budd, A.F., and Maier, K.L., This volume, An overview of the regional geology and stratigraphy of the Neogene deposits of the Cibao Valley, Dominican Republic, in: *Evolutionary Stasis and Change in the Dominican Republic Neogene*.
Nehm, R.H., 2001, Neogene Paleontology of the Dominican Republic: The genus *Prunum*, *Bull. Am. Paleontol.*, **359**:1–46.
Nehm, R.H., 2005, Patterns and processes of evolutionary stasis and change in *Eratoidea* (Gastropoda: Marginellidae) from the Dominican Republic Neogene, *Carib. J. Sci.*, **41**:189–214.
Nehm, R,H. and Geary, D.H., 1994, A gradual morphologic transition during a rapid speciation event in marginellid gastropod (Neogene: Dominican Republic), *J. Paleontol.*, **68**:787–795.
Pandolfi, J.M., 2007, A new, extinct Pleistocene reef coral from the Montastraea "annularis" species complex, *J. Paleontol.*, **81**:472–482.
Pandolfi, J.M., Lovelock, C.E., and Budd, A.F., 2002, Character release following extinction in a Caribbean reef coral species complex, *Evolution*, **56**:479–501.
Saunders, J.B., Jung, P., and Biju-Duval, B.,1986, Neogene paleontology in the northern Dominican Republic. 1. Field surveys, lithology, environment, and age, *Bull. Am. Paleontol.*, **89** (323):1–79, 9 pls.
Schultz, H.A. and Budd, A.F., This volume, Neogene evolution of the reef coral species complex *Montastraea "cavernosa"*, in: *Evolutionary Stasis and Change in the Dominican Republic Neogene*.
Smith, G.R., 1992, Introgression in fishes: significance for paleontology, cladistics, and evolutionary rates, *Syst. Biol.*, **41**:41–57.
Stanley, S.M. and Yang, X., 1987, Approximate evolutionary stasis for bivalve morphology over millions of years: a multivariate, multilineage study, *Paleobiology*, **13**:113–139.
Vaughan, T.W., 1919, Fossil corals from Central America, Cuba, and Porto Rico with an account of the American Tertiary, Pleistocene, and recent coral reefs, *U. S. Nat. Hist. Mus. Bull.*, **130**:189–524, pls. 68–152.
Vaughan, T.W. and Hoffmeister, J.E., 1926, Miocene corals from Trinidad. *Papers of the Department of Marine Biology, Carnegie Institution of Washington*, **23**:107–132, 7 pls.
Weil, E. and Knowlton, N., 1994, A multi-character analysis of the Caribbean coral *Montastraea annularis* (Ellis and Solander, 1786), and its two sibling species, *M. faveolata* (Ellis and Solander, 1786) and *M. franksi* (Gregory, 1895), *Bull. Mar. Sci.*, **55**:151–175.
Willis, B.L., van Oppen, M.J.H., Miller, D.J., Vollmer, S.V., and Ayre, D.J., 2006, The role of hybridization in the evolution of reef corals, *Ann. Rev. Ecol., Evol., Syst.*, **37**:489–517.
Wolstenholme, J.K., Wallace, C.C., and Chen, C.A., 2003. Species boundaries within the *Acropora humilis* group (Cnidaria; Scleractinia): a morphological and molecular interpretation of evolution, *Coral Reefs*, **22**:155–166.
Zelditch, M.L., Swiderski, D.L., Sheets, H.D., and Finks, W.L., 2004, *Geometric Morphometrics for Biologists*. Academic, London, 416 pp.

Chapter 6
Evolutionary Patterns Within the Reef Coral *Siderastrea* in the Mio-Pliocene of the Dominican Republic

Brian R. Beck[1,2] and Ann F. Budd[1]

Contents

6.1 Introduction

In 1972 Eldredge and Gould argued that the current evolutionary paradigm of the time (gradualism punctuated by stratigraphic gaps) did not explain patterns observed in the fossil record. They proposed an alternative hypothesis, punctuated equilibrium, which states that species lineages exhibit long periods of stasis and new species are formed during short periods of rapid evolutionary change (Eldredge and Gould, 1972; Gould, 2002). Since Eldredge and Gould's original studies, several additional studies have found examples of phyletic gradualism, punctuated equilibrium, and various combinations of the two modes (e.g., Cheetham, 1986; Stanley and Yang, 1987; Geary, 1990; Nehm and Geary, 1994; Erwin and Anstey, 1995; Jackson and Cheetham, 1999; Jablonski, 2000; Nehm, 2005; Cheetham et al.,

[1] Department of Geoscience, University of Iowa, Iowa City,
IA 52242, USA. Email: b.beck@uq.edu.au, ann-budd@uiowa.edu

[2] Centre for Marine Studies, University of Queensland, Brisbane, Queensland 4072, AUSTRALIA

R.H. Nehm, A.F. Budd (eds.) *Evolutionary Stasis and Change in the Dominican Republic Neogene,*

2007). The present study examines evolutionary patterns in the scleractinian reef coral *Siderastrea* (Blainville, 1830) to determine if the observed patterns follow phyletic gradualism or punctuated equilibrium.

Studies have been done in the Neogene of the Caribbean testing these alternative modes of evolution in various organisms (bryozoans, mollusks) (e.g., Cheetham, 1986; Nehm and Geary, 1994; Nehm, 2005; Cheetham et al., 2007). To date, only cursory work has been done studying the evolutionary patterns in corals (e.g., Potts et al., 1993; Budd et al., 1994; Budd and Klaus, this volume). Cheetham (1986), Cheetham et al. (2007), Nehm and Geary (1994), and Nehm (2005) all found morphological stasis, albeit sometimes oscillatory, but Nehm and Geary (1994) also found gradual change associated with speciation. The corals in our study are from the same localities in the Dominican Republic as the bryozoans and mollusks that were used in Cheetham (1986), Cheetham et al. (2007), Nehm and Geary (1994), and Nehm (2005). We test the hypothesis that the corals followed punctuated equilibrium in ways generally similar to that observed in bryozoans and mollusks.

Cheetham (1986), Cheetham et al. (2007), Nehm and Geary (1994), and Nehm (2005) used morphometrics to distinguish species and track evolutionary change over time. We also use morphometrics in our study of *Siderastrea*, but instead of basic linear measurements, we use two dimensional (2-D) geometric morphometrics to more effectively capture the changes within each species. With 2-D geometric morphometrics, landmarks are digitized on an image instead of taking traditional linear distance measurements. Two dimensional landmarks capture not only the traditional measurements of length, width, etc., but also other measurements not obtained by traditional morphometrics (e.g., angles) and ensure consistency by anchoring measures on spatially homologous points. The data are statistically analyzed to determine if species exhibited stasis or gradual evolutionary change over >5 million years of geologic time during the Mio-Pliocene. This period of time immediately preceded closure of the Central American Isthmus, which blocked the flow of ocean waters from the Atlantic to the Pacific. Closure is believed to have caused faunal turnover during the Plio-Pleistocene, in which many new species of reef corals originated and many species became extinct (Budd and Johnson, 1999). The present study will help to understand coral evolution just before this important event in Caribbean coral history.

We examine the morphometric data to determine whether the morphologic changes were evenly distributed through time or concentrated during certain intervals. If corals changed constantly through time that would indicate phyletic gradualism, but if periods of oscillatory change or no change at all are detected, that would support the hypothesis of punctuated equilibrium.

6.1.1 Taxonomy

Order: Scleractinia (Bourne, 1900)
Family: Siderastreidae (Vaughan and Wells, 1943)
Genus: *Siderastrea* (Blainville, 1830)

Fig. 6.1 Photographs of a colony of *Siderastrea* in hand sample (left) and corallite (right) views. Scale bar is 5 mm on the left, and 0.5 mm on the right. (Photos taken from NMITA < http://nmita.geology.uiowa.edu>)

Siderastrea (Fig. 6.1) is a zooxanthellate scleractinian coral of the family Siderastreidae, and although its modern geographic distribution is mostly Atlantic (Veron, 1995), some species of *Siderastrea* are found in the Pacific and Indian Oceans (Budd and Guzman, 1994). Colonies of *Siderastrea* are usually massive, have on average 48 septa per corallite, range widely in diameter (2.5–8 mm), vary in thickness of corallite walls, and are abundant in back reef environments. *Siderastrea* is cerioid (corallites are juxtaposed) and has extramural budding (new corallites bud outside the wall of the parent corallite). Along with numerous septa, *Siderastrea* also has many rings of synapticulae and a synapticulothecal wall (Veron, 1995; Budd and Guzman, 1994) (Table 6.1). There are five extant species recognized globally, two of which occur in the Caribbean (*S. radians* and *S. siderea*) (Veron, 1995), and there are four extinct species found in the Caribbean (*S. conferta*, *S. mendenhalli*, *S. pliocenica*, and *S. silecensis*) (Budd et al., 1994).

Siderastrea is a common coral used in many modern studies (Vermeij, 2005; Garcia et al., 2005; Forsman et al., 2005), but the fossil history of this coral is relatively unknown. The present study helps to further the paleontological knowledge of *Siderastrea* by examining its morphological history. Understanding how this reef coral has changed through geologic time will contribute to both paleontologic and neontologic studies of *Siderastrea* and reef corals in general.

Species: *Siderastrea siderea* (Ellis and Solander, 1786)

Siderastrea siderea forms colonies larger than most other living species of *Siderastrea*. The corallites, which are also larger than most other species of *Siderastrea*, average 4–5 mm in diameter, but can be as small as 3–3.5 mm in diameter. There are typically four complete septal cycles. The tertiary septa are fused to the

Table 6.1 Morphologic characters distinguishing eight Caribbean and two Pacific species of *Siderastrea* (Budd and Guzman, 1994)

Species	Distribution	Number of septa per corallite	Corallite diameter (mm)	Columella	Corallite wall
S. radians	Middle Pliocene to Recent; Caribbean, Bermuda, Brazil, W. Africa	30–40	2.5–3.5	Thick, solid; intermediate fossa depth	Thick, 2–3 synap. rings; septa usually continuous between calices
S. siderea	Early Miocene to Recent; Caribbean	44–50	3–5	Thin, papillose; deep fossa	Thin, 3–5 synap. rings; septa alternate between calices
S. stellata	Recent; Brazil	~48	~3	Thin papillose; very deep fossa	Thin, 3–4 synap. rings; septa usually continuous between calices
S. mendenhalli	Early Miocene to early Pliocene; Dominican Republic, California	48–54	3–5	Thick; shallow fossa	Thick, 3–4 synap. rings; septa continuous between calices
S. silecensis	early Miocene to early Pleistocene; Florida, Dominican Republic	48–60	>4.5	Intermediate thickness; deep fossa	Thin, 3–5 synap. rings; septa continuous between calices
S. pliocenica	Middle Pliocene to early Pleistocene; Florida	40–48	4.5–5	Thick, solid; shallow fossa	Thick, 4–5 synap. rings; septa usually continuous between calices
S. glynni	Recent; eastern Pacific	40–48	2.5–3.5	Intermediate thickness, papillose; shallow fossa	Intermediate thickness; 3–4 synap. rings; septa usually continuous between calices
S. savignyana	Recent; Red Sea, Indian Ocean	28–35	2.5–4	Thick, solid; intermediate fossa depth	Very thick, 2–3 synap. rings; septa continuous between calices

secondary septa, and the quaternary septa are fused to the tertiary septa closer to the wall than in *S. silecensis*. The columella is narrow in comparison to the other species of *Siderastrea* and is not as well defined. This species has a thin synapticulothecal wall and high calice relief.

Species: *Siderastrea silecensis* (Vaughan, 1919)

Siderastrea silecensis has a massive colony shape with a domed upper surface. The corallites are polygonal and the corallite wall is sometimes raised. An adult corallite is on average 5 mm in diameter, but oblong corallites can reach lengths of 7 mm and widths of 5 mm. The calice depth of most corallites of *S. silecensis* is 1.5 mm. This species has four complete septal cycles, but in larger corallites, a few quinary septa are often seen. The synapticulae are well developed and rather coarse. The columella width is average for species of *Siderastrea*. Its synapticulothecal wall is thin, and it has medium calice relief.

Species: *Siderastrea mendenhalli* (Vaughan, 1917)

Siderastrea mendenhalli forms massive colonies, which have a somewhat flat upper surface. Colonies are larger in size with diameters on average of 440 mm. Corallites are found to be polygonal and sometimes deformed. The diameter of corallites ranges from 4 mm to 7 mm with the more deformed and elongated corallites reaching the diameters around 7 mm. Four complete septa cycles occur in average sized corallites. The septa are thin and crowded within the corallite. The synapticulae are well developed and extend more than halfway from the wall of the corallite to the columella. The columella is thick. It has a thick synapticulothecal wall and low calice relief.

Additional taxonomic information, occurrence data, and photos are available in the NMITA (Neogene Marine Biota of Tropical America) database at http://nmita.geology.uiowa.edu (Budd et al., 2001).

6.1.2 Localities

The Dominican Republic is located in the Caribbean Sea and shares the island of Hispañola with Haiti. The sampling localities for our study are well-preserved river cut sections in the northern Dominican Republic. The river cut sections where we collected were Río Gurabo, Río Cana, Arroyo Bellaco, Cañada de Zamba, and Río Yaque del Norte. The sections consist of middle Miocene to lower Pliocene siliciclastic sediments and span a continuum of more than 5 million years. The original dating of the stratigraphic sections in the Dominican Republic was done using foraminifera and other microfossils (Saunders et al., 1986).

While the current stratigraphic dates are sufficient for this study, better chronostratigraphy is needed to expand this study further. McNeill et al. (this volume) have recently revised the age dates of these formations by integrating microfossil, paleomagnetic, and strontium isotope data, and are currently in the process of performing additional analyses to obtain a more accurate estimate of the timing of paleontological and sedimentological events in the sequence.

The stratigraphy of the northern Dominican Republic was originally studied by Saunders et al. (1986). On the basis of their fossil descriptions coupled with their stratigraphic columns and sedimentological interpretations, we assembled a list of potential collecting localities. These localities occurred in four stratigraphic formations:

the Baitoa Formation, Cercado Formation, Gurabo Formation, and Mao Formation (Table 6.2). These four formations are comprised mainly of silts and sands with conglomeratic beds near formation boundaries (Saunders et al., 1986).

The Baitoa Formation is composed of calcareous silts, which are occasionally separated by thin, sandy conglomerates. The sections of the Baitoa Formation that are dense with corals consist of a sandy limestone in which the coral heads are often in living position, or dense coral patches where many corals are in living position. The Cercado Formation, which lies above the Baitoa Formation, is similar in composition. The lower part of the formation is sandy with interbedded conglomerates. The sorting of these sands is poor and cross-bedding is often present. The upper part of the Cercado Formation is silty and thick with coral and algal debris. The Gurabo Formation is marked with a conglomerate at the base, and above the conglomerate are beds of calcareous and fossiliferous silts. The upper beds of these silts are extremely rich in coral colonies, and at the uppermost part of the coral rich beds of the Gurabo Formation are interbedded biostromal corals and coralliferous silts. The Mao Formation is composed of sandy silts with some sands and conglomerates. These beds are overlain by calcareous silts that have a significant

Table 6.2 A list of the localities sampled, the number of samples collected, the formation name, the river the samples were collected from, and the age of samples collected from those localities

Location	Number of samples collected	Formation	River	Age
15830	1	Mao	Gurabo	Pliocene
15845	2	Gurabo	Gurabo	Miocene
15846	1	Gurabo	Gurabo	Miocene
15848	2	Gurabo	Gurabo	Miocene
15859	5	Gurabo	Gurabo	Miocene
16815	2	Gurabo	Cana	Pliocene
16817	2	Gurabo	Cana	Pliocene
16818	24	Gurabo	Cana	Pliocene
16819	1	Gurabo	Cana	Pliocene
16883	1	Gurabo	Gurabo	Miocene
16884	10	Mao	Cana	Pliocene
16934	1	Gurabo	Gurabo	Miocene
16937	4	Baitoa	Yaque del Norte	Miocene
16939	3	Baitoa	Yaque del Norte	Miocene
16943	8	Baitoa	Yaque del Norte	Miocene
16944	5	Baitoa	Yaque del Norte	Miocene
17273	1	Cercado	Yaque del Norte	Miocene
17282	1	Baitoa	Yaque del Norte	Miocene
17289	1	Baitoa	Yaque del Norte	Miocene
CCE 4	1	Cercado	Arroyo Bellaco	Pliocene
CCE 5	1	Cercado	Arroyo Bellaco	Pliocene
CCE 12	23	Cercado	Arroyo Bellaco	Pliocene

amount of clay in them. Above the silts is a series of massive limestones. Corals are common in these upper limestones (Saunders et al., 1986).

The corals used in our study were collected from areas in the silty layers that were packed with coral debris. The overall depositional environment of these beds is shallow marine, but the depth and exact water conditions vary. The changes in shallow marine conditions were interpreted on the basis of overall faunal changes (Saunders et al., 1986), and have recently been modified on the basis of various sedimentary features (McNeill et al., this volume). The energy levels of the depositional environments range from high energy levels represented by the conglomeratic formation boundaries to low energy levels represented by the calcareous clays (Saunders et al., 1986).

6.2 Methods and Materials

6.2.1 Sampling

Siderastrea is a relatively uncommon coral in the Dominican Republic sequences. We searched for colonies throughout the exposed sequence and made collections at all localities where *Siderastrea* was found by Saunders et al. (1986). At each locality, all colonies of *Siderastrea* were collected unless we found more than 20 colonies. Collecting colonies of *Siderastrea* within the rock sequence involved removing them from the face of the outcrop and packaging them for shipping back to the University of Iowa. For each sample, we recorded data on geographic location, stratigraphic position, and the interpreted geologic age (Table 6.2). After the samples were shipped back to Iowa, we cleaned off the excess dirt and prepared one thin section per colony to a thickness of 30 μm, and mounted them on 1½" × 3" slides.

With the exception of the Baitoa Formation, the Dominican Republic sections are continuous, with little or no missing time gaps. Table 6.2 shows a list of the localities that were sampled and how many samples were taken. It also indicates the river section, the formation, and the age of each locality. Even though the sampling was not even through the section due to occasional large samples in the Gurabo Formation and the Cercado Formation, samples were collected throughout the sequence.

In addition, we also obtained supplemental collections from the Natural History Museum (NMB) in Basel, Switzerland, which were made through the same sections in the late 1970s. These collections were used to increase sample size and to extend the temporal range of the analysis. They increased the temporal range because that they were the only samples from the Baitoa Formation. Since the time they were collected, the Baitoa section has become covered in water due to the construction of a dam near Santiago.

The specimens collected in this study are listed in the Appendix and deposited in the University of Iowa Paleontology Repository.

6.2.2 Morphometrics

We used 2-D geometric morphometric analysis to study the morphologic differences among colonies, and group them into species. Twenty-eight landmarks (Fig. 6.2) were digitized using Image J (ver. 1.31, 2004; available at http://rsb.info.nih.gov/ij/) on six corallites per colony. Due to the symmetry of corallites, landmarks were only taken from an area extending from one primary septum counter-clockwise to a secondary septum within each corallite (Fig. 6.2). Landmarks were selected to capture all of the measurements that were traditionally obtained by using linear measurements (e.g., corallite diameter, length of primary septa, etc.), along with other measurements not easily obtained with linear measurements (e.g., septum length ratios, septum width ratios, etc.). The landmarks were also chosen to capture information about distance between septa, characteristics of the synapticulae, and septal angles. Traditionally species of *Siderastrea* have been distinguished mainly using number of septa, corallite diameter, and thickness of corallite wall (Table 6.1).

After species were determined using 2-D geometric morphometrics, these traditional measures were compared among species to better understand differences among species.

Using the computer program CoordGen6f in the IMP software series (Integrated Morphometrics Package, 2004, written by H. David Sheets, available at http://www2.canisius.edu/~sheets/morphsoft.html), Bookstein coordinates (Bookstein,

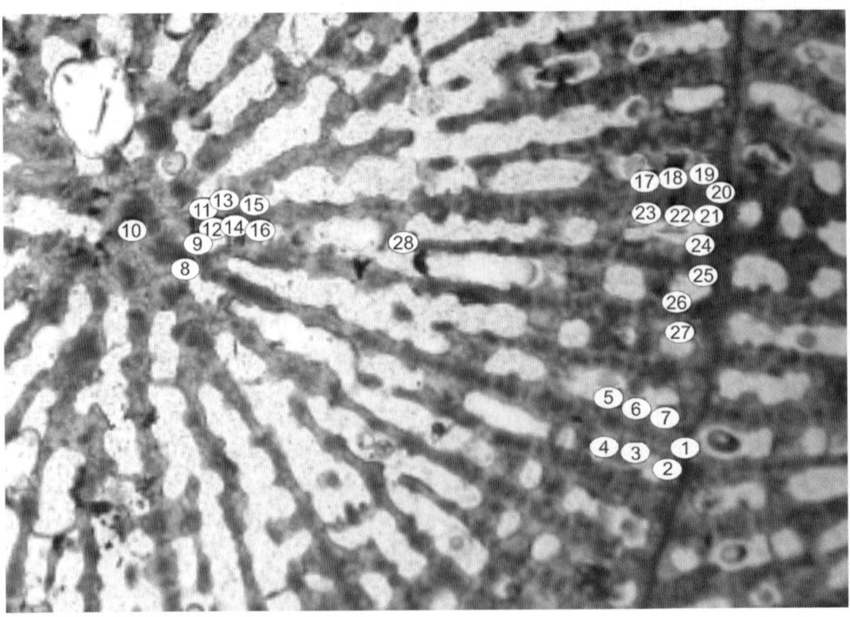

Fig. 6.2 Transverse sections of a corallite of *Siderastrea* showing the landmark scheme used in the 2-D geometric morphometric analysis

1991) were calculated using a baseline consisting of landmarks 1 and 8 (Fig. 6.2). Centroid size was also calculated using the same program.

We used SPSS (Statistical Package for Social Science, ver. 11.0.0, 2001) to analyze the measurements statistically. Because they were used as the baseline, landmarks 1 and 8 were normalized to 0, 0 and 1, 0 and left out of the analysis. All of the x and y coordinates for the other 26 landmarks were included in the analysis. In combination with c-size (centroid size) and number of septa, a total of 54 variables were entered into the analysis.

We performed a series of iterative canonical discriminant analyses within each formation in order to recognize morphospecies. The iterative procedure was the same as that used by Cheetham et al. (2007) and Schultz and Budd (this volume). The initial *a priori* groups consisted of colonies, and their differences were evaluated using F-values corresponding with Mahalanobis distances between colonies. If two colonies were not statistically significantly different (having a p-value of 0.05 or greater), we would group them together and rerun the analysis. This process was continued until all groups were found to be statistically different from one another.

After we determined the groups within each formation (= "fossil clusters" in Budd and Klaus, this volume), all of the groups were then combined into one large global dataset. We then performed canonical discriminant analysis comparing all of the previously recognized groups using the same iterative procedure as described above. Groups that were not statistically significantly different were combined and the analyses rerun. Morphospecies were defined on the basis of the final groups. To aid in identification of these morphospecies, we added three colonies of modern *Siderastrea siderea* and three colonies of *Siderastrea radians* collected in Jamaica to the global dataset and reran the canonical discriminant analysis using the four fossil morphospecies and two modern species as *a priori* groups.

In order to examine change over time within each of the morphospecies, we compared the morphospace occupation among formations for each species. If morphologic stasis occurred, the morphospace covered by each formation would be expected to completely overlap. If evolutionary change occurred, the morphospace covered by each formation would not be expected to completely overlap, and a directional change would be exhibited. We also created boxplots and performed nonparametric Kruskal-Wallis tests comparing formations. We used the most significant variables in the discriminant analyses as well as the discriminant functions in these analyses. Significant variables were determined by examining the structure matrix (Table 6.3).

We compared traditional morphometrics and 2-D geometric morphometrics using the resulting morphospecies. This comparison was performed to determine if 2-D geometric morphometrics is a better identification tool for species, or if traditional measures are equally effective. Traditional measurements were calculated using TMorphGen (ver. 6c, 2004) in the IMP software series (Integrated Morphometrics Package, 2004, written by H. David Sheets, available at http://www2.canisius.edu/~sheets/ morphsoft.html). Corallite radius was calculated by using the distance from landmark 1 to landmark 10 (Fig. 6.2). After calculating corallite radius and making

Table 6.3 Structure matrix for the canonical discriminant analysis displaying the significant variables for each discriminant function. Values are Pearson's correlation coefficients with each function

	Function		
Landmark	1	2	3
x20	−0.070(*)	−0.047	−0.028
x26	−0.065(*)	−0.029	−0.051
x19	−0.060(*)	−0.042	−0.055
x21	−0.059(*)	−0.038	−0.058
x15	−0.054(*)	−0.025	.000
x27	−0.054(*)	−0.020	−0.040
y2	0.052(*)	0.008	−0.039
x13	−0.052(*)	−0.007	0.010
x12	−0.038(*)	−0.021	0.024
y4	0.047	0.093(*)	−0.029
y15	0.015	0.080(*)	−0.037

*Relatively highly correlated variable

boxplots for number of septa and corallite radius, nonparametric Mann-Whitney U tests were performed to test for differences between species.

6.3 Results

6.3.1 Species Recognition

The canonical discriminant analyses resulted in three groups within each of the individual four formations (Baitoa, Cercado, Gurabo, Mao), and a total of four species in the global dataset (Fig. 6.3). There was no overlap between any of the four species. Discriminant function 1 explained 78.9% of the variation, discriminant function 2 explained 14.5% of the variation, and discriminant function 3 explained 6.7% of the variation. Of these three functions, the first two were found to be significant with p-values for Wilks Lambda below 0.05, and the third function was found to be less significant with a p-value of 0.052. The three landmarks that were most influential in discriminant function 1 are x20, x26, and x19 (Table 6.3). These three variables represent the outermost point of the secondary septa (x20), the right junction of the secondary septa with the corallite wall (x19), and the right junction of the tertiary septa with the corallite wall (x26).

6.3.2 Comparison of Fossil and Modern Species

When the modern samples of *S. siderea* and *S. radians* were added to the fossil data set, the results support the morphospecies defined based solely on the fossil data. As

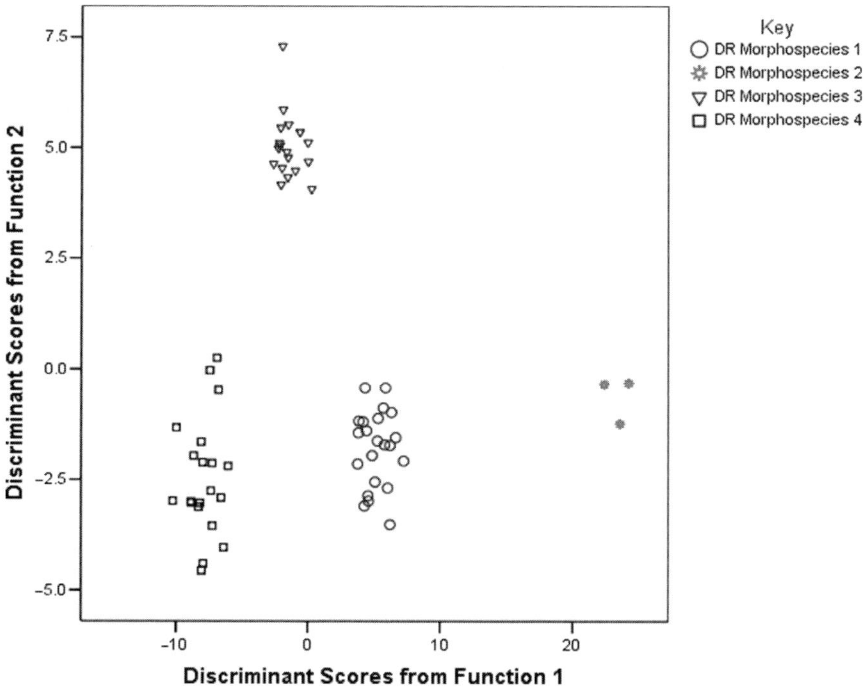

Fig. 6.3 Scatterplot of discriminant function 1 vs. discriminant function 2 of the global canonical discriminant analysis comparing DR morphospecies 1–4. Each point represents one colony. Colonies are labelled by species. The four species are clearly distinct

seen in Fig. 6.4, *S. siderea* groups with DR morphospecies 3, allowing us to interpret DR morphospecies 3 as being the same species as *S. siderea. S. radians* did not group with any fossil group, which was expected due to the fact that *S. radians* has no previously known fossil record in the Dominican Republic and its corallites are considerably reduced in size (Table 6.1) in comparison with the other species of *Siderastrea* in the analysis. The two modern species are clearly distinct using 2-D geometric morphometrics, but they are also easy to separate using traditional methods. Both corallite diameter and number of septa differ between the two modern species.

The remaining two abundant DR morphospecies (DR morphospecies 1 and 4) can be classified respectively as *S. silecensis* and *S. mendenhalli*. These identifications are based upon the species described using traditional characters in Budd et al. (1994). However, not all of the characters in Budd et al. (1994) exactly match. DR morphospecies 1 (Fig. 6.5) has a thick columella and a thin wall, and DR morphospecies 4 (Fig. 6.5) has a thick wall and a thin to intermediate columella. *S. silecensis* is described as having an intermediate columella width and a thin wall, and *S. mendenhalli* as having a thick columella and thick corallite wall.

In sum, of these four fossil species in this study, three (DR morphospecies 1, 3, and 4) belong to previously known fossil species (Fig. 6.5) and one (DR morphospecies 2) is a possible new species (Fig. 6.5).

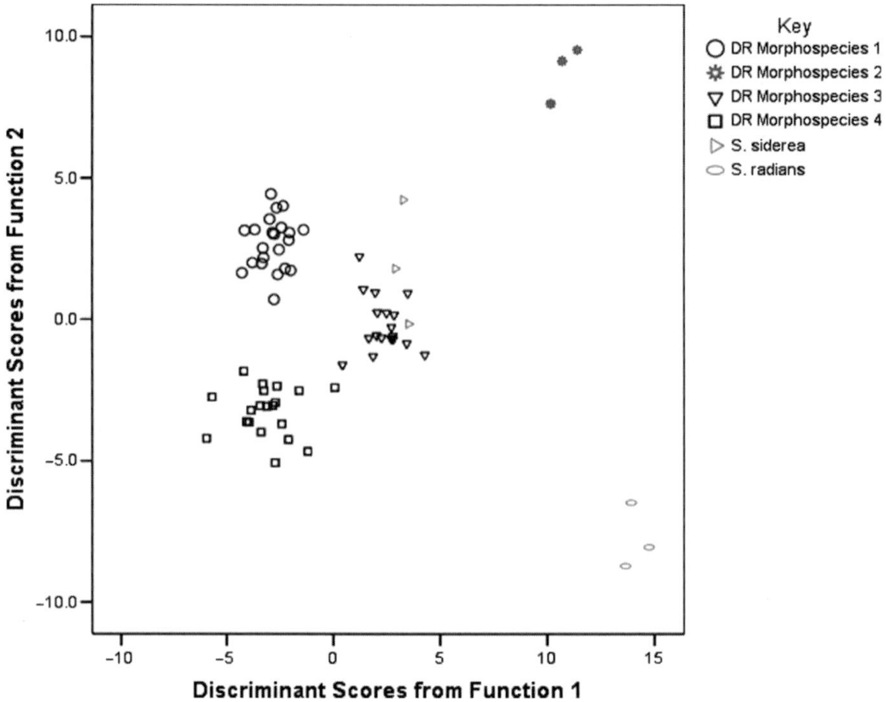

Fig. 6.4 Scatterplot of discriminant function 1 vs. discriminant function 2 of a canonical discriminate analysis comparing DR morphospecies 1–4, modern *S. siderea,* and modern *S. radians*. Each point represents one colony. Colonies are labelled by species

6.3.3 *Stratigraphic Ranges*

Figure 6.6 shows the same plot as Fig. 6.3 but with the markers set to display the formation in which each colony was sampled. The sampling in the four formations covers >5 million years of geologic time. Of the three known species (*S. siderea, S. mendenhalli, S. silecensis*), the >5 million year period that was sampled covers the entire known range of *S. mendenhalli* and >50% of the total range of *S. siderea* and *S. silecensi*. These three species of *Siderastrea* are known from Florida to the Dominican Republic to Costa Rica (Budd et al., 1994).

6.3.4 *Morphologic Changes Through Time*

Figure 6.7 shows boxplots of DR morphospecies 1, 3, and 4 (species 2 was left out because it is only found in a single horizon) examining the change of discriminant function 1. Kruskal-Wallis tests show the p-values of discriminant function 1,

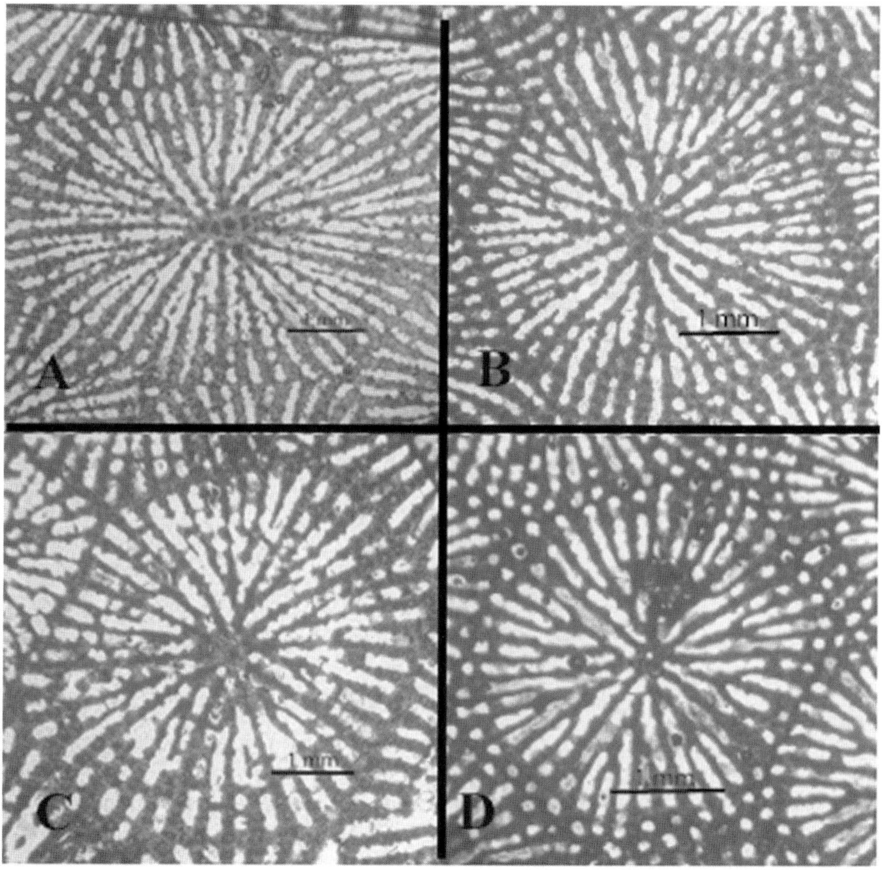

Fig. 6.5 Photographs of transverse sections through corallites of DR morphospecies 1 (A), DR morphospecies 2 (B), DR morphospecies 3 (C), and DR morphospecies 4 (D)

discriminant function 2, x20, x19, x26, and y4 throughout the formations for DR morphospecies 1, 3, and 4 to be over 0.05, indicating that there was no significant change in any of these variables between any of the formations. Combined with Fig. 6.6, which shows that the formations almost completely overlap within each species group, the results indicate that the different coral species occupied the same morphospace during the >5 million year time period represented in this study.

6.3.5 Comparisons Using Traditional Measures

The two traditional features used for identifying species of *Siderastrea* are corallite diameter and number of septa. Number of septa and corallite radius were determined

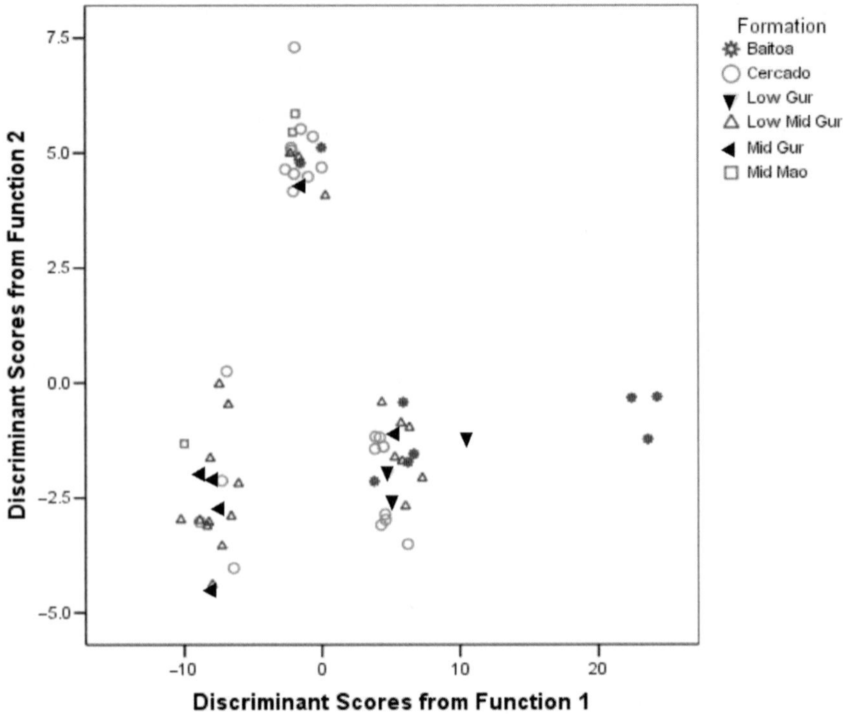

Fig. 6.6 Scatterplot of discriminant function 1 vs. discriminant function 2 of a canonical discriminant analysis comparing DR morphospecies 1–4 (same analysis as Fig. 6.3). Each point represents one colony. Colonies are labelled by formation. Three of the four fossil species extend though numerous horizons

for each species using 2-D geometric morphometrics, and boxplots were constructed comparing species. No significant differences were found when comparing the number of septa between the four DR morphospecies, but when comparing coral radius, a significant difference (p-value < 0.05) was found between DR morphospecies 1 and DR morphospecies 4. The difference in sizes found between DR morphospecies 1 and 4 helps to further differentiate the 2 species. *Siderastrea mendenhalli* has a smaller corallite size in comparison to *Siderastrea silecensis* (Budd et al., 1994).

6.4 Discussion

The scatterplot showing the fossil species of corals only (Fig. 6.3) clearly outlines the three fossil species that were expected to be found in the Dominican Republic sections. DR morphospecies 1, 3, and 4 have absolutely no overlap and have significant gaps between them. If overlap was observed, it would make it difficult to correctly parse out the species and even more difficult to correctly observe how the species changed morphometrically through time.

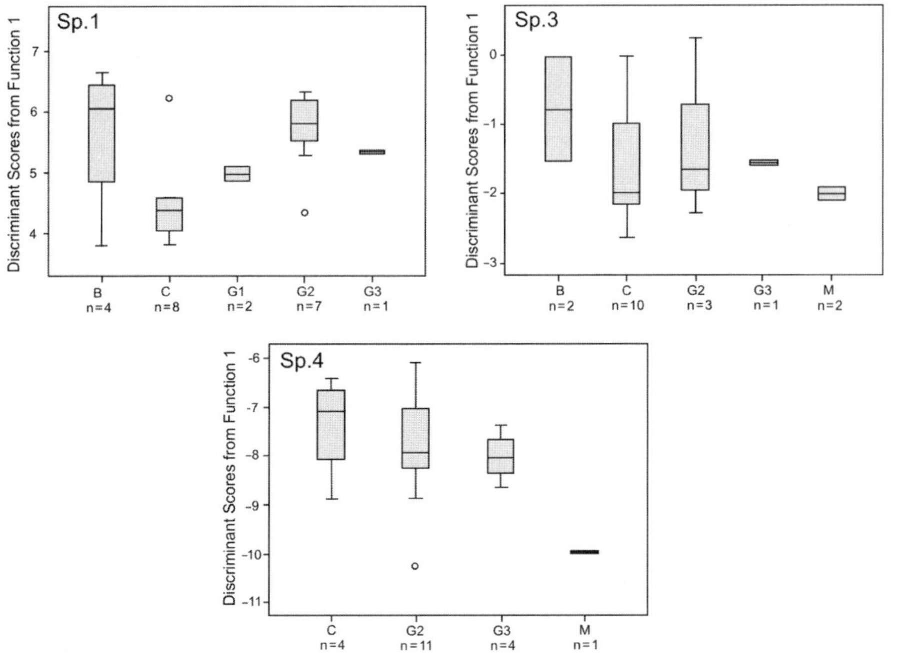

Fig. 6.7 Boxplots illustrating the variation of discriminant function 1 (y-axis) throughout the sampled formations for DR morphospecies 1, 3, and 4 (x-axis). Kruskal-Wallis tests of the data for each graph give a p-value above 0.05 indicating no difference between the formations. B, Baitoa; C, Cercado; G1, lower Gurabo; G2, lower to middle Gurabo; M, Mao

Along with the three fossil species that were expected to be seen in the Dominican Republic formations, a fourth fossil species was also found (DR morphospecies 2). Only three colonies of this fourth fossil species were found and analyzed. These three colonies only occurred in the Baitoa Formation, which was sampled by Saunders et al. (1986). The Baitoa localities that contain colonies of *Siderastrea* are now under water due to dam construction near Santiago. The identification of this species will be explored in future studies.

After species groupings were established and the separation found to be well-defined, it is possible to look at the morphological evolutionary history of each species. Due to the fact that DR morphospecies 2 is only found in the Baitoa Formation, it is not possible to obtain data on how it changed over time so it is excluded from further temporal analyses. The first test of morphometric evolution was to label cases by formation on the same scatterplot that was initially labelled with fossil species. Figure 6.6 shows that the formations completely cover the morphospace range of each fossil species (with the exception of DR morphospecies 2). If these species of *Siderastrea* changed morphometrically through time, you would expect to see a directional shift in the area that each formation covered. Instead we see that each formation completely covers the range of each species. Combined with the fact that there was significant environmental change with the depositional environments of the Cercado and Gurabo Formations, with it deepening up section (McNeill et al., this volume), this result supports the hypothesis that these species of *Siderastrea* display morphologic stasis during the Mio-Pliocene in the Dominican Republic.

To further test the hypothesis of stasis, a Kruskal-Wallis was performed to determine if discriminant function 1, which explained 78.9% of the total variation, showed any significant change between any of the formations in which species of *Siderastrea* were sampled. Along with discriminant function 1, Krustal-Wallis tests were also performed on discriminant function 2, variable x20, variable x19, variable x26, and variable y4. In all of the Kruskal-Wallis tests, the p-values were greater than 0.05 (Table 6.4), which signifies that there is no significant difference in discriminant function 1 throughout the formations sampled. Boxplots showing discriminant function 1, discriminant function 2, variable x20, and variable y4 versus the formations sampled were also constructed to illustrate the patterns examined using the Kruskal-Wallis tests (Fig. 6.7).

By adding the modern species of *Siderastrea*, it not only aided in species identification, but it gave some insight concerning the evolutionary history of *S. siderea* from the Pliocene until the Recent. In Fig. 6.4, the modern *S. siderea* plots at the upper right edge of the species boundary of the fossil *S. siderea*. This could indicate that from the Pliocene to the Recent, *S. siderea* has undergone some change and was no longer in stasis. This potential morphologic change could be attributed to environmental factors associated with the closing of the Central American Isthmus and Plio-Pleistocene reef coral turnover (Budd and Johnson, 1999). These differences could also be caused by the geographic difference between the Dominican fossil morphospecies and the Jamaican Recent coral. In future studies, colonies of *S. siderea* from the Pleistocene and Holocene, as well as other geographic locations, should be added to those in this study to better understand the nature of morphologic change.

Another line of evidence that helps to support the hypothesis of evolutionary stasis for the species of *Siderastrea* is the apparent environmental change through part of the section. Nehm and Geary (1994) show that the water depth of the depositional environment of the Cercado and Gurabo Formation was deepening rapidly from the Cercado Formation through the Gurabo Formation (see also McNeill et al., this volume). The morphologic stasis of *Siderastrea* during an obvious environmental change through this period of time further supports the hypothesis that *Siderastrea* was in morphologic stasis.

While it was found that traditional morphometric techniques alone were not sufficient to parse out all of the species, with no differences between number of

Table 6.4 Results of the Kruskal-Wallis test comparing formations within the three known DR morphospecies

DR morph.		y4_1	x19_1	x20_1	x26_1	Discriminant function 1	Discriminant function 2
1.00	p-value	0.100	0.659	0.666	0.714	0.163	0.410
3.00	p-value	0.212	0.318	0.273	0.365	0.711	0.252
4.00	p-value	0.442	0.174	0.190	0.238	0.373	0.722

All p-values are above 0.05 indicating that there is no significant change in any of the variables analyzed over the period of time sampled septa and only one difference between corallite radius, these measures are still useful as a supplement to the 2-D geometric morphometric data. With both the 2-D geometric morphometric data and the tests done on the traditional measurements, we were able to identify DR morphospecies 1 as being *Siderastrea silecensis*, DR morphospecies 3 as *Siderastrea siderea*, and DR morphospecies 4 as *Siderastrea mendenhalli*.

6.5 Conclusions

Using iterative canonical discriminant analyses, three known DR morphospecies were found in the Mio-Pliocene of the Dominican Republic. These three known species were identified as *S. siderea*, *S. mendenhalli*, and *S. silecensis*. A fourth unknown species was also found in the Baitoa Formation. The successful parsing and identification of these species of *Siderastrea* supports the further use of 2D geometric morphometric methods for similar studies distinguishing species.

Several lines of evidence support the theory that these species of *Siderastrea* underwent evolutionary stasis during the Mio-Pliocene. Based on the scatterplot with markers set by formation (Fig. 6.6), the formations cover the entire morphospace of the three known species. If there was directional morphologic change, one would expect to see linear change in the area covered by each formation. The boxplots of the discriminant functions and most heavily weighted variables (Fig. 6.7) show that there is no significant difference in any of the variables among formations within each species. Nonparametric Kruskal-Wallis tests support these interpretations (Table 6.4).

Sedimentological studies (McNeill et al., this volume) indicate that the water depth of the depositional environment of the Cercado and Gurabo Formation was increasing through time. During a period of non-evolutionary stasis, this environmental change would be expected to be reflected in coral morphology. This is not the case in this study and so it is likely that *Siderastrea* was in evolutionary stasis during the >5 million year time period of this study.

Traditional measures have been used to identify colonies of *Siderastrea* in the past. While some distinctions were found involving corallite radius, all of the species were not parsed out. Also, there were no differences found between species when number of septa was analyzed. These results suggest that these traditional measures are not the most accurate methods to identify species. The observed similarity of these measures may have caused the misidentification of many colonies of *Siderastrea* in other work.

Acknowledgments We are grateful to the following people for their help during this research: Dr. Jonathan Adrain and Dr. Christopher Brochu for helpful comments and suggestions during the editorial process; Kay Saville for help with thin-sections; Tiffany Adrain (SUI) and Arne Ziems (NMB) for help with museum specimens. BRB would like to thank the University of Iowa Department of Geoscience and the Littlefield Fund for support of fieldwork. Additional financial support was provided by a NSF grant (DEB-0102544) to AFB.

Appendix

List of specimens analyzed in morphometric analyses. SUI = University of Iowa Paleontology Repository; NMB = Natural History Museum, Basel, Switzerland

Colony #	Formation	Locality	# of septa	Museum	Catalog #	Morphospecies
1	Low mid Gurabo	16818A	48	SUI	102554	4
2	Cercado	CCE-12A	44	SUI	102555	1
3	Cercado	CCE-12B	46	SUI	102556	3
4	Cercado	CCE-12C	48	SUI	102557	3
5	Cercado	CCE-12D	50	SUI	102558	3
6	Cercado	CCE-12E	50	SUI	102559	1
7	Mid Gurabo	15845A	48	SUI	102560	4
8	Mid Gurabo	15845B	50	SUI	102561	1
9	Low mid Gurabo	16818B	44	SUI	102562	4
10	Low mid Gurabo	16819	52	SUI	102563	3
11	Cercado	CCE-12F	48	SUI	102564	1
12	Cercado	CCE-12G	34	SUI	102565	3
13	Low mid Gurabo	16817A	44	SUI	102566	1
14	Low Gurabo	16837A	48	SUI	NA	1
15	Mid Mao	16884A	46	SUI	102567	3
16	Baitoa	16937B	50	NMB	D5781	1
17	Baitoa	16937C	56	NMB	D5781	1
18	Baitoa	16943A	60	NMB	D5782	2
19	Baitoa	16943B	50	NMB	D5782	2
20	Baitoa	16943C	48	NMB	D5783	3
21	Baitoa	16943D	56	NMB	D5784	2
22	Baitoa	16944A	58	NMB	D5785	1
23	Baitoa	16944B	62	NMB	D5786	1
24	Baitoa	17273	52	NMB	SH Norte7	3
25	Mid Gurabo	15846	46	SUI	102568	4
26	Mid Gurabo	15859A	38	SUI	102569	4
27	Mid Gurabo	15859B	68	SUI	102570	3
28	Mid Gurabo	15859C	48	SUI	102571	4
29	Low Gurabo	15885	58	SUI	102572	1
30	Low mid Gurabo	16817B	50	SUI	102573	4
31	Low mid Gurabo	16818C	46	SUI	102574	1
32	Low mid Gurabo	16818D	46	SUI	102575	1
33	Low mid Gurabo	16818E	48	SUI	102576	1
34	Low mid Gurabo	16818F	48	SUI	102577	3
35	Low mid Gurabo	16818G	40	SUI	102578	3
36	Low mid Gurabo	16818H	44	SUI	102579	4
37	Low mid Gurabo	16818I	50	SUI	102580	1
38	Low mid Gurabo	16818J	48	SUI	102581	4
39	Low mid Gurabo	16818K	48	SUI	102582	4
40	Low mid Gurabo	16818L	48	SUI	102583	4
41	Low mid Gurabo	16818M	98	SUI	102584	4
42	Low mid Gurabo	16818N	34	SUI	102585	4
43	Low mid Gurabo	16818O	56	SUI	102586	4
44	Low mid Gurabo	16818P	50	SUI	102587	4

(continued)

Appendix (continued)

Colony #	Formation	Locality	# of septa	Museum	Catalog #	Morphospecies
45	Low mid Gurabo	16818Q	46	SUI	102588	1
46	Low mid Gurabo	16818R	46	SUI	102589	1
47	Mid Mao	16884B	48	SUI	102590	4
48	Mid Mao	16884C	44	SUI	102591	3
49	Cercado	CCE4	48	SUI	102592	1
50	Cercado	CCE5	52	SUI	102593	3
51	Cercado	CCE12H	46	SUI	102594	3
52	Cercado	CCE12I	44	SUI	102595	3
53	Cercado	CCE12J	32	SUI	102596	4
54	Cercado	CCE12K	48	SUI	102597	4
55	Cercado	CCE12L	50	SUI	102598	3
56	Cercado	CCE12M	54	SUI	102599	1
57	Cercado	CCE12N	62	SUI	102600	1
58	Cercado	CCE12O	44	SUI	102601	1
59	Cercado	CCE12P	48	SUI	102602	3
60	Cercado	CCE12Q	52	SUI	102603	3
61	Cercado	CCE12R	45	SUI	102604	4
62	Cercado	CCE12S	46	SUI	102605	4
63	Cercado	CCE12T	50	SUI	102606	1

References

Blainville, H.M. de., 1830, Zoophytes, *Dictionnaire des Sciences Naturelles*, Paris, v. 60.

Bookstein, F.L., 1991, *Morphometric Tools for Landmark data*. Cambridge University Press, Cambridge, 435 pp.

Bourne, G.C., 1900, Anthozoa, in: *Treatise on Zoology*, II (E. R. Lankester, ed.), Adam and Black, London.

Budd, A.F., Foster, C.T., Dawson, J.P., and Johnson, K.G., 2001, The Neogene Marine Biota of Tropical America ("NMITA") Database: accounting for Biodiversity in Paleontology, *J. Paleontol.*, **75**:743–751.

Budd, A.F. and Guzman, H.M., 1994, *Siderastrea glynni*, a new Scleractinian coral (Cnidaria: Anthozoa) from the Eastern Pacific, *Proc. Biol. Soc. Wash.*, **107**:591–599.

Budd, A.F. and Johnson, K.G., 1999, Origination preceding extinction during Late Cenozoic turnover of Caribbean reefs, *Paleobiology*, **25**:188–200.

Budd, A.F. and Klaus, J.S., This volume, Early evolution of the *Montastraea "annularis"* species complex (Anthozoa: Scleractinia): evidence from the Mio-Pliocene of the Dominican Republic, in: *Evolutionary Stasis: Species and Communities Through Geologic Time* (R.H. Nehm, and A.F. Budd, eds.), Kluwer/Plenum, New York.

Budd, A.F., Stemann, T.A., and Johnson, K.G., 1994, Stratigraphic distributions of genera and species of Neogene to Recent Caribbean reef corals, *J. Paleontol.*, **68**:951–977.

Cheetham, A.H., 1986, Tempo of evolution in a Neogene bryozoan: rates of morphometric change within and across species boundaries, *Paleobiology*, **12**:190–202.

Cheetham, A.H., Sanner, J., and Jackson, J.B.C., 2007, *Metrarabdotos* and related genera (Bryozoa: Cheilostomata) in the late Paleogene and Neogene of Tropical America. *Paleontol. Soc. Mem.*, **67**:1–96.

Eldredge, N. and Gould, S.J., 1972, Punctuated equilibria: an alternative to phyletic gradualism, in: *Models in Paleobiology* (T. J. M. Schopf, ed.), Freeman, Cooper, San Francisco, pp. 82–115.

Ellis, J. and Solander, D., 1786, *The Natural History of Many Curious and Uncommon Zoophytes*, White and Son, London, 208 pp., 63 pls.

Erwin, D.H. and Anstey, R.L., 1995, Speciation in the fossil record, in: *New Approaches to Speciation in the Fossil Record* (D.H. Erwin and R.L. Anstey, eds.), Columbia University Press, New York, pp. 11–38.

Forsman, Z.H., Guzman, H.M., Chen, C.A., Fox, G.E., and Wellington, G.M., 2005, An ITS region phylogeny of *Siderastrea* (Cnidaria: Anthozoa): is *S. glynni* endangered or introduced?, *Coral Reefs*, **24**:343–347.

Garcia, E., Ramos, R., and Bastidas, C., 2005, Presence of cytochrome P450 in the Caribbean corals *Siderastrea siderea* and *Montastraea faveolata, Ciencias Marinas*, **31**:23–30.

Geary, D.H., 1990, Patterns of evolutionary temp and mode in the radiation of *Melanopsis* (Gastropoda; Melanopsidae), *Paleobiology*, **16**:492–511.

Gould, S.J., 2002, *The Structure of Evolutionary Theory*. Belknap Press of Harvard University Press, Cambridge, MA, 1433 pp.

Jablonski, D., 2000, Micro- and macroevolution: scale and hierarchy in evolutionary biology and paleontology, in: *Deep Time: Paleobiology's Perspective* (D.H. Erwin and S.L. Wing, eds.), *Paleobiology* (Suppl.) **26**:15–52.

Jackson, J.B.C. and Cheetham, A.H., 1999, Tempo and mode of speciation in the sea. *Trends Ecol. Evol.*, **14**:72–77.

McNeill, D.F., Klaus, J.S., Evans, C.C., Budd, A.F., and Maier, K.L., This volume, An overview of the regional geology and stratigraphy of the Neogene deposits of the Cibao Valley, Dominican Republic, in: *Evolutionary Stasis and Change in the Dominican Republic Neogene.*

Nehm, R.H., 2005, Patterns and processes of evolutionary stasis and change in *Eratoidea* (Gastropoda: Marginellidae) from the Dominican Republic Neogene, *Carib. J. Sci.*, **41**:189–214.

Nehm, R.H. and Geary, D., 1994, A gradual morphologic transition during a rapid speciation event in marginellid gastropods (Neogene; Dominican Republic), *J. Paleontol.*, **68**:787–795.

Potts, D.C., Budd, A.F., and Garthwaite, R.L., 1993, Soft tissue vs. skeletal approaches to species recognition and phylogeny reconstruction in corals, *Courier Forschungsinst. Senckenberg*, **164**:221–231.

Saunders, J.B., Jung, P., and Biju-Duval, B.,1986, Neogene paleontology in the northern Dominican Republic. 1. Field surveys, lithology, environment, and age, *Bull. Am. Paleontol.*, **89**(323):1–79, 9 pls.

Schultz, H.A. and Budd, A.F., This volume, Neogene evolution of the reef coral species complex *Montastraea "cavernosa"*, in: *Evolutionary Stasis and Change in the Dominican Republic Neogene.*

Stanley, S.M. and Yang, X., 1987, Approximate evolutionary stasis for bivalve morphology over millions of years: a multivariate, multilineage study, *Paleobiology*, **13**:113–139.

Vaughan, T.W., 1917, The reef-coral fauna of Carrizo Creek, Imperial County, California and its significance, *U.S. Geol. Surv. Prof. Paper* **98T**:355–386, pls. 92–102.

Vaughan, T.W., 1919, Fossil corals from Central America, Cuba, and Porto Rico with an account of the American Tertiary, Pleistocene, and recent coral reefs, *U. S. Nat. Hist. Mus. Bull.*, **130**:189–524, pls. 68–152.

Vaughan, T.W. and Wells, J.W. 1943. Revision of the suborders, families, and genera of the Scleractinia, *Geol. Soc. Am. Spec. Pap.*, **104**:363 pp., 51 pls.
Vermeij, M.J.A., 2005, Substrate composition and adult distribution determine recruitment patterns in a Caribbean brooding coral, *Mar. Ecol. Prog. Ser.*, **295**:123–133
Veron, J.E.N., 1995, Corals in Space and Time: the Biogeography and Evolution of the Scleractinia, UNSW Press, Sydney, 321 p.

Chapter 7
Neogene Evolution of the Reef Coral Species Complex *Montastraea "cavernosa"*

Holly A. Schultz[1,2] and Ann F. Budd[1]

Contents

7.1 Introduction

Species are the fundamental unit of evolution for most studies in paleontology; however, what constitutes a species is often disputed. Controversy stems from the differences between various concepts and their applications. According the biological species concept, species are defined as "groups of actually or potentially interbreeding populations which are reproductively isolated from other such groups" (Mayr, 1963). This concept is difficult to apply to corals due to the potential for hybridization and asexual reproduction. According to the phylogenetic species concept, species are defined as "the smallest aggregation of populations (sexual) or lineages (asexual) diagnosable by a unique combination of character states in comparable individuals" (Nixon and Wheeler, 1990). This concept

[1]Department of Geoscience, University of Iowa, Iowa City, IA 52242, USA. Email: schultz@geology.ucdavis.edu, ann-budd@uiowa.edu

[2]Current address: Department of Geology, University of California at Davis, One Shields Avenue, Davis, CA 95616, USA.

R.H. Nehm, A.F. Budd (eds.) *Evolutionary Stasis and Change in the Dominican Republic Neogene,*

emphasizes a common origin for members of a species (Cracraft, 1987), but is even more difficult to apply to corals because it assumes a hierarchical pattern and transmission of discrete characters by inheritance. It therefore completely excludes transmission of characters by processes such as hybridization (Veron, 1995). Many coral workers lean toward the biological species concept, and integrate multiple lines of evidence (e.g., genetics, morphology, reproduction) in making interpretations (Knowlton and Weigt, 1997; Willis et al., 2006). Identifying extinct coral species can therefore be especially problematic because the only available data source for distinguishing fossil species is their morphology and most characters are continuous. This problem is magnified by the high amount of environmentally-induced morphological variation found in many coral species (Budd and Pandolfi, 2004; Fukami et al., 2004a).

Species delineation is a particularly central issue in studies of evolutionary stasis and change, the theme of this volume. To determine how, or even if, a species is changing through time, it is necessary first to accurately distinguish species. If intraspecific variation is not recognized and species are oversplit, their evolution could be greatly misconstrued. The converse, lumping together species that form complexes, is equally problematic (see Jackson and Cheetham, 1999, for discussion).

Many genera of coral, including *Montastraea* and *Acropora*, are thought to form species complexes (Knowlton and Weigt, 1997; Knowlton and Budd, 2001; Willis et al., 2006), which consist of numerous genetically distinct species or lineages that periodically split and/or fuse as they extend through time. During splitting or fusing, morphologic intermediates form and species overlap (Budd and Pandolfi, 2004). Species complexes in corals are a relatively new discovery, and the dynamics of speciation, extinction, and hybridization within complexes over long periods of time have not been studied (Odorico and Miller, 1997; Budd and Pandolfi, 2004). One complex that is currently under investigation is the *Montastraea "annularis"* species complex (see Budd and Klaus, this volume). The three members of the complex are ecologically dominant, broadly sympatric, and overlap in distribution on many tropical reefs in the western Atlantic. The members were once thought to represent one morphologically variable species, but in reality they differ in many attributes including growth rate, stable isotope geochemistry, behavior, and life history (Weil and Knowlton, 1994; Knowlton and Budd, 2001; Fukami et al., 2004a). Recent research indicates that the members of the complex differ genetically (Fukami et al., 2004a). However, the complex has also been found to undergo hybridization in experimental fertilizations in the lab, and hybridization has been inferred on Pleistocene-age reefs of the Bahamas (Budd and Pandolfi, 2004).

Overall variation within the *Montastraea "cavernosa"* species complex is even more extensive and complicated than in the *M. "annularis"* complex (Foster, 1985; Budd, 1990). An early qualitative monograph contains identification keys based on traditional morphological characters for various *M. "cavernosa"*-like corals, such as corallite diameter and septal number (Vaughan, 1919). One of the first quantitative studies analyzing these traditional characters was performed by Budd (1991), and was based solely on linear measurements and counts made on corallite architecture. The present study expands on the work done in the latter monograph.

It examines the complex during the Mio-Pliocene in the northern Dominican Republic, over approximately 5 million years of geologic time, and uses a combination of traditional measurements of corallite architecture and geometric morphometrics. The sample size of the present work is also greatly expanded over previous work. Here, 95 fossil colonies are analyzed, fifteen of which (out of a total of 41) were also analyzed by Budd (1991). Several questions are addressed in the present study regarding the *M. "cavernosa"* complex.

1. How many species constitute the *Montastraea "cavernosa"* species complex in the Mio-Pliocene of the Dominican Republic?
2. How do these species differ morphologically from one another?
3. How did the diversity of the complex vary over time?
4. How does the current research compare with that of Budd (1991)?

7.2 Geologic Setting

The Dominican Republic occupies the eastern two-thirds of the island of Hispañola, which it shares with the country of Haiti. Hispañola is occupied by several mountain ranges, which trend from northeast-southwest (Lewis, 1980; Mann et al., 1991). The largest is the Cordillera Central, which forms the backbone of the island. The Cibao Valley is formed by a large synclinal basin (Mann et al., 1991) that is bordered to the south by the Cordillera Central and to the north by the Cordillera Septrionale. Several rivers cut across the basin and expose an approximately 5,000 meter sequence of sediments (Mann et al., 1991), which range in age from Eocene to Pliocene. The majority of these sediments comprises the Yaque Group, a northward prograding wedge of sediment eroded from the Cordillera Central. The Yaque Group is composed of the Cercado, Gurabo, and Mao Formations, which are Miocene to Pliocene in age (Mann et al., 1991).

The Cercado Formation is the oldest of the three formations, and based on planktic foraminifera and nannofossil zones, paleomagnetic data, and strontium isotope data, it is estimated to be late Miocene in age (approximately 6.5–5.6 Ma). The reef at Arroyo Bellaco is approximately 6.2 Ma (McNeill et al., this volume). Throughout the Cibao Valley, the Cercado Formation is approximately 145 meters thick, and dips to the north at about ten degrees (Evans, 1986). The lower part of the formation consists of sands with interbedded conglomerates deposited in a shallow marine to brackish environment, while the upper portion contains coral thickets with massive heads and algal mounds.

Coral-rich units within the Cercado Formation are exposed along Arroyo Bellaco, a tributary of the Río Cana, and are approximately nineteen meters thick and extremely well preserved. Based on the types of coral present, seven reef zones have been recognized: thicket, mixed coral thicket, small branched coral zone, pillar zone, head coral zone, interlayered pillar and stick zone, and reef flat (Evans, 1986). Samples for the present study were collected from the head coral zone using maps from Klaus and Budd's (2003) study of coral reef communities. The reef area located at Bel-1 and Bel-2 consists of the genera *Agaricia*, *Dichocoenia*, *Gardinoseris*, *Goniopora*, *Montastraea "annularis"* complex, *Pocillopora*, *Porites*, *Siderastrea*, *Solenastrea*, *Stephanocoenia*, and *Stylophora*. Other areas of the reef along Arroyo Bellaco contain the same genera, but they also include *Madracis*, *Favia*, and *Montastraea "cavernosa"* complex (Klaus and Budd, 2003).

The Gurabo Formation is late Miocene to early Pliocene in age (approximately 5.6–4.5 Ma) based on calcareous nannofossils, foraminifera, paleomagnetic data, and strontium isotope data (McNeill et al., this volume). It is approximately 425 m thick (Saunders et al., 1986), consists of a basal conglomerate whose clasts are well rounded but poorly sorted, and contains poorly preserved mollusks and corals. Above this conglomerate are calcareous silts containing numerous mollusks and corals. Corals are concentrated in horizons, two of which were collected in the present study. The first horizon is approximately 300 m up the section (localities NMB 16933, 15847), and is composed of both massive heads and branching corals that have been transported but are still in life position (Saunders et al., 1986). Genera found in this area include *Dichocoenia, Hadrophyllia, Leptoseris, Madracis, Montastraea "annularis"* complex, *Montastraea "cavernosa"* complex, *Placocyathus,* and *Stephanocoenia* (see the Neogene Marine Biota of Tropical America (NMITA) Website, http://nmita.geology.uiowa.edu). The second horizon (localities NMB 15808, 15837, 16817, and 16818; approximately 400 m up in the sequence) is composed of calcareous silts and sandy silts, and are also rich in corals. Corals occur in beds and as separate, scattered heads (Saunders et al., 1986). Some corals are in growth position while others have been transported. Genera include *Diploria, Gardineroseris, Madracis, Montastraea "annularis"* complex, *Montastraea "cavernosa"* complex, *Manicina, Meandrina, Undaria, Siderastrea, Stylophora,* and *Trachyphyllia* (see the NMITA Website, http://nmita.geology.uiowa.edu). Throughout the Gurabo Formation, there is a general trend of increasing water depth with corals absent near the top of the formation (Saunders et al., 1986).

The Mao Formation rests atop the Gurabo Formation. It is early Pliocene to middle Pliocene in age (approximately 4.5–3.4 Ma), based on calcareous nannofossils and foraminifera, paleomagnetic data, and strontium isotope data (McNeill et al., this volume; Saunders et al., 1986). The Mao Adentro Limestone, the unit from which all of the samples from the Mao Formation were collected, is thought to be approximately 3.48 Ma in age (McNeill et al., this volume). The Mao Formation is approximately 600 m thick beginning with a base of sandy silts, overlain by coarse sands and conglomerates, followed by calcareous silts. Numerous corals, mollusks and microfauna are found throughout. The massive Mao Adentro Limestone (loc. NMB 16884) forms a cap, and contains a rich assemblage of corals, most of which are not in life position (Saunders et al., 1986). The corals include the genera *Agaricia, Dichocoenia, Diploria, Favia, Hadrophyllia, Isophyllia, Madracis, Manicina, Montastraea "annularis"* complex, *Montastraea "cavernosa"* complex, *Mussismilia, Pavona, Placocyathus, Porites, Siderastrea, Solenastrea, Stephanocoenia, Stylophora,* and *Undaria* (see the NMITA Website, http://nmita.geology.uiowa.edu).

7.3 Sampling

A large collection of *Montastraea "cavernosa"* – like corals is needed to examine the variation and changes in the complex over time. We analyzed fifteen specimens in a previous collection made by Saunders et al. (1986), which is deposited at the Natural History Museum (NMB) in Basel, Switzerland. These specimens were also studied in the previous monograph (Budd, 1991). In addition, we analyzed 80 specimens that we collected ourselves during the summers of 2003 and 2004, making a grand total of 95 colonies in the analyses (see Appendix 1 for complete list of specimens).

Using the maps of Saunders et al. (1986) and Klaus and Budd (2003), we collected samples along the Arroyo Bellaco, Río Cana, and Río Gurabo from the Cercado, Gurabo, and Mao Formations. These new samples have greatly enhanced the existing collection of *Montastraea "cavernosa"* – like corals made by Saunders et al. (1986) and analyzed in Budd's (1991) monograph. Not only are the overall sample sizes higher, but also their distribution through the section is more even.

7.4 Study Taxa

The genus *Montastraea* (Blainville, 1830) is characterized by having colonies that are massive, encrusting, or subfoliaceous (Budd, 1991). Colonies are plocoid and formed by extratentacular budding; corallite walls are septothecate with dentate margins, the columella is trabecular, and costae are well developed (Wells, 1956; Budd, 1991). *Montastraea "cavernosa"* – like corals were initially distinguished using traditional morphologic features (Vaughan, 1919), including a septal number of 32–60 and a corallite size that ranges from five to nine millimeters. Definitions and illustrations of these characteristics are available on the Neogene Marine Biota of Tropical America Website, http://nmita.geology.uiowa.edu. Recent molecular data have shown that the *Montastraea "cavernosa"* species complex is not related to the *Montastraea "annularis"* species complex indicating that the genus *Montastraea* is polyphyletic (Fukami et al., 2004b).

Currently five species are assigned to the *Montastraea "cavernosa"* species complex during the Mio-Pliocene (Budd, 1991). Those species are *Montastraea brevis* (Duncan, 1864), *Montastraea canalis* (Vaughan, 1919), *Montastraea cylindrica* (Duncan, 1863), *Montastraea cavernosa* (Linnaeus, 1767), and *Montastraea endothecata* (Duncan, 1863). All members of the complex are common to abundant in reef zones ranging from backreef across the reef crest to deeper areas.

Montastraea brevis has small colonies, small to intermediate corallites, thin walls and a narrow columella. It is most similar morphologically to *Montastraea cylindrica.* It is early Pliocene in age and is found in the Gurabo Formation along the Río Cana (loc. NMB 16881) and the Gurabo Formation along the Río Gurabo (loc. NMB 15807, 17837, 15838, 15839, 15841, 15846, 15847, 15850, 15851, 15858, 16883, 16921, 16934).

Montastraea canalis has thicker walls, smaller corallites, and equal costae. It is most similar morphologically to *M. cavernosa. Montastraea canalis* ranges from late Oligocene to the Pliocene. It is found in the Gurabo Formation (loc. NMB 16814, 16815, 16817, 16881) and Mao Formation (loc. NMB 16875, 16876, 16877) along the Río Cana, the Mao Formation (loc. NMB 15830) on the Río Gurabo, and the Baitoa Formation (loc. NMB 16943, 16944, 17279, 17289) along the Río Yaque del Norte.

Montastraea cavernosa is highly variable. It has an intermediate corallite size and subequal costae. It is found from the Pliocene to Recent. In the Dominican Republic it occurs along the Río Cana in the Mao Formation (loc. NMB 16884), along the Río Gurabo in the Gurabo Formation (loc. NMB 15808, 15836, 15847, 15858, 15893, 16184, 16921), along the Río Mao in the Cercado Formation (loc. NMB 16908) and the Gurabo Formation (loc. NMB 16911), and along the Río Yaque del Norte in the Baitoa Formation (loc. NMB 16944, 17279).

Montastraea cylindrica has large, plate-like colonies with widely spaced calices that are slightly elevated. This species generally has fewer septa. It is found from the late Miocene to the late Pliocene and occurs along the Río Cana in the Mao Formation (loc. NMB 16876, 16884) and along the Río Gurabo in the Gurabo Formation (loc. NMB 15838, 15841, 15846, 16921).

Lastly, *Montastraea endothecata* has large calices, thick walls, strong costae, and a strongly whirled columella. It is found from the Oligocene to the late Pliocene. It occurs in the Gurabo Formation along the Río Cana (loc. NMB 16817, 16818), the Gurabo Formation along the Río Gurabo (loc. NMB 16933), and in the Gurabo Formation along the Río Mao (loc. NMB 16911). See Budd (1991) for more information and photographs of each member of the complex.

Montastraea "cavernosa" – like corals have been present in the Caribbean since the Eocene (Budd et al., 1992). Budd et al. (1992) named three new species (*Montastraea nodosa, Montastraea prima,* and *Montastraea? rotunda*) from the Eocene of Panama. Based on corallite diameter and number of septa, these species could be members of the *Montastraea "cavernosa"* species complex. *Montastraea "cavernosa"* – like corals are present in late Oligocene sediments as well (Budd et al., 1994; Vaughan, 1919). From the early Miocene through to the present, *M. "cavernosa"* – like corals have been found in almost every part of the Caribbean, from Florida and Jamaica in the north to Panama and Brazil in the south.

Ecological studies have shown that modern *M. "cavernosa" s.l.* is not a dominant reef coral and is more abundant in intermediate to deeper depths rather than shallower depths such as the reef crest (Goreau, 1959; Rützler and Macintyre, 1982; Burke, 1982). Experiments have shown that colonies are especially effective at removing sediment (Lasker, 1980). Although it is widely believed that modern *Montastraea "cavernosa"* consists of a complex of several species, very few studies have attempted to distinguish and characterize species within the complex. Lasker (1976, 1979, 1980, 1981; Budd, 1993) recognized two distinct feeding morphs of *Montastraea cavernosa*: a "diurnal" morph with smaller corallites that is expanded both day and night, and a "nocturnal" morph with larger corallites that is only expanded at night and is more effective at capturing zooplankton. Ongoing research on the two morphs suggests that they may be sibling species, with the "nocturnal" morph found in deeper waters and the "diurnal" morph found in shallower waters (Ruiz Torres, 2004). Studies of the diurnal morph have shown that it varies in corallite size, septal number, thecal thickness, and coenosteal porosity among habitats (Foster, 1985; Budd, 1990).

7.5 Geometric Morphometrics

Transverse thin sections were prepared from each specimen and corallites were photographed on each thin section. Seventeen landmarks were digitized on six mature corallites per colony (Fig. 7.1). These landmarks are spatially homologous points, which include juxtaposition of skeletal structures and maxima of curvature (i.e., Types 1 and 2 of Bookstein, 1991). They were chosen to reflect shape of the corallite wall and costosepta (see Appendix 2).

Traditionally, species have been distinguished based on corallite diameter and number of septal cycles (Vaughan, 1919, pp. 363–364). Budd (1991) distinguished species based on five broad categories of characteristics: (1) corallite size and spacing; (2) septal number and length; (3) columella (and associated paliform lobes) width and porosity; (4) septum, theca, and costa thickness; and (5) development of the coenosteum. Landmark data can produce some of these traditional measurements as well as other features that are important for identifying *Montastraea* species, such as wall thickness and costae extensions (Appendix 3). Using the computer program CoordGen6f in the IMP software series (Integrated Morphometrics Package, 2004, written by H. David Sheets, available at http://www2.canisius.edu/~sheets/morphsoft.html), centroid size and Bookstein shape coordinates were calculated using points nine and ten as the baseline. By registering two points on a common baseline, shape coordinates contain all the information about an object's shape, independent of the object's size (Bookstein, 1991). Bookstein coordinates are used instead of other methods; because they can be utilized in both graphical displays and statistical analyses, they are biologically meaningful, and they are suitable for subsequently identifying unknowns in canonical discriminant analyses (Zelditch et al., 2004). In previous work (Budd, 1991), species within the complex were found to differ primarily using measurements made parallel to the baseline (not perpendicular to it). To focus on this variation and reduce noise in the present study, only x-coordinates of shape coordinates were used in the analysis (Appendix 3). Centroid size is calculated by summing the squared distances from each of the seventeen landmarks to a common centroid.

In addition to landmark data, two linear distance measurements were taken for each corallite, corallite and columella diameter. These were studied because they have traditionally been key distinguishing characteristics for members of the species complex (Vaughan, 1919). The number of septa in each corallite was also counted.

To distinguish species, all samples were subdivided into three time bins that correspond with the geologic formations (Table 7.1). Bin one contains a total of thirty-two samples collected from the Cercado Formation (loc. Bel-1, Bel-2, Bel-3, Bel-4, Bel-10, Bel-5, Bel-6). Bin two contains fifty-four colonies collected from the Gurabo Formation

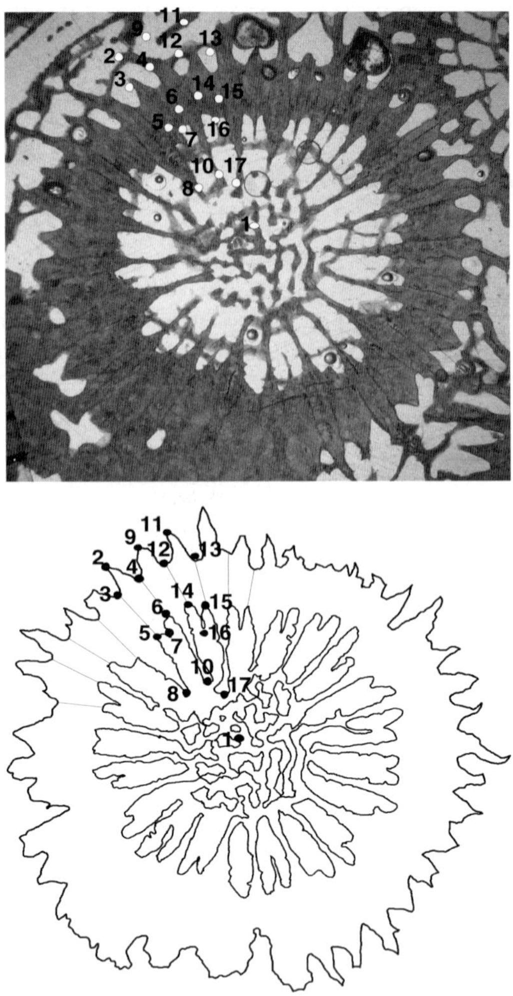

Fig. 7.1 Two-dimensional Cartesian coordinates collected for 17 landmarks on transverse thin-sections of corallites. Points 9 and 10 were used as the baseline. Linear measurements were taken from point 1 to point 17 and point 1 to point 5. Line drawing is courtesy of Reggie Schreiber

(NMB loc. 15808, 15837, 15855, 15894, 16817, 16818, 16859, 16933, 16911). Bin three contains 14 samples collected from the Mao Formation (NMB 15830, 16884).

Canonical discriminant analyses were first performed separately for each time bin using 15 shape coordinates, number of septa, centroid size, and two linear measurements as variables, and colonies as groups, following the iterative approach of Cheetham et al. (2007). The same approach was used by Beck and Budd (this volume). P-values based on the F-values associated with Mahalanobis distances were used to determine whether colonies were significantly different from one

Table 7.1 Collecting localities and number of colonies used in this analysis and previous monograph (Budd, 1991)

		Number of colonies measured	
Formation	Locality no.	Current study	Budd (1991)
Tabera	NMB 17279	0	3
Baitoa	NMB 16944	1	1
Cercado	Bel-1, 2	18	
Cercado	Bel-3, 4, 10	10	0
Cercado	Bel-5, 6	1	0
Cercado	TU 1422	0	1
Gurabo	TU 1405	0	1
Gurabo	NMB 15808	10	1
Gurabo	NMB 15837	2	1
Gurabo	NMB 15838	1	3
Gurabo	NMB 15841	0	3
Gurabo	NMB 15847	1	1
Gurabo	NMB 15855	2	0
Gurabo	NMB 15893	0	1
Gurabo	NMB 15894	2	0
Gurabo	NMB 16815	0	1
Gurabo	NMB 16817	5	0
Gurabo	NMB 16818	24	2
Gurabo	NMB 16859	1	0
Gurabo	NMB 16881	0	2
Gurabo	NMB 16911	1	0
Gurabo	NMB 16921	0	1
Gurabo	NMB 16933	1	2
Gurabo	NMB 16934	0	2
Gurabo	TU 1231	0	1
Gurabo	TU 1215	0	1
Gurabo	TU 1246	0	1
Gurabo	TU 1208	0	1
Mao	TU 1344	0	1
Mao	NMB 15830	1	2
Mao	NMB 16876	0	1
Mao	NMB 16877	0	1
Mao	NMB 16884	12	5
		Total: 93	41

another. If colonies were not significantly different (p-value greater than 0.05), they were grouped together. After successive iterations in which colonies that were not significantly different from one another were lumped into the same group (="fossil cluster" in Budd and Klaus, this volume), the number of groups found within each bin was determined. For example, a canonical discriminant analysis is performed on all of the colonies from one formation using SPSS, and a matrix of p-values is produced comparing each colony with every other colony in the formation. Colonies that are not significantly different from each other (p-value greater than 0.05) are combined into one group, and the analysis is run again. This process continues until all of the groups are significantly different from one another.

After each bin was analyzed, a global analysis was performed using the groups found within each time bin. The same iterative grouping procedure was followed combining groups whose Mahalanobis distances were not statistically significant. Species (= "morphospecies" in Budd and Klaus, this volume) are defined using the final groups, and the differences between species are then examined multivariately using canonical discriminant analysis and univariately using one-way analyses of variance. The results are used to determine the total number of species found in this study and their stratigraphic ranges.

Finally, modern specimens of *Montastraea cavernosa* were measured using the same landmark scheme in order to both compare the morphology of the Neogene members of the complex with extant specimens and to study the typical amount of morphological variation in the species complex. All nine modern *M. cavernosa* specimens used in the analysis were determined to be the diurnal morph (morph 2) by Budd (1993) and were collected from three different environments. Samples were collected from the lagoon, the sand channel, and the reef environments (SUI 48759, SUI 48764, SUI 48767, SUI 48771, SUI 48773, SUI 48777, SUI 48754, SUI 48752, SUI 48755).

7.6 Results

7.6.1 Cercado Formation

Discriminant analyses of the samples from the Cercado Formation indicate five distinct groups in the thirty-two samples (Fig. 7.2). Ninety-five percent of the total variation is explained by the first three discriminant functions. Function one (70%) is most strongly correlated with the number of septa and centroid size. Function two (20%) is correlated with the number of septa. Function three (7%) is correlated with centroid size and the size of the corallite.

The most strongly correlated variables were further examined using one-way analyses of variance and a Tukey HSD analyses ($p < 0.05$). The results show that numbers of septa are low in group 43, intermediate in groups 60 and 63, and high in group 68 (Fig. 7.3A). Corallite diameter and centroid size are low in group 63, intermediate in group 60, and high in groups 59 and 68 (Fig. 7.3B, C). Using Bookstein coordinates for wall thickness (x3, x4), thickness is low for group 57, intermediate for group 68, and high for group 59. Though similar in corallite diameter, centroid size, and number of septa, groups 60 and 63 are separated from each other based on wall variables.

7.6.2 Gurabo Formation

Discriminant analyses on samples from the Gurabo Formation produced four significantly different final groups (Fig. 7.2). Ninety-three percent of the total variation is explained by the first three discriminant functions. Function one (54%) is strongly

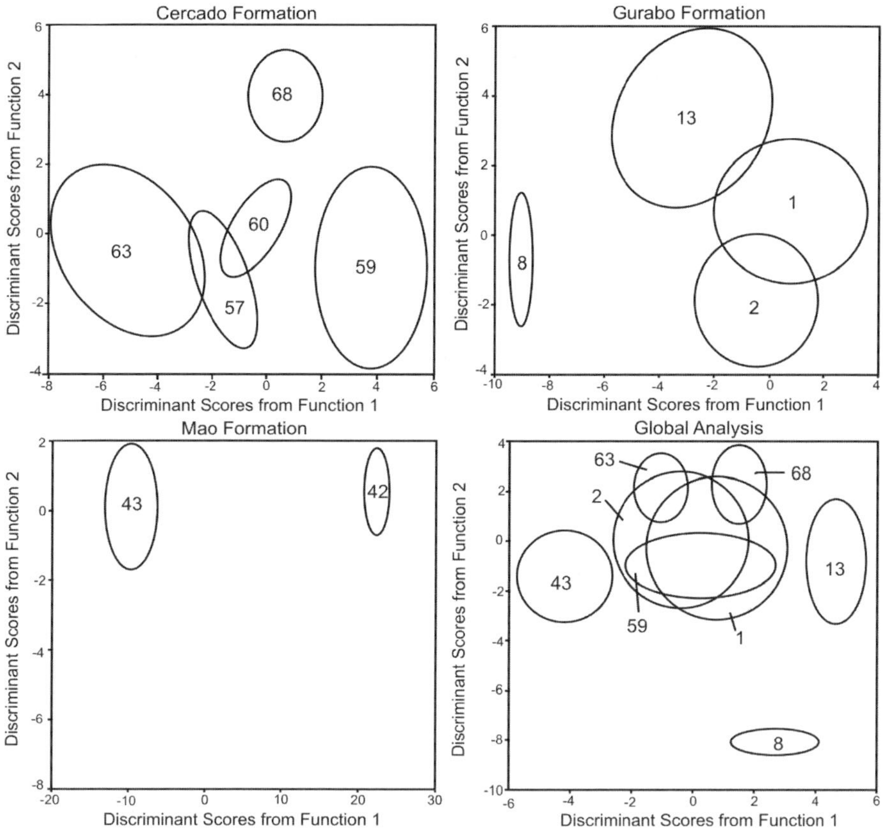

Fig. 7.2 Plots of scores on the first two canonical variables of the discriminant analyses used to distinguish species of *M. "cavernosa"* for each of the formations and the global analysis of fossil species. Ellipses indicate maximum variation within each final group; ellipse labels correspond with group numbers described in the text

correlated with wall thickness, function two (26%) is strongly correlated with number of septa, and function three (14%) is strongly correlated with centroid size.

The most strongly correlated variables were further examined using one-way analyses of variance and Tukey HSD analyses ($p < 0.05$). The results show that group 13 has a significantly higher number of septa than groups 1, 2, and 8 (Fig. 7.3A). Group 2 has a smaller corallite diameter than group 13 (Fig. 7.3C). Centroid size is small in group 2, intermediate in groups 1 and 13, and large in group 8 (Fig. 7.3B). When compared to group 8, groups 1, 2, and 13 have significantly thinner walls. Group 1 has significantly thicker walls than group 2 when variables associated with wall thickness are examined (x3 and x4). The walls of groups 1 and 13 are also significantly different from one another.

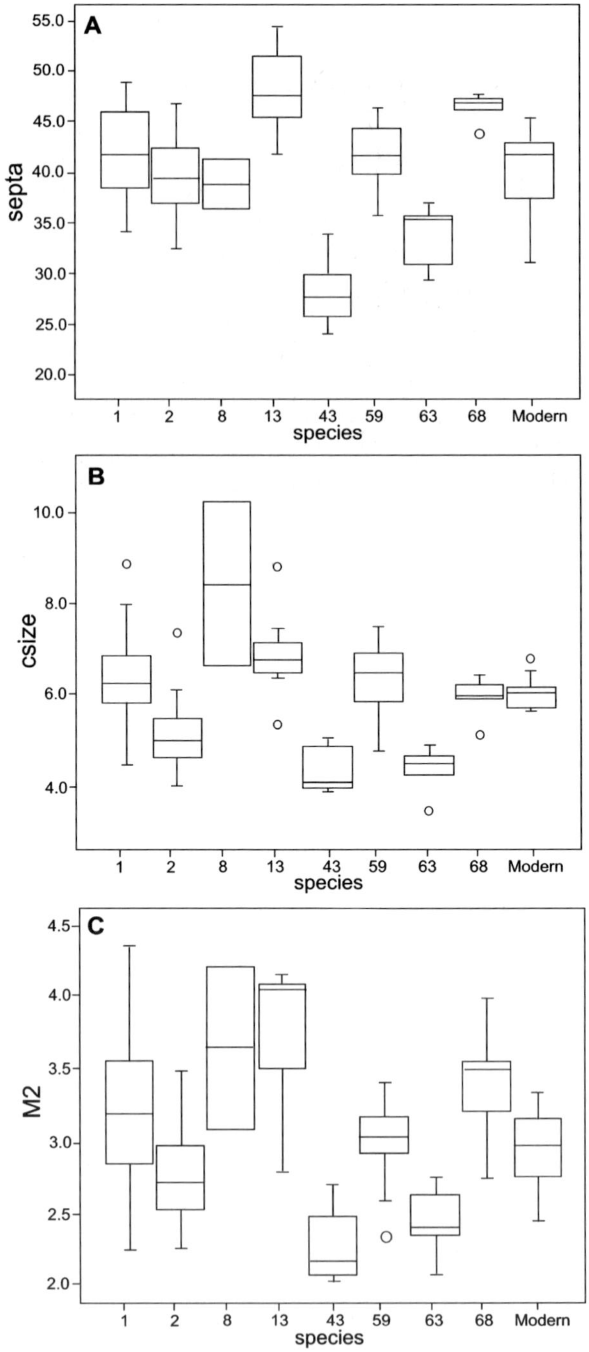

Fig. 7.3 Boxplots showing differences among species of modern and fossil *Montastraea "cavernosa"* like corals. Number of septa, centroid size, and corallite diameter (M2) showed the greatest differences among species in the discriminant analysis. Species also differ significantly in various combinations of wall characters. Each box represents the interquartile range containing 50% of the values; the line across each box represents the median. The whiskers represent the highest and lowest values, excluding outliers (indicated by circles)

7.6.3 Mao Formation

Discriminant analysis of the samples from the Mao Formation resulted in two significantly different final groups (Fig. 7.2). In the analysis some colonies were difficult to classify due to the low sample size. These colonies are not assigned to a species at the present time. Approximately 97% of the total variation is along function one, which is strongly correlated with shape coordinates and linear measurements (M3) associated with corallite diameter.

When number of septa, centroid size, corallite diameter, and wall thickness were examined using one-way analysis of variance ($p < 0.05$), the groups clearly differed. Group 43 has far fewer septa than group 42, and is significantly smaller in corallite diameter and centroid size (Fig. 7.3).

Only one variable associated with wall thickness (x16) is significantly different between the groups. Group 43 has a thicker wall than group 42.

7.6.4 Global Analysis

A global canonical discriminant analysis using the 11 final groups indicated a total of eight species (Fig. 7.2). Group 57 from the Cercado Formation and group 1 from the Gurabo Formation are not significantly different and were therefore combined to form species 1. Group 60 from the Cercado Formation, group 2 from the Gurabo Formation, and group 42 from the Mao Formation were combined to form species 2. In the analysis, four discriminant functions are needed to explain the variation. Function one (55%) is strongly correlated with number of septa and corallite diameter. Function two (27%) is strongly correlated with shape coordinates related with corallite diameter. Function three (12%) is correlated with centroid size and shape coordinates associated with wall thickness. Function four (6%) is correlated with length of quaternary septa (Appendix 4).

The most heavily weighted variables from the canonical discriminant analysis were examined using a one-way analysis of variance and a Tukey HSD analysis. Number of septa is low in species 43; intermediate in species 63; high in species 1, 2, 8, 68; and very high in species 13 (Fig. 7.3A). Species that were not significantly different in the number of septa are clearly different using either shape coordinates or linear measurements. Differences can be seen especially in corallite diameter (M2). Corallite diameters are small in species 43 and 63, intermediate in species 2 and 59, and large in species 1, 8, 13 and 63 (Fig. 7.3C). Generally in this analysis, species that are significantly different in corallite diameter are also significantly different in centroid size (Fig. 7.3B). The exception is species 8, which differs from species 1, 2, and 68 in centroid size. Species 63 and 43, though similar in septal number and size, are significantly different in shape coordinates related to wall thickness. Species 43 has a thicker wall (x14, x15) than species 63. Species 8 also has a significantly thicker wall than all other species.

Only two species cross formation boundaries (1 = 57; 2 = 42 = 60), indicating that most of the species are restricted to a single formation. Species 59, 63, and 68

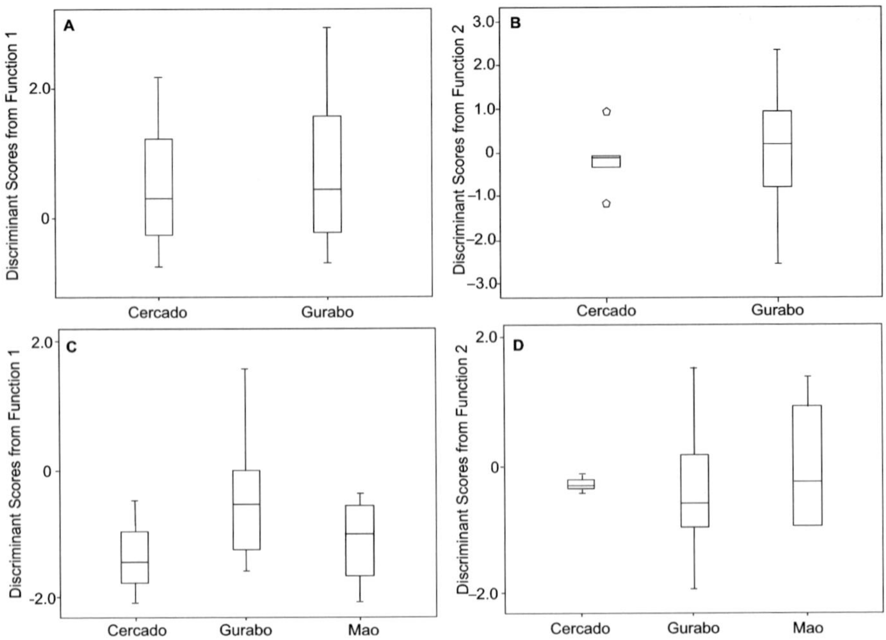

Fig. 7.4 (A and B) Boxplots showing morphologic stasis of species 1 as it spans the Cercado and Gurabo Formations. (C and D) Boxplots showing morphologic stasis of species 2 as it spans all three formations. Each box represents the interquartile range containing 50% of the values; the line across each box represents the median. The whiskers represent the highest and lowest values, excluding outliers (indicated by circles)

are only found in the Cercado Formation. Species 8 and 13 are only found in the Gurabo Formation and species 43 is unique to the Mao Formation.

The species that spanned more than one formation (species 1 and species 2) were further analyzed for evolutionary stasis or change. Variation within the species was examined among formations using nonparametric analyses of variance (Kruskal-Wallis tests). Discriminant functions were used as variables in these analyses, to emphasize morphological features that differed among species. Morphologic stasis was found in both species (Fig. 7.4).

7.6.5 Comparisons with Previous Work

The larger sample size of this study has allowed more variation within the complex to be observed and three new species to be recognized (Table 7.2). Geometric morphometrics allowed certain aspects of the corallite morphology to be studied that would have been difficult using linear measurements, such as costa length (x2, x11) and wall thickness (x3, x4, x14, x15). For example, species 13 and 43 have very thin, elongate

Table 7.2 Comparison between classifications of specimens used in current analysis and Budd (1991). All specimens are deposited at the Natural History Museum in Basel, Siwtzerland (Budd, 1991)

	Identification	
Specimen from Budd (1991)	Current study	Budd (1991)
16944, D5740	sp. 68	*M. canalis*
16817, D5616	sp. 13	*M. endothecata*
16818, D5618	sp. 1	*M. endothecata*
16818, D5619	sp. 2	*M. endothecata*
16818, D5620	sp. 1	*M. endothecata*
16818, D5621	sp. 13	*M. endothecata*
15830, D5551	sp. 43	*M. canalis*
15838, D5557	sp. 2	*M. cylindrica*
15847, D5579	sp. 2	*M. cavernosa*
16884, D5647	sp. 43	*M. cylindrica*
16884, D5648	sp. 2	*M. cavernosa*
16884, D5649	sp. 43	*M. cylindrica*
16888, D5659	sp. 43	*M. cylindrica*
16911, D5702	sp. 13	*M. endothecata*
16933, D5718	sp. 1	*M. endothecata*

costae but the wall is much thinner in species 43, while species 59 has short, thick costae. The landmark scheme is better able to capture this variation than traditional linear measurements because shape is measured independent of size using geometric morphometrics. Equally, this study did not examine some aspects of coral morphology that Budd (1991) studied. Septum thickness, costa thickness and corallite spacing were all measured by Budd (1991) and not examined in the current study.

Because the current study has a larger sample size and is based on geometric morphometrics, rather than traditional measurements, some of the species of Budd (1991) appear to represent more than one species. Budd's original *Montastraea endothecata* species separates into at least two distinct species in this analysis (species 1 and 13; Table 7.2). Both species 1 and 13 have a large corallite diameter and many septa, but species 1 has a thicker wall. Specimens of Budd's original *Montastraea canalis* are identified as species 43 and 68 in the current study, however, sample sizes are too low to be certain (Table 7.2). Species 43 appears to be *M. cylindrica*, and species 8, 59, and 63 appear to be new species (Table 7.3).

7.6.6 Comparisons with Modern Specimens

When fossil specimens were compared with modern *Montastraea cavernosa* specimens using the same procedure as above, modern *Montastraea cavernosa* colonies (diurnal morph) were significantly different from all fossil species at

Table 7.3 Distinguishing morphological characteristics of the eight species of *Montastraea "cavernosa"* – like corals from current analysis

Sp.	Wall (mm)	Corallite (mm)	Septal number	Relative septal thickness	Columella size (mm)	Relative costa length
Sp. 1	1.0–1.5	8.0–11.0	36–48	Equal	3.0–4.0	Equal
Sp. 2	0.2–0.7	5.0–8.0	34–46	Unequal	2.0–3.0	Unequal
Sp. 8	2.0–3.0	7.0–10.0	38–42	Unequal	2.0–3.0	Equal
Sp. 13	0.2–0.5	8.0–11.0	44–56	Equal	2.0–3.0	Equal
Sp. 43	0.3–0.5	4.0–5.0	24–36	Unequal	1.0–2.0	Unequal
Sp. 59	1.5–2.0	8.0–10.0	36–46	Equal	2.0–3.0	Equal
Sp. 63	0.2–0.5	5.0–6.0	30–36	Equal	1.0–2.0	Equal
Sp. 68	0.5–1.0	8.0–10.0	48–52	Unequal	3.0–4.0	Unequal

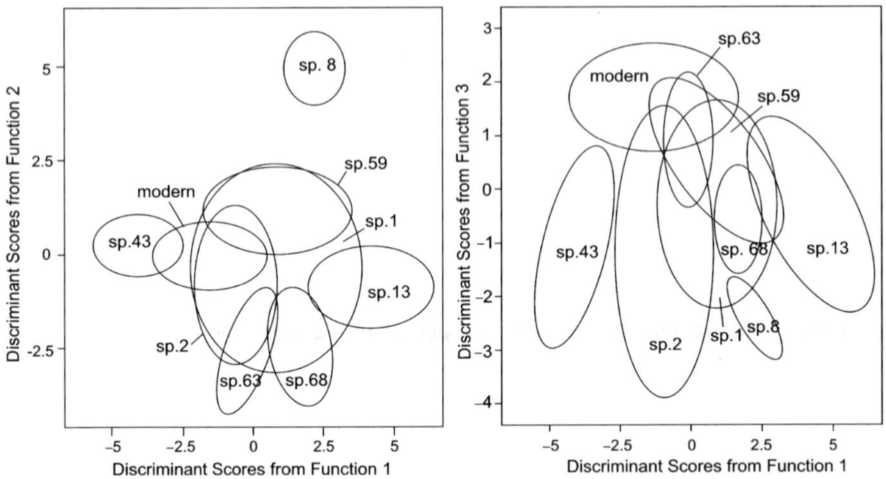

Fig. 7.5 Plots of scores on the first two canonical variables of the discriminant analyses used to distinguish species of *M. "cavernosa"* (one modern species and 8 fossil species). Ellipses indicate maximum variation within each species

$p < 0.05$. The modern species differed from the fossil species in the same morphological characteristics as those that differed among fossil species. Septal number, corallite diameter, and wall thickness were the primary characteristics that differed among species (Fig. 7.3). Modern specimens of *M. cavernosa* have an intermediate number of septa, an intermediate centroid size, and an intermediate corallite diameter. Modern specimens consistently grouped together to the exclusion of the fossil species in discriminant analyses, though some overlap was seen (Fig. 7.5).

Visual examination of the morphospace occupied by modern *M. cavernosa* and each of the fossil species (Fig. 7.5) indicates that roughly the same range of morphological variation is represented by each species.

7.7 Discussion and Conclusion

The results of the present study show that at least eight species existed within the *Montastraea "cavernosa"* complex in the Neogene of the Dominican Republic (Fig. 7.6; Table 7.3). This number is twice as high as that found by Budd (1991). Previously recognized species, such as *Montastraea endothecata*, are found to be more than one species, and three of the species discovered in the present study are likely to be new (species 8, 59, and 63). The increase in total number of species is due in part to an increased number of colonies used in these analyses and more localities having been represented. Many localities that are not represented in previous studies were represented in the current study, such as localities along Arroyo Bellaco (Bel-1, 2, 3, 4, 5, 6, 10) and several NMB localities (NMB 15894, 16911). In addition, geometric morphometrics has facilitated more refined characterization of the corallite wall.

Diversity decreased through time. Six of the eight species were short-lived and did not span more than one formation. The Cercado Formation contained five species (species 1, 2, 59, 63, 68); the Gurabo Formation contained four (species 1, 2, 8, 13); and the Mao Formation contained only two species (species 2, 43). However, in contrast, visual examination of Fig. 7.2D suggests that disparity was lower in the Cercado Formation and higher in the Gurabo and Mao Formations.

Studies such as this one are a critical first step in analyses of evolutionary stasis and change. Identifying the number of species in a given time interval and understanding the range of morphological variation within those species is necessary before testing for evolutionary change. Six of the eight species in the present study were short-lived and did not span more than one formation. The other two species were longer lived and experienced evolutionary stasis.

This study also paves the way for further work on the long-term evolution of the *M. cavernosa* species complex. It seems unlikely that these eight species represent all of the species within the *Montastraea "cavernosa"* species complex during the Mio-Pliocene. Additional specimens from the Dominican Republic and from other Neogene sequences in the Caribbean need to be included in the analyses, and additional landmarks involving the primary and secondary septa need to be added to the landmark scheme. Moreover, morphological characters need to be delineated for use in a phylogenetic analysis of the complex. Recent work (Budd and Smith, 2005) has shown that continuous morphologic variables, such as corallite diameter, can be used in a phylogenetic analysis using the step-matrix gap-weighting technique (Wiens, 2001).

Acknowledgments We would like to thank Dana Geary for reviewing this chapter. Special thanks to Kay Saville for lab assistance, Jonathan Adrain and Chris Brochu for helpful comments during the writing process, and Reggie Schreiber for help with figures. This research was supported by the U.S. National Science Foundation Grants DEB-0102544 and DEB-0343208 and the University of Iowa, Department of Geoscience.

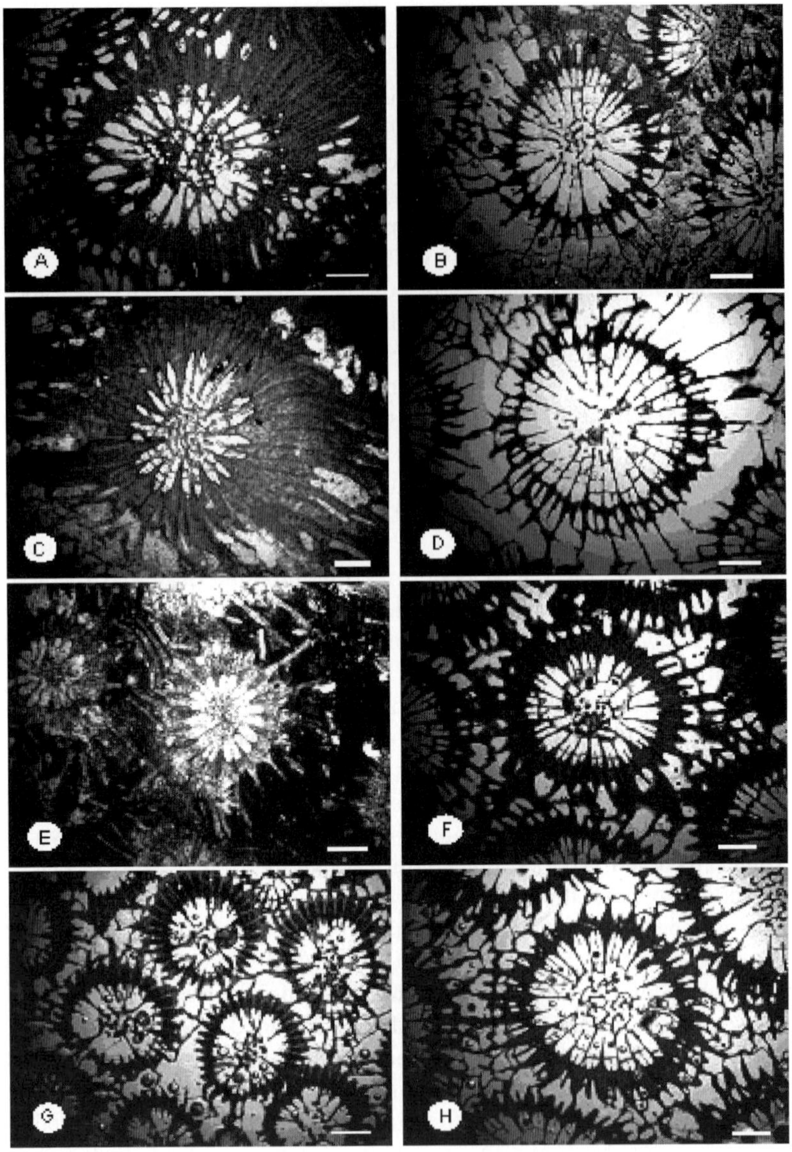

Fig. 7.6 Transverse thin-sections of representative corallites of the eight species of Montastraea "cavernosa" – like corals. (A) species 1, SUI 101386, locality NMB 16817, Río Cana, Gurabo Formation, early Pliocene; (B) species 2, SUI 101411, locality NMB 15855, Río Gurabo, Gurabo Formation, late Miocene; (C) species 8, SUI 101422, locality NMB 15808, Río Gurabo, Gurabo Formation, late Miocene; (D) species 13, SUI 101424, locality NMB 16817, Río Cana, Gurabo Formation, early Pliocene; (E) species 43, SUI 101434, locality NMB 16884, Río Cana, Mao Formation, early Pliocene; (F) species 59, SUI 101435, locality Bel-1&Bel-2, Río Cana, Cercado Formation, late Miocene; (G) species 63, SUI 101446, locality Bel-1 & Bel-2, Río Cana, Cercado Formation, late Miocene; (H) species 68, SUI 101451, locality Bel-1&Bel-2, Río Cana, Cercado Formation, late Miocene

Appendix 1

List of fossil specimens analyzed in morphometric analysis

Species #	Specimen #	Locality No.	Formation	Geologic Age
1	SUI 101373	Bel-1&Bel-2	Cercado	l. Miocene
1	SUI 101374	Bel-1&Bel-2	Cercado	l. Miocene
1	SUI 101375	Bel-1&Bel-2	Cercado	l. Miocene
1	SUI 101376	Bel-1&Bel-2	Cercado	l. Miocene
1	SUI 101377	Bel-1&Bel-2	Cercado	l. Miocene
1	SUI 101378	15808	Gurabo	l. Miocene
1	SUI 101379	15808	Gurabo	l. Miocene
1	SUI 101380	15808	Gurabo	l. Miocene
1	SUI 101381	15808	Gurabo	l. Miocene
1	SUI 101382	15837	Gurabo	l. Miocene
1	SUI 101383	15855	Gurabo	l. Miocene
1	SUI 101384	15894	Gurabo	l. Miocene
1	SUI 101385	15894	Gurabo	l. Miocene
1	SUI 101386	16817	Gurabo	e. Pliocene
1	SUI 101387	16817	Gurabo	e. Pliocene
1	SUI 101388	16817	Gurabo	e. Pliocene
1	SUI 101389	16818	Gurabo	e. Pliocene
1	SUI 101390	16818	Gurabo	e. Pliocene
1	SUI 101391	16818	Gurabo	e. Pliocene
1	SUI 101392	16818	Gurabo	e. Pliocene
1	SUI 101394	16818	Gurabo	e. Pliocene
1	SUI 101395	16818	Gurabo	e. Pliocene
1	SUI 101396	16818	Gurabo	e. Pliocene
1	SUI 101397	16818	Gurabo	e. Pliocene
1	SUI 101398	16818	Gurabo	e. Pliocene
1	SUI 101399	16818	Gurabo	e. Pliocene
1	SUI 101400	16818	Gurabo	e. Pliocene
1	SUI 101401	16818	Gurabo	e. Pliocene
1	NMB D5618	16818	Gurabo	e. Pliocene
1	NMB D5620	16818	Gurabo	e. Pliocene
1	SUI 101402	16859	Gurabo	e. Pliocene
1	NMB D5718	16933	Gurabo	e. Pliocene
2	SUI 101403	Bel-1&Bel-2	Cercado	l. Miocene
2	SUI 101404	Bel-1&Bel-2	Cercado	l. Miocene
2	SUI 101405	Bel–3, 4, 10	Cercado	l. Miocene
2	SUI 101406	15808	Gurabo	l. Miocene
2	SUI 101407	15808	Gurabo	l. Miocene
2	SUI 101408	15808	Gurabo	l. Miocene
2	SUI 101409	15808	Gurabo	l. Miocene
2	SUI 101410	15837	Gurabo	l. Miocene
2	NMB D5557	15838	Gurabo	l. Miocene
2	NMB D5579	15847	Gurabo	l. Miocene
2	SUI 101411	15855	Gurabo	l. Miocene
2	SUI 101412	15894	Gurabo	l. Miocene
2	SUI 101413	15894	Gurabo	l. Miocene
2	SUI 101414	16818	Gurabo	e. Pliocene
2	SUI 101415	16818	Gurabo	e. Pliocene

(continued)

Appendix 1 (continued)

Species #	Specimen #	Locality No.	Formation	Geologic Age
2	SUI 101416	16818	Gurabo	e. Pliocene
2	SUI 101417	16818	Gurabo	e. Pliocene
2	NMB D5619	16818	Gurabo	e. Pliocene
2	SUI 101418	16884	Mao	e. Pliocene
2	SUI 101419	16884	Mao	e. Pliocene
2	SUI 101420	16884	Mao	e. Pliocene
2	SUI 101421	16884	Mao	e. Pliocene
8	SUI 101422	15808	Gurabo	l. Miocene
8	SUI 101423	15808	Gurabo	l. Miocene
13	SUI 101424	16818	Gurabo	e. Pliocene
13	NMB D5616	16817	Gurabo	e. Pliocene
13	SUI 101425	16818	Gurabo	e. Pliocene
13	SUI 101426	16818	Gurabo	e. Pliocene
13	SUI 101427	16818	Gurabo	e. Pliocene
13	NMB D5621	16818	Gurabo	e. Pliocene
13	NMB D5702	16911	Gurabo	l. Miocene
43	SUI 101428	16884	Mao	e. Pliocene
43	SUI 101430	16884	Mao	e. Pliocene
43	SUI 101431	16884	Mao	e. Pliocene
43	SUI 101433	16884	Mao	e. Pliocene
43	SUI 101434	16884	Mao	e. Pliocene
43	NMB D5647	16884	Mao	e. Pliocene
43	NMB D5551	15830	Mao	e. Pliocene
59	SUI 101435	Bel-1&Bel-2	Cercado	l. Miocene
59	SUI 101436	Bel-1&Bel-2	Cercado	l. Miocene
59	SUI 101437	Bel-1&Bel-2	Cercado	l. Miocene
59	SUI 101438	Bel-1&Bel-2	Cercado	l. Miocene
59	SUI 101439	Bel-3, 4, 10	Cercado	l. Miocene
59	SUI 101440	Bel-3, 4, 10	Cercado	l. Miocene
59	SUI 101441	Bel-1&Bel-2	Cercado	l. Miocene
59	SUI 101442	Bel-1&Bel-2	Cercado	l. Miocene
59	SUI 101443	Bel-3, 4, 10	Cercado	l. Miocene
59	SUI 101444	Bel-3, 4, 10	Cercado	l. Miocene
59	SUI 101445	Bel-5&Bel-6	Cercado	l. Miocene
63	SUI 101446	Bel-1&Bel-2	Cercado	l. Miocene
63	SUI 101447	Bel-3, 4, 5	Cercado	l. Miocene
63	SUI 101448	Bel-3, 4, 5	Cercado	l. Miocene
63	SUI 101449	Bel-3, 4, 5	Cercado	l. Miocene
63	SUI 101450	Bel-3, 4, 5	Cercado	l. Miocene
63	NMB D5738	16944	Baitoa	m. Miocene
68	SUI 101451	Bel-1&Bel-2	Cercado	l. Miocene
68	SUI 101452	Bel-1&Bel-2	Cercado	l. Miocene
68	SUI 101453	Bel-1&Bel-2	Cercado	l. Miocene
68	SUI 101454	Bel-1&Bel-2	Cercado	l. Miocene
68	SUI 101455	Bel-3, 4, 10	Cercado	l. Miocene

Appendix 2

Landmarks on transverse thin-sections of corallites of *Montastraea "cavernosa"*-like corals

Number	Type	Description
1	na	Center of corallite
2	2	Outermost point on quaternary costa
3	1	Outer left junction of quaternary costoseptum and wall dissepiment
4	1	Outer right junction of quaternary costoseptum and wall dissepiment
5	1	Inner left junction of quaternary costoseptum and wall dissepiment
6	1	Inner right junction of quaternary costoseptum and wall dissepiment
7	2	Innermost point of quaternary septum
8	2	Innermost point of primary septum
9	2	Outermost point of tertiary costa
10	2	Innermost point of tertiary septum
11	2	Outermost point of quaternary costa
12	1	Outer left junction of quaternary costoseptum and wall dissepiment
13	1	Outer right junction of quaternary costoseptum and wall dissepiment
14	1	Inner left junction of quaternary costoseptum and wall dissepiment
15	1	Inner right junction of quaternary costoseptum and wall dissepiment
16	2	Innermost point of quaternary septum
17	2	Innermost point of secondary septum

1 = juxtaposition of structures; 2 = maxima of curvature

Appendix 3

Shape coordinates used in discriminant analyses of 2 dimensional landmark data collected on corallites in transverse thin-section

Shape coordinates	Definition
x1	Center of corallite
x2	Extension of quaternary costa
x3	Wall thickness
x4	Wall thickness
x5	Corallite diameter
x6	Corallite diameter
x7	Length of quaternary septum
x8	Length of primary septum
x9	Baseline for analysis
x10	Baseline for analysis
x11	Extension of quaternary costa
x12	Wall thickness
x13	Wall thickness
x14	Corallite diameter
x15	Corallite diameter
x16	Length of quaternary septum
x17	Length of secondary septum
m1	Columella diameter (linear measurement)
m2	Corallite diameter (linear measurement)

Appendix 4

Results of the global canonical discriminant analysis

(A) Eigenvalues

Function	Eigenvalue	% of Variance	Cumulative %	Canonical Correlation
1	3.746	55.1	55.1	0.888
2	1.815	26.7	81.8	0.803
3	0.798	11.7	93.6	0.666
4	0.438	6.4	100.0	0.552

(B) Wilks Lambda

Test of function	Wilks Lambda	Chi-square	Deg. of freedom	Significance
1 Through 4	0.029	309.909	40	0.000
2 Through 4	0.137	173.637	27	0.000
3 Through 4	0.387	83.078	16	0.000
4	0.696	31.759	7	0.000

(C) Structure matrix

	Function			
Shape coordinate	1	2	3	4
No. of septa	0.583*	0.316	−0.013	−0.046
M2	0.427*	0.030	0.278	0.072
M1	0.349*	0.023	0.264	0.012
X2	0.243*	0.162	0.128	0.019
X15	−0.081	0.358*	−0.209	0.057
X14	−0.015	0.320*	−0.227	0.108
X6	−0.008	0.313*	−0.197	0.168
X5	−0.030	0.280*	−0.196	0.134
X17	−0.074	0.270*	0.202	0.022
X11	0.199	0.267*	0.133	−0.005
X8	−0.025	0.192*	0.169	0.165
Centroid size	0.463	−0.062	0.550*	−0.185
X3	0.108	0.095	0.358*	−0.094
X4	0.100	0.170	0.340*	−0.154
X13	0.061	0.230	0.330*	−0.260
X12	0.069	0.167	0.317*	−0.193
X1	−0.066	0.120	0.243*	−0.055
X7	−0.218	0.193	−0.024	0.316*
X16	−0.221	0.217	−0.074	0.239*

*Largest absolute correlation between each variable and any discriminant function

References

Blainville, H.M. de, 1830, Zoophytes, *Dictionnaire des Sciences Naturelles*, Paris, v. 60.

Beck, B.R. and Budd A.F., This volume, Evolutionary patterns within the reef coral *Siderastrea* in the Mio-Pliocene of the Dominican Republic, in: *Evolutionary Stasis: Species and Communities Through Geologic time* (R.H. Nehm and A.F. Budd, eds.), Kluwer/Plenum, New York.

Bookstein, F.L., 1991, *Morphometric Tools for Landmark Data*. Cambridge University Press, Cambridge, 435 pp.

Budd, A.F., 1990, Long term patterns of morphological variation within and among species of reef-corals and their relationship to sexual reproduction, *Syst. Bot.*, **15**:150–165.

Budd, A.F., 1991, Neogene paleontology in the Northern Dominican Republic. 11. The Family Faviidae (Anthozoa: Scleractinia), *Bulls. Am. Paleontol.*, **101**:5–83.

Budd, A.F., 1993, Variation within and among morphospecies of *Montastraea*, *Courier Forschungsinst. Senckenberg*, **164**:241–254.

Budd, A.F. and Klaus, J.S., This volume, Early evolution of the *Montastraea "annularis"* species complex (Anthozoa: Scleractinia): Evidence from the Mio-Pliocene of the Dominican Republic, in: *Evolutionary Stasis and Change in the Dominican Republic Neogene.*

Budd, A.F. and Pandolfi, J.M., 2004, Overlapping species boundaries and hybridization within the *Montastraea "annularis"* reef coral complex in the Pleistocene of the Bahama Islands. *Paleobiology*, **30**:396–425.

Budd, A.F. and Smith, N., 2005, A new clade of Atlantic reef corals: Paleontological Society Papers, v. 11, pp. 103–128.

Budd, A.F., Stemann, T.A., and Johnson, K.G., 1994, Stratigraphic distributions of genera and species of Neogene to Recent Caribbean reef corals, *J. Paleontol.*, **68**:951–977.

Budd, A.F., Stemann, T.A., and Stewart, R.H., 1992, Eocene Caribbean reef corals: a unique fauna from the Gatuncillo Formation of Panama, *J. Paleontol.*, **66**:570–594.

Burke, R.B., 1982, Reconnaissance study of the geomorphology and Benthic communities of the outer barrier reef platform, Belize, in: *The Atlantic Barrier Reef Ecosystem at Carrie Bow Cay, Belize, I. Structure and Communities* (K. Rützler and I.G. Macintyre, eds.), Smithsonian Institution Press, Washington, DC, pp. 509–526.

Cheetham, A.H., Sanner, J., and Jackson, J.B.C., 2007, *Metrarabdotos* and related genera (Bryozoa: Cheilostomata) in the late Paleogene and Neogene of Tropical America, *Paleontol. Soc. Mem.*, **67**:1–96.

Cracraft, J., 1987, Species concepts and the ontology of evolution, *Biol. Philos.*, **2**:63–80.

Duncan, P.M., 1863, On the fossil corals of the West Indian Islands, Part I, *Quart. J. Geol. Soc. Lond.*, **19**:406–458.

Duncan, P.M., 1864, On the fossil corals of the West Indian Islands, Part II, *Quart. J. Geol. Soc. Lond.*, **20**:20–44.

Evans, C.C., 1986, *A Field Guide to the Mixed Reefs and Siliciclastics of the Neogene Yaque Group, Cibao Valley, Dominican Republic*, University of Miami, RSMAS, Fisher Island Station, 98 pp.

Foster, A.B., 1985, Variation within coral colonies and its importance for interpreting fossil species, *J. Paleontol.*, **59**:1359–1383.

Fukami, H., Budd, A.F., Levitan, D.R., Jara, J., Kersanach, R., and Knowlton, N., 2004a, Geographic differences in species boundaries among members of the *Montastraea annularis* complex based on molecular and morphological markers, *Evolution*, **58**:324–337.

Fukami, H., Budd, A.F., Pauley, G., Solé-Cava, A., Chen, C.A., Iwao, K., and Knowlton, N., 2004b, Conventional taxonomy obscures deep divergence between Pacific and Atlantic corals, *Nature*, **427**:832–835.

Goreau, T.F., 1959, The ecology of Jamaican coral reefs, Part I. Species composition and zonation, *Ecology*, **40**:67–90.

Jackson, J.B.C., and Cheetham, A.H., 1999, Tempo and mode of speciation in the sea, *Trends Ecol. Evol.*, **14**:72–77.

Klaus, J.S. and Budd., A.F., 2003, Comparison of Caribbean coral reef communities before and after Plio-Pleistocene faunal turnover: analyses of two Dominican Republic reef sequences, *Palaios*, **18**:3–21.

Knowlton, N. and Budd, A.F., 2001, Recognizing coral species past and present, in: *Evolutionary Patterns: Growth, Form, and Tempo in the Fossil Record* (J.B.C. Jackson, S. Lidgard, and F. K. McKinney, eds.), University of Chicago Press, Chicago, IL, pp. 97–119.

Knowlton, N. and Weigt, L.A., 1997, Species of marine invertebrates: a comparison of the biological and phylogenetic species concepts, in: *Species – The Units of Biodiversity* (M. F. Claridge, H.A., Dawah, and M.R. Wilson, eds.), Chapman & Hall, London, pp. 199–219.

Lasker, H.R., 1976, Intraspecific variability of zooplankton feeding in the hermatypic coral *Montastrea cavernosa*, in: *Coelenterate Ecology and Behavior* (Mackie, G.W., ed.), Plenum, New York, pp. 101–109.

Lasker, H.R., 1979, Light dependent activity patterns among reef corals: *montastrea cavernosa*, *Biol. Bull.*, **156**:196–211.

Lasker, H.R., 1980, Sediment rejection by reef corals: the roles of behavior and morphology in *Montastrea Cavernosa* (Linnaeus), *J. Exp. Mar. Biol. Ecol.*, **47**:77–87.

Lasker, H.R., 1981, Phenotypic variation in the coral *Montastrea cavernosa* and its effects on colony energetics, *Biol. Bull.*, **160**:292–302.

Lewis, J.F., 1980, Resumé of the geology of Hispañola, in: *Field Guide to the 9th Caribbean Geological Conference, Santo Domingo Dominican Republic*, Amigo del Hogar, Santo Domingo, pp. 5–31.

Linnaeus, L., 1767, *Madrepora*, Systema Naturae, Holmiae, Editio Duodecima, t.1, pt. 2, pp. 1272–1282.

Mann, P., Grenville, D., and Lewis, J.F., 1991, An overview of the geologic and tectonic development of Hispañola, in: *Geologic and Tectonic Development of the North American-Caribbean Plate Boundary in Hispañola* (P. Mann, D. Grenville, and J.F. Lewis, eds.), *Geol. Soc. Am. Spec Paper*, **262**:1–28.

Mayr, E., 1963, *Animal Species and Evolution*, Harvard University Press, Cambridge, MA, 811 p.

McNeill, D.F., Klaus, J.S., Evans, C.C., Budd, A.F., and Maier, K.L., This volume, An overview of the regional geology and stratigraphy of the Neogene deposits of the Cibao Valley, Dominican Republic, in: *Evolutionary Stasis and Change in the Dominican Republic Neogene*.

Nixon, K.C. and Wheeler, Q.D., 1990, An amplification on the phylogenetic species concept, *Cladistics*, **6**:211–223.

Odorico, D. and Miller, D.J., 1997, Variation in the ribosomal internal transcribed spacers and 5.8S rDNA among five species of Acropora (Cnidaria: Scleractinia): patterns of variation consistent with reticulate evolution, *Mol. Biol. Evol.*, **14**:465–473.

Ruiz Torres, H.J., 2004, Morphometric examination of corallite and colony variability in the Caribbean coral *Montastraea cavernosa*. M.S. thesis, University of Puerto Rico.

Rützler, K. and Macintyre, I.G., 1982, The habitat distribution and community structure of the barrier reef complex at Carrie Bow Cay, Belize, in: *The Atlantic Barrier Reef Ecosystem at Carrie Bow Cay, Belize, I. Structure and Communities* (K. Rützler and I.G. Macintyre, eds.), Smithsonian Institution Press, Washington, DC, pp. 9–45.

Saunders, J.B., Jung, P., and Biju-Duval, B., 1986, Neogene paleontology in the northern Dominican Republic. 1. Field surveys, lithology, environment, and age, *Bull. Am. Paleontol.*, **89** (323):1–79, 9 pls.

Vaughan, T.W., 1919, Fossil corals from Central America, Cuba, and Porto Rico with an account of the American Tertiary, Pleistocene, and recent coral reefs, *U. S. Nat. Hist. Mus. Bull.*, **130**:189–524, pls. 68–152.

Veron, J.E.N., 1995, *Corals in Space and Time: The Biogeography and Evolution of the Scleractinia*, UNSW Press, Sydney, 321 pp.

Weil, E. and Knowlton, N., 1994, A multi-character analysis of the Caribbean coral *Montastraea annularis* (Ellis and Solander, 1786), and its two sibling species, *M. faveolata* (Ellis and Solander, 1786) and *M. franksi* (Gregory, 1895), *Bull. Mar. Sci.*, **55**:151–175.

Wiens, J.J., 2001, Character analysis in morphological phylogenetics: problems and solutions, *Syst. Biol.*, **50**:689–699.

Wells, J.W., 1956, Scleractinia, in: *Treatise on Invertebrate Paleontology, Coelenterata* (R. C. Moore, ed.), Geological Society of America and University of Kansas Press, Lawrence, KS, pp. F328–440.

Willis, B.L., van Oppen, M.J.H., Miller, D.J., Vollmer, S.V., and Ayre, D.J., 2006, The role of hybridization in the evolution of reef corals, *Annu. Rev. Ecol. Evol. Syst.*, **37**:489–517.

Zelditch, M.L., Swiderski, D.L., Sheets, H.D., and Finks, W.L., 2004, *Geometric Morphometrics for Biologists*. Academic, London, 416 pp.

Chapter 8
The Dynamics of Evolutionary Stasis and Change in the '*Prunum maoense* Group'

Ross H. Nehm

Contents

8.1 Introduction

Causal explanations for evolutionary patterns in the fossil record have long oscillated between intrinsic and extrinsic mechanisms (Gould, 1977). Early work on evolutionary stasis, for example, favored intrinsic mechanisms such as genetic and developmental constraints and internal homeostatic mechanisms as important causal factors in the production of patterns of morphological stasis through geological time (Eldredge and Gould, 1972). These intrinsic explanations were endorsed because alternative hypotheses focusing on extrinsic causes, such as faunal tracking, stabilizing selection, and the absence of abiotic change, often failed to sufficiently explain particular evolutionary scenarios (Eldredge and Gould, 1972; Lieberman et al., 1996). In recent years, however, the conceptual pendulum has swung towards extrinsic causal explanations for evolutionary stasis, specifically factors relating to population structure and the spatial dynamics of species (Eldredge et al., 2005; Gould, 2002). This shift in perspective is in line with Gould's (2002) recent reconsideration of the importance of developmental constraints as causes of stasis.

The Ohio State University, Columbus, OH, USA. Email: nehm.1@osu.edu

R.H. Nehm, A.F. Budd (eds.) *Evolutionary Stasis and Change in the Dominican Republic Neogene,*

Many empirical studies have also contributed to the weakening of arguments for the dominance of intrinsic factors as causes of stasis. These studies have demonstrated that stasis is unlikely to be caused by the lack of production of phenotypic novelty because high levels of genetic and morphological variation, and the general absence of severe genetic or developmental constraints on the production of variation, are ubiquitous characteristics of extant populations. Rapid evolutionary transformations, possible only through the availability of genetic and developmental variation, are quite common in many lineages and also question the dominance of intrinsic constraints as causes of stasis (Grant, 1999; Eldredge et al., 2005). Much like the rapid morphological transformations frequently observed in extant populations (e.g., Grant, 1999), the fossil record reveals numerous cases of short-term reversals, oscillations, or brief evolutionary diversions within episodes of "stasis" (Nehm, 2005). Such studies support the perspective that the production of phenotypic novelty may not be the primary limiting factor in evolutionary change (Erwin and Anstey, 1995; Nehm, 2001a, 2005).

Given these recent perspectives on evolutionary stasis, the major questions facing paleobiologists are: Is the production of morphological novelty though time and space in fact common within species displaying stasis? And, if it is: Why does such novelty fail to become established beyond its site of origin, thus leading to patterns of stasis in the fossil record (Eldredge et al., 2005)? Finally, if the production of phenotypic novelty is not exclusively associated with speciation, how can fossil species be reliably recognized?

I investigate these questions in an extensively studied and sampled sequence of gastropods (the *Prunum maoense* group) from the Dominican Republic (DR) Neogene. The DR Neogene is a geological system containing the largest number of well-established cases of evolutionary stasis within invertebrate species (See Nehm and Budd, this volume) and thus serves as a useful context for exploring whether or not the production of phenotypic novelty is commonplace within fossil lineages that have been argued to display stasis. Additionally, it serves as a useful system for exploring the complexities of species delineation and the fate of novelty over macroevolutionary timescales.

8.2 Materials and Methods

Within the clade *Prunum* + *Volvarina* (Nehm, 2001c), I focus on morphological variation and evolution within a small but temporally long-ranging clade referred to here as the "*Prunum maoense* group." This group is distinguished from other *Prunum* and *Volvarina* species by several conchological features and color patterns: (1) A large posterior aperture margin callus that is thicker than the body whorl shell layers (2) a posterior lip indentation and (3) three stripes of color on the body whorl. These character states are not known to collectively occur in any other living or fossil marginellids.

The *P. maoense* clade consists of three species: *P. maoense* (Maury), *P. latissimum* (Dall), and *P. dasum* (Gardner). The clade ranges temporally from the middle Miocene Chipola Formation of Florida to the lower Pliocene Mao Formation of the Dominican Republic (Fig. 8.1). Although phylogenetic analysis appears to support

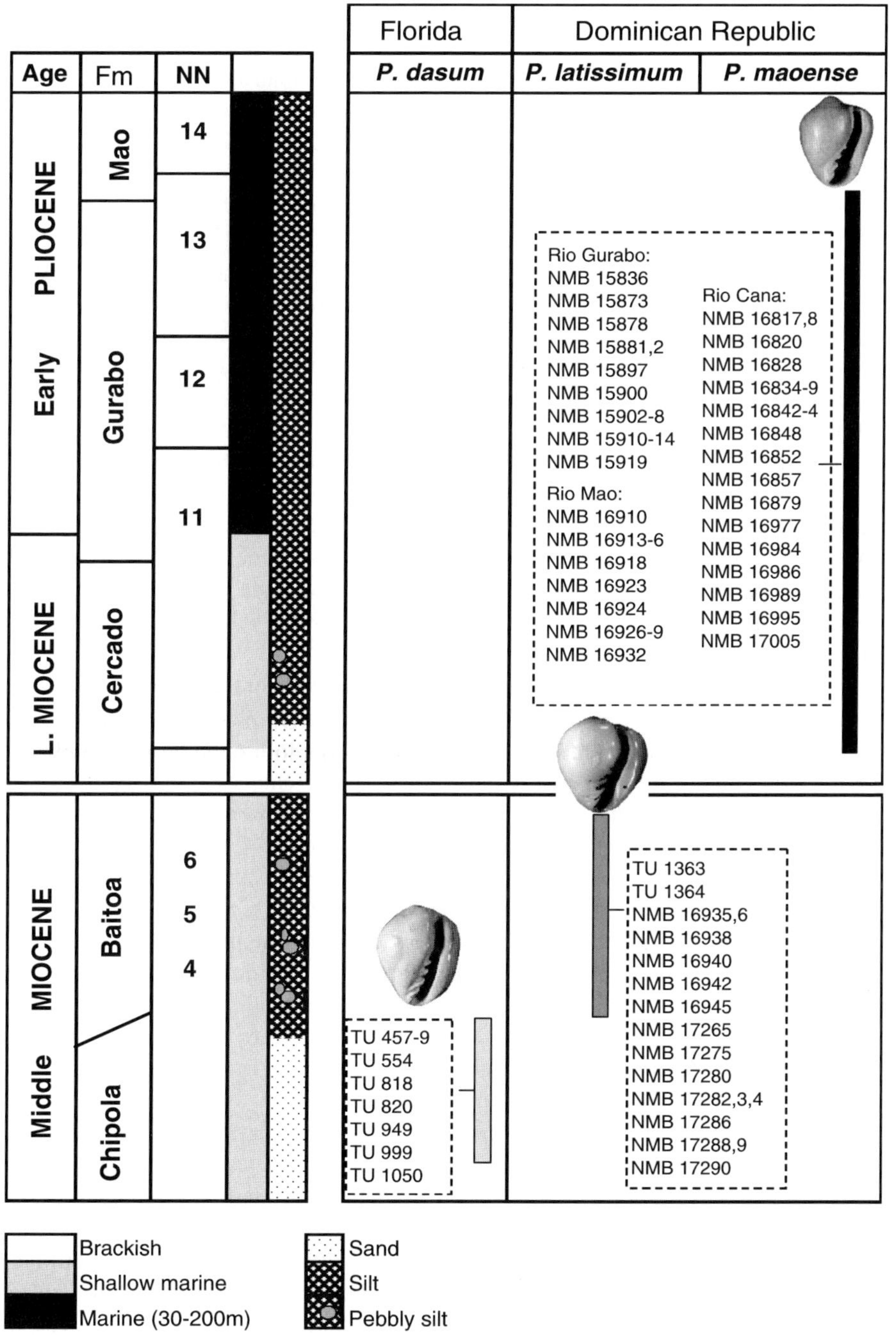

Fig. 8.1 Sample data for the *P. maoense* group in tropical America

uniting these species as a clade, as I will demonstrate below, evolutionary patterns within this clade are complex.

The three species in the *P. maoense* group are abundant and well-preserved in the Neogene of tropical America. Approximately 1,600 specimens were used to establish the stratigraphic and geographic ranges of species (Fig. 8.1). *P. dasum* was studied in 191 specimens from 9 Tulane University (TU) samples. It occurs in the lower to middle Miocene Chipola Formation of Florida. *P. latissimum* was studied in 703 specimens from 18 Naturhistorisches Museum Basel [NMB] samples. It is restricted to the middle Miocene Baitoa Formation of Río Yaque del Norte (Lopez section). *P. maoense* is the most abundant Dominican *Prunum* species (represented by 704 specimens from 51 NMB samples) and occurs in the upper Miocene to lower Pliocene Cercado and Gurabo Formations of Río Cana and Río Gurabo, and the Cercado Formation of Río Mao. More detailed information may be found in Nehm (2001a).

Each sample used in this study was categorized lithologically (e.g., sand, pebbly silt, or silt) and paleoenvironmentally (e.g., brackish, shallow marine, marine, deep marine) in order to examine the relationship of morphological variability with lithology and paleoenvironment. These seven categorical designations were made from observations in the field and/or paleoecological data from Saunders et al. (1982, 1986), Van den Bold (1988), Vokes (1979, 1989), Nehm and Geary (1994), Anderson (1994), and Anderson et al. (1992).

Brackish-water paleoenvironments were identified by brackish-restricted ostracode species (see Van den Bold, 1988) and the *Larkinia* (= *Anadara* [*Grandiarca*]), *Mytilus*, and *Melongena* mollusk assemblage (Saunders et al., 1986). Shallow marine paleoenvironments (<30 m paleodepth) were characterized by a lack of planktonic foraminifera and the presence of the shallow marine and intertidal mollusks *Anadara, Tellina, Strombina*, and *Pachycrommium* (Saunders et al., 1986; Nehm and Geary, 1994), benthic foraminifera (e.g., soritids, miliolids, and *Amphistegina*), and ostracods (e.g., *Cytherella, Randimella, Caudites, Proteoconcha, Loxoconcha*, and *Paracytheridea*). Moderately deep marine paleoenvironments (30 to 100 m paleodepths) were characterized by common marine mollusks (e.g., *Oliva, Prunum, Lyria*, and *Polystira*) and corals (Saunders et al., 1982). Deep marine paleoenvironments (exceeding paleodepths of 100 m) were characterized by rich assemblages of planktonic foraminifera and size patterns in the deep-water ostracode *Krithe* (Van den Bold, 1988; Saunders et al., 1986).

Marginellid gastropods are appropriate morphological systems for the study of variation and evolution because they (1) preserve a complete record of shell ontogeny and (2) shells record the termination of growth associated with sexual maturity. This allows the recognition of juvenile and adult shells. The description and study of ontogenetic variability is an essential component of systematic and evolutionary research (Raup and Stanley, 1978:55), yet its explicit consideration is often absent from contemporary paleobiological studies of molluscs. An understanding of ontogenetic variability is necessary for recognizing members of the same species at different ages, and comparing morphological differences among individuals of different ages. Specifically, an understanding of *Prunum* ontogeny to the species level is essential because (1) species-diagnostic characters (e.g., callus morphology, shape of the outer lip, presence of an external varix, and the presence and morphology of denticulations) are variable through ontogeny and (2) juvenile stratigraphic and geographic distributions are not completely concordant with adult distributions; thus, identifying juveniles to species is necessary for accurately determining temporal and geographic ranges of taxa.

Separation of marginellid specimens into juvenile and adult classes is possible because living and fossil species develop unique morphological features at adulthood when growth in body size and shape stops (the callus, however, may continue to thicken to a small degree). In gastropods, determinate growth is recognized by the development of a lip varix, internal lip thickening, an ascending suture, and apertural callusing patterns (Vermeij and Signor, 1992; Papadopoulos et al., 2004); *Prunum* displays these four morphological changes at the termination of growth that appear to be correlated with sexual maturity. Similar morphological and maturational patterns have been observed in the laboratory in the closely related *P. apicinum* (Nehm, personal observations).

An important issue in studies that examine morphological patterns in gastropods—and many other clades—is the ability to accurately detect significant morphological differences (or a lack thereof) in samples from different times, places, environments, populations, and species. The study of any biological pattern requires the comparison of biologically meaningful, standardized, and equivalent units. Many, if not most, mollusk species display significant morphological changes through ontogeny (Vermeij, 1995). Thus, studies of living and fossil mollusks must account for ontogenetic variation prior to making between-sample comparisons. Unfortunately, many have not done so, resulting in the potential interpretative conflation of size-frequency change with evolutionary change. The approach adopted here to avoid such comparative errors was to initially differentiate juvenile and adult age classes (discussed above) and subsequently analyse size and shape variation in adults only though time and space.

Morphometric analyses were used to quantify morphological change in relation to time, lithology, and paleoenvironment and compare morphological evolution quantitatively within and among species. All intact juvenile and adult shells from each stratigraphic sample were mounted on cardboard trays in apertural view with the columellar axis oriented horizontally to the surface (orientation shown in Fig. 8.2). Images of the shells were captured using a video camera mounted to a dissecting microscope and manipulated in Scion Image™ to increase the clarity of morphological features prior to measurement. Ten variables were measured on 520 adult specimens: six distances, three areas, and one angle (Fig. 8.5A). Fifteen landmarks were identified adult specimens from each species in order to perform geometric morphometric analyses of shell morphology (Fig. 8.5B; see also Bookstein, 1996; Rohlf and Marcus, 1993). Raw measurements were imported and analyzed in SYSTAT 7.0 (SPSS, Inc. 1997). Differences in these measured variables are traditionally used by marginellid systematists to diagnose recent and fossil marginellid species.

Variables were tested for significant differences among species using an analysis of variance (ANOVA). In cases of significant differences in morphometric variables

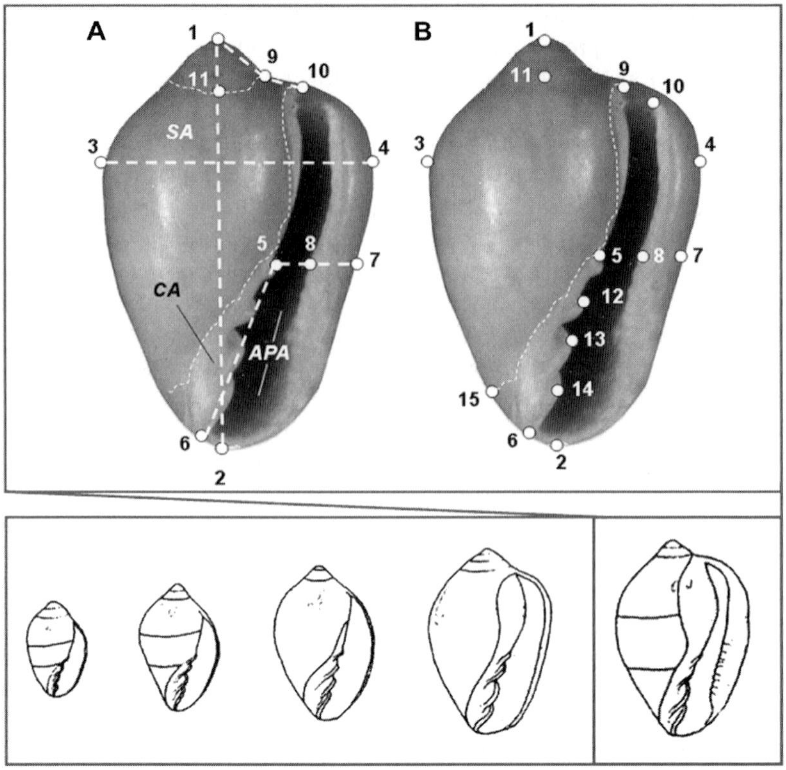

Fig. 8.2 Morphometric template for adult *Prunum*. A. Measured distance variables on adults: Shell height (landmark 1 to 2); Shell width (landmark 3 to 4); Spire height (landmark 1 to 11); Columellar fold height (landmark 5 to 6); 5). Lip width (landmark 7 to 8); Aperture width (landmark 5 to 8). Measured variables (areas): Aperture area (APA); Columellar callus area (CA); Shell area (SA). Measured variables (angle): landmarks 1 to 10. (B). Landmarks for geometric morphometric analyses

among species, post-hoc tests were used to determine which species pairs were significantly different.

Principal Component Analyses (PCAs) of character correlation matrices were also performed in order to summarize patterns of morphological variability among morphometric variables with respect to species, lithology, paleoenvironment and geography. Log-transformed measurements of shell height, shell width, aperture area, columellar fold height, lip thickness, spire height, aperture width, spire angle, and callus area were taken from 520 adult specimens and used to generate correlation matrices. PCAs of correlation and covariation matrices were performed in SYSTAT 7.0.

Relative Warps Analysis (RWA) was also used to determine if the shape changes captured using traditional morphometric analyses paralleled those documented using geometric morphometrics. Unlike PCA using linear distances, RWA is capable of unambiguously separating variation in size and shape. RWA is also unique in that it separates shape variation at different spatial scales (global and local) (see Bookstein, 1996; Zelditch and Fink, 1996, for a detailed explanation of the advantages and disadvantages of RWA). RWA characterizes shape change as a deformation of landmarks from a "reference" form into each specimen under analysis. The reference form or tangent configuration used here was generated by calculating a generalized least squares consensus of all specimens in TPSRELW ver. 1.13 (Rohlf, 1998, 2001). Shape variation among specimens was separated into uniform and nonuniform shape change using TPSRELW (alpha set to 0). Variation in shape space is presented as a series of bivariate plots comparing the relative warps (RWs) that explain the most variation in each RWA. Several authors have warned against exclusive use of RWA because of the complexity of the analyses and difficulties in interpreting results. Thus, I present RWA results in parallel to traditional multivariate results in order to increase interpretability.

8.3 Results

8.3.1 Interspecific Adult Morphological Differences

ANOVAs of adult distance measurements produced significant differences in all cases (Table 8.1). Adult *P. dasum* and *P. latissimum* are large, obovate, and heavily callused, whereas adult *P. maoense* are small, cylindrical, and less callused. Post-hoc tests revealed the location of significant morphological differences between species pairs (Table 8.1). The only variables that did *not* differ significantly between species were (1) shell shape (height/width) and callus area between *P. dasum* and *P. latissimum*, and (2) aperture width and spire angle between *P. dasum* and *P. maoense* (Table 8.1). Thus, the three species were significantly different in most measured variables.

A PCA of adult specimens also produced separation of species, although considerable overlap is also apparent (Fig. 8.3a). The first two principal components explained more than 78% of the variance in the morphometric variables used in the analysis. The first principal component (PC1) explained 61% of the variance and represented variation in size and size-correlated shape, as indicated by the high loadings of size variables, such as shell height, width, and columellar fold height. Nearly all variables were strongly and positively correlated with the first axis, suggesting size-correlated shape variation. The second principal component (PC2) explained about 17% of the variance, and represented variation in spire morphology, as indicated by high loadings for spire height and spire angle. Mean values for principal component scores on PC1 and PC2 were significantly different in all species comparisons (Table 8.1) and corroborate univariate conclusions that adults of all three species are significantly different from one another.

Relative Warp Analyses of landmarks from *P. dasum*, *P. latissimum*, and *P. maoense* produced results similar to the PCA analyses of distance measures: while the three species generally occupy different regions of morphospace, considerable overlap is also apparent (Fig. 8.3b). The RWA represents both uniform and non-uniform shape

Table 8.1 ANOVAs of adult morphometric variables. See text for morphometric variable description

Variables	n	SS	d.f.	Mean square	F-ratio	P	Post-Hoc test
Shell height	520	448.07	2	224.03	301.75	0.000	1**,2**,3**
Shell width	520	394.51	2	197.25	528.92	0.000	1**,2**,3**
Aperture area	520	951.91	2	475.95	209.48	0.000	1**,2**,3**
Columellar fold height	520	74.10	2	37.05	188.47	0.000	1**,2**,3**
Height/Width	520	1.56	2	0.78	127.88	0.000	2**,3**
Lip thickness	520	24.85	2	12.42	321.68	0.000	1**,2**,3**
Spire height	520	4.77	2	2.38	54.55	0.000	1**,2**,3**
Aperture width	520	1.99	2	0.99	80.87	0.000	1**,2**
Spire angle	518	1614.87	2	807.43	45.89	0.000	1**,2**
Callus area	520	7932.19	2	3966.09	181.51	0.000	2**,3**
PC1	518	305.91	2	152.95	373.17	0.000	1**,2**,3**
PC2	518	62.48	2	31.24	35.40	0.000	1**,2**,3**

*p<.05, **p<.01
(1) *P. dasum-P. latissimum*
(2) *P. latissimum-P. maoensis*
(3) *P. dasum-P. maoensis*

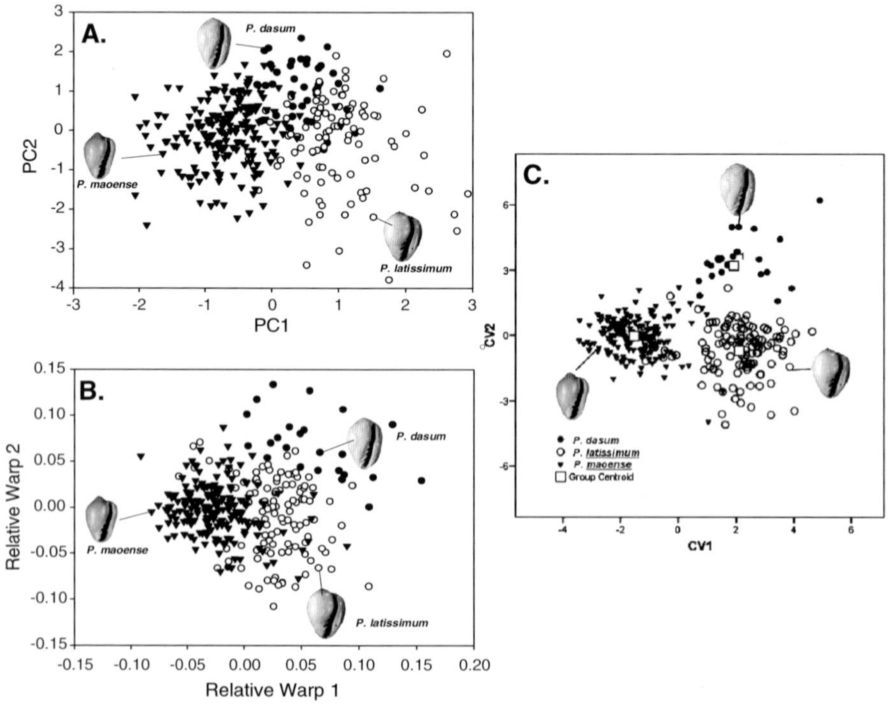

Fig. 8.3 A. plot of the first and second principal component scores (representing size and size-correlated shape) of *adult* specimens of *P. dasum*, *P. latissimum*, and *P. maoense* derived from distance measurements. B. Relative Warp Analyses of landmarks illustrating uniform and non-uniform shape differences among adult specimens. C. Canonical variates analysis of relative warp scores of adult specimens

differences among species. RW1 explained 25% of the shape variation, RW2 explained 15% of the shape variation, and RW3 explained 14% of the shape variation. Each remaining RW explained less than 8% of the shape variation. A plot of RW1 and RW2 displays good separation between *P. maoense* and the other two species along RW1, whereas *P. dasum* and *P. latissimum* are not well-differentiated from one another along this axis. RW2 scores poorly separated specimens assigned to different species. A plot of RW2 and RW3 (not shown) produced good separation of *P. dasum* from the other two species along RW3, whereas specimens assigned to *P. dasum* and *P. maoense* overlapped. A canonical variates analysis (CVA) of landmarks produced the best separation of groups (Fig. 8.3C), although several individuals from each species cluster unambiguously within the shape space of other species.

8.3.2 Adult Morphological Variation in Different Palaeoenvironments, Lithologies, and Sections

A plot of adult PC1 scores (representing size and size-correlated shape) by paleoenvironmental category illustrates morphological overlap between adult individuals of *P. latissimum* from the shallow marine Baitoa Formation and adult individuals of *P. maoense* from brackish, shallow marine, and deeper marine paleoenvironments (Fig. 8.4, middle panels). Adult *P. maoense*, however, displayed generally similar patterns of morphological variation among different paleoenvironments. A plot of PC2 scores (representing variation in spire morphology) display nearly complete overlap between *P. latissimum* and *P. maoense*. Overall, adult morphological variation does not appear to be correlated with paleoenvironment in *P. maoense*, and morphological overlap does occur between adults of *P. latissimum* and *P. maoense*.

A plot of adult PC1 scores by lithological category produced similar patterns as paleonvironmental comparisons: there is some overlap between adult individuals of *P. latissimum* from the pebbly silts of the Baitoa Formation and adult individuals of *P. maoense* collected from silts and pebbly silts (Fig. 8.4, top panels). However, adult *P. maoense* display similar patterns of morphological variation in samples containing silts and pebbly silts. Plots of PC2 scores produced nearly complete overlap between adult specimens of *P. latissimum* and *P. maoense* in samples with different lithologies. Overall, adult morphological variation does not appear to be correlated with lithology in *P. maoense*, and morphological overlap occurs between adults of *P. latissimum* and *P. maoense* in all of the categories available for analysis.

Like the paleoenvironmental and lithological comparisons, a plot of PC1 scores of adult specimens by river section indicated some overlap between adult individuals of *P. latissimum* from the Baitoa Formation at Lopez and adult individuals of *P. maoense* from the Río Cana, Río Gurabo, and Río Mao sections (Fig. 8.4, lower panels). However, adult individuals of *P. maoense* from the Río Cana, Río Gurabo, and Río Mao sections display similar patterns of morphological variation. A plot of PC2 scores illustrates much greater overlap between *P. latissimum* and *P. maoense*.

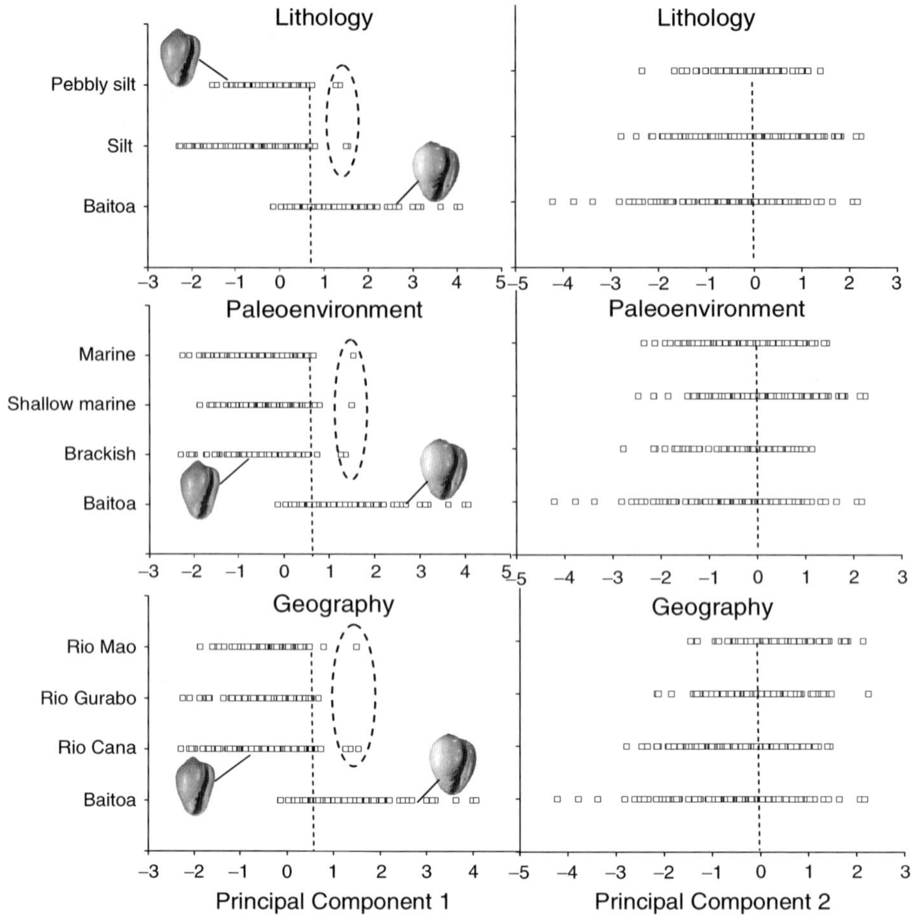

Fig. 8.4 Morphological variation in *P. latissimum* and *P. maoense*. Note that *P. latissimum* in these plots is restricted to the Baitoa section. (top) A lithological category plot of the first principal component scores of adult *P. maoense* from silts and pebbly silts compared to *P. latissimum* from the Baitoa section; (Middle) A paleoenvironment category plot; (bottom) a category plot by river section (Baitoa, Río Cana, Río Gurabo, and Río Mao)

In summary, significant geographic differentiation is not apparent in adult *P. maoense*, and morphological overlap occurs between *P. latissimum* and *P. maoense*.

A UPGMA cluster analysis was used to summarize the morphological similarities among samples (Fig. 8.5). This analysis reveals that (1) samples of *P. dasum* cluster with samples of *P. latissimum*; (2) Two clusters of *P. latissimum* were produced; (3) NMB samples 15913 from the Río Gurabo section and NMB sample 16818 from the Río Cana section were initially assigned to *P. maoense,* but cluster analysis places them with samples of *P. latissimum*; and (4) Two clusters of *P. maoense* were produced, but these clusters were not unique in terms of geography or paleoenvironment.

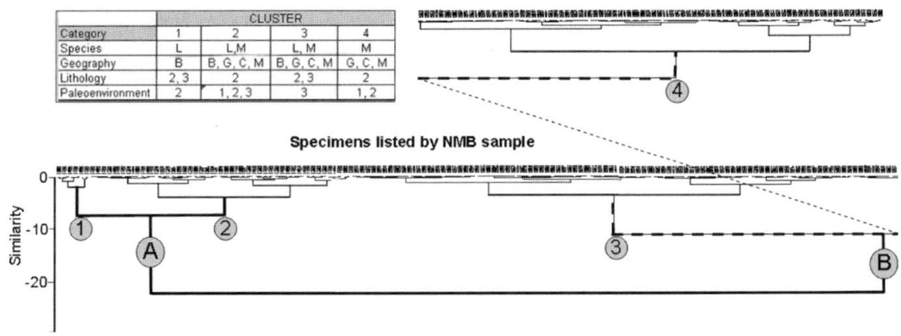

Fig. 8.5 Cluster analysis of all specimens. Four major groups were produced: (1) *P. dasum*; (2) primarily *P. latissimum*; (3) primarily *P. maoense* group A; and (4) primarily *P. maoense* group B. L = *P. latissimum*; M = *P. maoense*; B = Baitoa; G = Gurabo; C = Cana, and M = Mao. Lithology: 2 = silts; 3 = pebbly silts. Paleoenvironment: 1 = brackish; 2 = shallow marine; 3 = marine; 4 = deep marine

In summary, morphometric analyses indicate that morphological variation is not clearly associated with environment, lithology, or geography.

8.4 Discussion

In this chapter I performed a fine-grained analysis of morphological variation in the *Prunum maoense* group from the Caribbean Neogene. The biological attributes of this lineage and the geological context in which it lived together comprise a research system that meets nearly all of the systematic, morphological, and geological prerequisites for rigorous studies of species-level patterns in the fossil record. The *Prunum maoense* group is comprised of abundant, well-preserved, and temporally and geographically widespread species within the Cibao Valley basin. The unique morphological attributes of this lineage permit ontogenetic separation and morphometric comparison of biologically equivalent stages through time and space. Previous studies of the lithological and palaeoenvironmental context of these samples also permit the exploration of potential correlates of morphological variation. Thus, overall, this system has many advantages for species-level research in the fossil record.

As many chapters in this volume have discussed, morphometric methodology may have a significant effect on species discrimination and subsequent analyses of morphological change through time and space. Beck and Budd (this volume) specifically found that whereas exclusively traditional (i.e., distance) morphometric measures were not sufficient to parse out all of the species of Dominican *Siderastrea*, landmark-based morphometric measures were able to clearly delineate additional clusters interpreted to be new morphospecies. Similarly, Budd and Klaus (this volume), using geometric morphometric methods, uncovered eight previously unrecognized species of *Montastraea*, although they note that increased sample sizes may also have contributed to this finding. In their comparison of patterns derived from geometric and traditional-morphometric analyses, Budd and Klaus also found that geometric morphometric analysis is more sensitive to environmentally induced variation, whereas traditional measures are more effective at tracing the distribution of lineages through time (that is, comparing fossil clusters from among different time intervals). Finally, Schultz and Budd (this volume), using geometric morphometrics and larger sample sizes than previous investigations of *Montastraea*, found that some of the species of Budd (1991) appear to represent more than one species. In summary, both morphometric methodology and sample size have been shown to have significant effects on both species discrimination and subsequent interpretations of stasis in Dominican invertebrate lineages (see also Cheetham et al., 2007).

Study of Dominican *Prunum* produced somewhat different conclusions about morphometric methodology than those documented in Beck and Budd, Budd and Klaus, and Schultz and Budd (all in this volume). Prior to comparing these studies, it is important to point out several differences between these unique data sets: (1) Less species occur in the *P. maoense* group than in any of the coral lineages analyzed; (2) the gastropod data were age-standardized prior to analysis (that is, only adults were compared through time and space) whereas this is not possible with the coral data; (3) within-colony variation was used in the coral morphometric studies; and (4) morphological data from genetically-differentiated extant taxa were included in several of the coral morphometric analyses.

Unlike the coral studies, the study of *Prunum* using multivariate analyses based on distance measurements vs. landmarks did not lead to different interpretations of the number of species and did not lead to appreciably different patterns of variation. Canonical Variates Analysis unsurprisingly generated more discrete species groupings than Principal Components Analysis, but both data types (distances and landmarks) and both analysis methods (PCA and CVA) produced outliers in the Cercado and Gurabo Formations that were phenotypically similar to, and in some cases indistinguishable from, specimens of *P. latissimum* from the Baitoa Formation. Thus, unlike the Dominican corals discussed in this volume, morphometric methodologies do not appear to have a major impact on species delineation or patterns of morphological variation through time and space in the *Prunum maoense* group.

Nehm and Geary (1994) and Nehm (2001a) quantitatively evaluated Maury's (1917a,b) qualitative determination that *P. maoense* and *P. latissimum* were separate morphospecies. The traditional and geometric-morphometric analyses performed in Nehm (2001b) and the present chapter generally corroborate this conclusion: statistically significant between-group differences were found in nearly every measured variable. The weak association of phenotypic differences with palaeoenvironment, lithology, and geography suggests that the morphological differences between species were not ecophenotypic. Additionally, the strong association of phenotypic differences with time (i.e., nearly all *P. latissimum* occur in the Miocene, and nearly all *P. maoense* occur in the Pliocene) also provides evidence in support of separate species designations. Patterns of phenotypic overlap in *P. maoense* and *P. latissimum*, notable in the morphometric plots (Fig. 8.3), warrant greater scrutiny, however, and raise several questions about the strength of the above interpretations. Additionally, the occasional recurrence of "*P. latissimum*" phenotypes in the Cercado and Gurabo Formations, at least 5 million years after their abundant representation in the Baitoa section, encourages the consideration of alternative explanatory models.

The range of phenotypic variation in Baitoa samples of *P. latissimum* overlaps with a portion of the variation observed in the Cercado Fm. and Gurabo Fm. samples of stratigraphically younger *P. maoense* (Fig. 8.6). The converse is true of *P. maoense*. One could argue that this pattern of overlap supports the interpretation that *P. latissimum* and *P. maoense* represent the endpoints along a continuum of variation, and that they are arbitrary designations along a single and morphologically variable anagenetic lineage. Consequently, the designation of separate species would be unwarranted (Fig. 8.7A).

Alternatively, one could argue that the phenotypic variants identified in morphometric analyses as "outliers" of *P. latissimum* are in fact individuals of *P. maoense*, and likewise the outliers of *P. maoense* are individuals of *P. latissimum*. This argument is supported by the observation that the outliers of *P. maoense* from the Cercado and Gurabo Formation are phenotypically indistinguishable from *P. latissimum*, despite millions of years of separation. In this model, the outliers of each species would be rejected as such, and the two species would be recognized as having almost completely overlapping stratigraphic ranges (Fig. 8.7C). Reciprocal abundance patterns (i.e., when *P. latissimum* is abundant *P. maoense* is very rare, and *vice versa*) would characterize the two species through time in this model.

The CVA unambiguously classifies "*P. latissimum*" specimens from the Cercado and Gurabo Formations with *P. latissimum* from the Baitoa Formation even though these specimens are separated in time by a minimum of 5 million years. Paleobiological studies employing morphometric data often accept such CVA classifications, revise *a priori* species designations to reflect these classifications, and subsequently plot stratigraphic ranges. In contrast, other studies have relied more heavily on a 'literal' interpretation of the stratophenetic pattern and endorsed alternative explanations. Gould's (1969) case study of land snails from Bermuda illustrates some aspects of this situation.

Gastropod specialists, notably those working with *Poecilozonites*, have encountered similar stratophenetic complexities as those found in the *P. maoense* group when attempting to interpret the evolutionary history of forms from Bermuda (Gulick, 1904; Verill, 1905; Gould, 1969, Fig. 8.8). Interestingly, the *Prunum* and *Poecilozonites* case studies also share developmental-evolutionary features (Nehm, 2001b; Gould, 1969). Briefly, Gould (1969) argued that four independent episodes of paedomorphosis characterized the evolutionary history of *Poecilozonites* in the Pleistocene of Bermuda. He rejected Peile's (1926) suggestion that two stratigraphically overlapping taxa persisted through time in the region and offered detailed morphometric data as his primary (although not exclusive) justification (Gould, 1969, Figs. 8.17–8.19). Gould's (1969) study concluded that (1) the

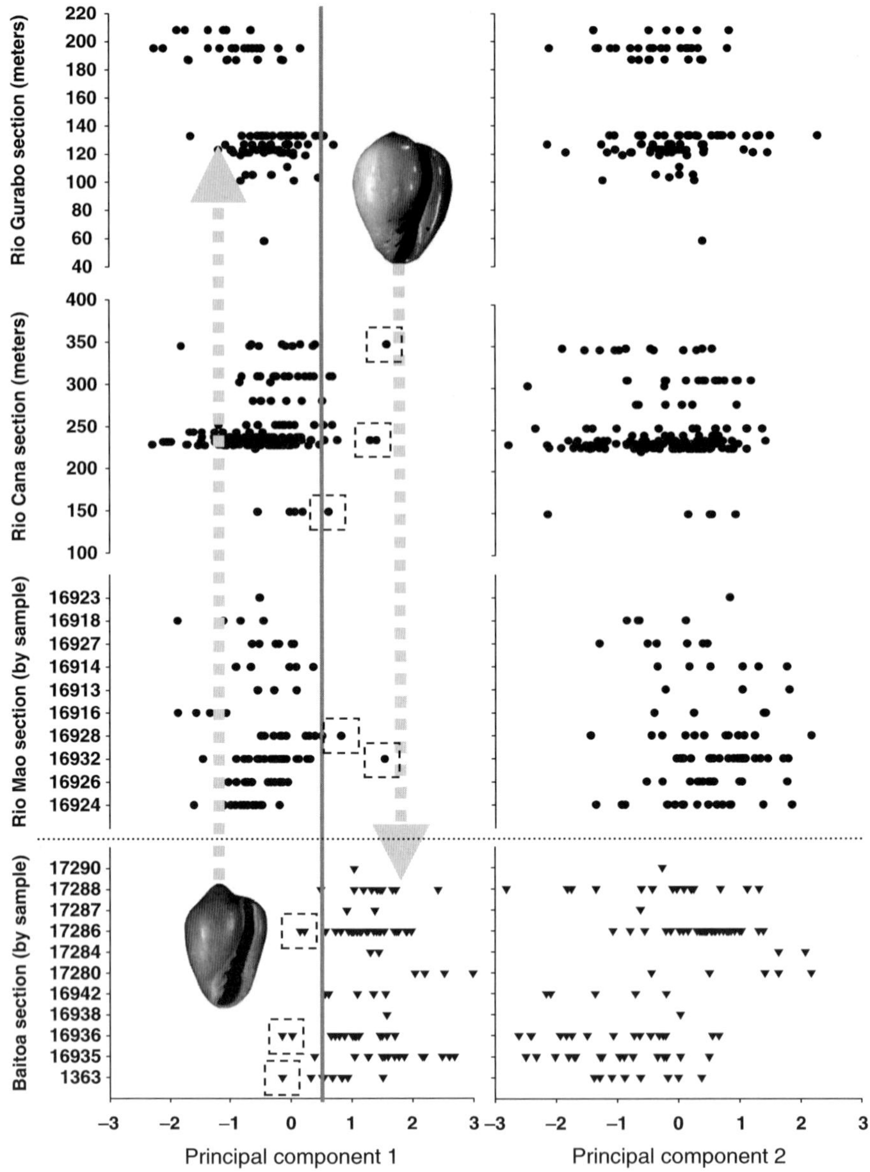

Fig. 8.6 Spatiotemporal patterns of multivariate variation in the *P. maoense* group; left = *P. maoense*, right = *P. latissimum*. Note the repeated appearance of "*P. latissimum*" phenotypes in different places and times. Dashed boxes encompass individuals of interest. The dotted line between the Baitoa and subsequent sections represents a temporal gap of at least 5 my. See Fig. 8.1 for the temporal arrangements of each section in time

repeated evolution of suites of morphological (paedomorphic) features in separate lineages characterizes *Poecilozonites* in Bermuda and (2) the appearance and disappearance of these lineages through time is not exclusively a result of sampling bias,

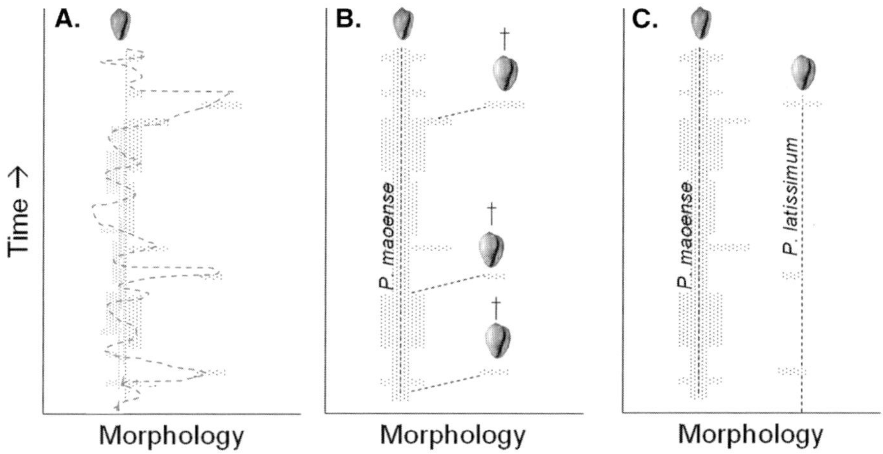

Fig. 8.7 Models of species-level change in the *P. maoense* group and their stratigraphic manifestations. A. Highly variable single-species model; B. Iterative production of short-lived "*P. latissimum*" phenotypes; C. Overlapping two-species model

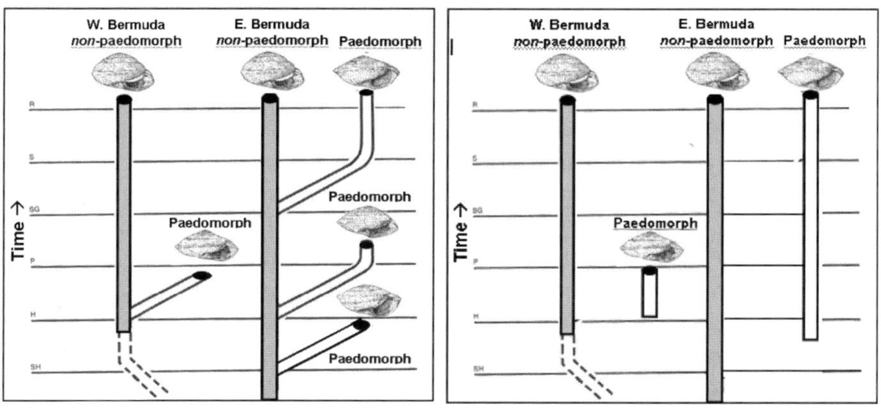

Fig. 8.8 Iterative progenesis in the Pleistocene gastropod *Poecilozonites* from Bermuda. A. Iterative origins of paedomorphic forms in time and space. (Modified from Gould 1977, P. 276); B. Alternative interpretation, similar to some extent to Peile, 1926)

but reflects the repeated origin and selective persistence of paedomorphs in geologically ephemeral red soil habitats characterized by limited calcium.

8.4.1 *Parallelisms and the Problem of Stasis*

Eldredge et al. (2005) argue that a strength of paleobiological data is that "within-population variation can be compared over time as well as space, allowing analysis of the importance of spatial structure of species…". While most workers would accept this point as uncontroversial, species delineation is a complex and difficult prerequisite to taking advantage of such data. Indeed, this is the crux of the problem faced in the present study: How should the "*P. latissimum*" outliers from the Cercado and Gurabo Formations be treated in our analyses? On the one hand, should they be included in the within-population variation of *P. maoense*, or, on the other hand, should they be treated as holdovers of *P. latissimum* from the Baitoa Formation (i.e., a separate species from *P. maoense*)?

The documentation and interpretation of patterns of morphological variation though time and space may be considered among the most basic tasks of paleobiological research. Given that the fossil record affords rich opportunities for documenting morphological variation at vastly different temporal and phylogenetic scales (from within populations to among clades), paleobiologists have unsurprisingly been confronted with the simple question of what processes may link frequently observed morphological recurrences among disparate evolutionary levels (Schindewolf, 1950; Gould, 2002). Numerous empirical studies have documented parallel evolutionary patterns across temporal and phylogenetic scales in a diverse array of taxa, including angiosperms (Rudall and Bateman, 2004), ammonites (Landman, 1989; Jacobs et al., 1994; Schlögl et al., 2006), foraminifera (Norris, 1991; Gill and Kelly, 2003), gastropods (Gould, 1969; Geary et al., 2002) trilobites (Fortey and Owens, 1990; Lauridsen and Nielsen, 2005), graptolites (Rickards and Wright, 2003) and mammals (Martin and Meehan, 2001).

Morphological 'parallelisms' have likewise been of considerable interest to neontologists (e.g., Sanderson and Hufford, 1996, references therein). Wake (1996) for example has argued that studies of variant traits within populations should be of particular value to evolutionary biologists because such traits "frequently duplicate conditions fixed in other taxa" and provide insights into the origins and persistence of morphological traits. Perhaps the clearest case of parallelism between intraspecific morphological variation and its evolutionary fixation in extant taxa was documented by Shubin et al. (1995). These authors documented the limited domains of phenotypic expression in limb elements, and illustrated how patterns of intraspecific variation paralleled changes in salamander phylogeny (Shubin et al., 1995: 881–882). Such studies underscore the clear connections between micro- and macroevolutionary patterns.

Genetic and developmental processes are known to be responsible for evolutionary 'parallelisms' (Sucena et al., 2003; Shapiro et al., 2006; Pelosi et al., 2006). The repeated evolution of toxin resistance in prokaryotes, for example, has been shown in some cases to be a result of the same genetic causes (Woods et al., 2006). More recent work has begun to tease apart the genetic and developmental contributions to morphological parallelisms in eukaryotic metazoans. Notable is Yoon and Baum's (2004) study of plant flowering. They found that three independent origins of rosette flowering were caused by changes in the same developmental regulatory program—but by different genetic changes. Developmental models have long predicted similar mechanisms (e.g., Oster et al., 1988) and paleobiologists have considered similar causes to account for particular parallel patterns (e.g., Nehm, 2001c).

If parallelisms are in fact commonplace in the history of life, which empirical studies suggest, then they may introduce serious complications for studies of evolutionary stasis and change in the fossil record. Specifically, the repetition of morphologies *via* parallelism in *Prunum* would likely confound commonly used morphometric classification procedures such as CVA, which in turn would produce a number of potential 'downstream' errors such as inflation of the number of co-existing species, extension of species' stratigraphic ranges, and the deflation of measures of intraspecific variation and morphological novelty (Fig. 8.7). These complications are much less likely to be significant in other Dominican invertebrate groups, notably colonial bryozoans and corals (e.g., Budd and Klaus, this volume, Cheetham et al., 2007), because genetic data have been used to successfully corroborate species designations based on morphology, and within-colony variation has been used to tease apart genetic and environmentally induced morphological variation.

8.4.2 Patterns and Processes of Stasis and Change

Eldredge et al. (2005) provide a useful framework for thinking about the causes of evolutionary stasis. Their framework can be formulated as three related questions that serve to divide and conquer the problem of stasis: (a) Is stasis caused by constraints on the origin of phenotypic variation? (b) If not (a), is stasis caused by the failure of novel phenotypes to become established in local populations? And (c) If not (a) or (b), is stasis caused by the failure of novelties to spread throughout a species' geographic range? Unfortunately, the answers to all of these questions return us to the interpretive problems outlined in the present case study of the *Prunum maoense* group.

Eldredge et al. (2005), for example, ask: "Do novelties arise only at speciation, or do they arise but are typically not conserved throughout the history of the species…" If, for example, the outliers of *P. maoense* in the Cercado and Gurabo Formations are treated as such, we would likely conclude that the production of novel phenotypes are relatively common in the evolutionary history of this species, but that they appear to not have been successful except in isolated populations. Thus, the lack of production of variation would not explain stasis, and morphological innovation would not be restricted to speciation. Only if these iteratively produced parallelisms were sequestered through reproductive isolation would they be secured as independent lineages (cf. Futuyma, 1987). Thus, the interpretation of data in this case would appear to support Eldredge et al.'s (2005) claim that spatially induced hurdles, rather than the production of novelty, may be the most potent evolutionary forces maintaining stasis. Alternatively, if we were to accept the CVA classification (i.e., the outliers in the Cercado and Gurabo Formation are in fact holdouts of *P. latissimum*), then we would document long-ranging species displaying limited morphological variation, a lack of production of novelty, and clear cases of stasis.

This study illustrates how a relatively small number of phenotypic outliers can play central roles in the interpretation of nearly every issue that is of concern in the interpretation of patterns and processes of stasis and change in the fossil record. Most geological research systems are not characterized by the extensive spatial and temporal sampling as the DR Neogene, and most palaeobiological studies have not focused their attention on phenotypic outliers. While frustratingly complex, such minutiae likely hold the key to understanding the causes of stasis and change in the fossil record.

8.5 Conclusions

In this chapter I performed a fine-grained analysis of morphological variation in the *Prunum maoense* group from the Caribbean Neogene. The unique morphological attributes of this lineage permit ontogenetic separation and morphometric comparison of biologically equivalent stages through time and space. Statistically significant between-group differences in nearly every measured variable, regardless of data type (distances or landmarks) or analysis method (PCA or CVA) and the weak association of phenotypic differences with environment, lithology, and geography suggests that three species occur in this lineage and that these morphological differences are not ecophenotypic. Additionally, the strong association of phenotypic differences with time (i.e., nearly all *P. latissimum* occur in the Miocene, and nearly all *P. maoense* occur in the Pliocene) provides evidence in support of separate species designations. The spatial and temporal patterns of phenotypic overlap in *P. maoense* and *P. latissimum*, and the repeated recurrence of "*P. latissimum*" phenotypes at least 5 million years after the Baitoa Formation, raise questions about the strength of the above interpretations. The repeated production of "*P. latissimum*" morphologies *via* parallelism may explain the complex stratophenetic patterns documented in this study.

Acknowledgments I thank Jonathan Baez, Brian Beck, Novia Jarrett, Victor Matos, and Moziah Saad for assistance in the field, Nancy Budd for reviews of the manuscript, and the National Science Foundation for financial support. The material for this study was provided by Peter Jung and René Panchaud of the Naturhistorisches Museum, Basel, Switzerland; Jack and Winifred Gibson-Smith of Surrey, England; Emily Vokes of Tulane University; Tom Waller and Warren Blow of the Smithsonian Institution; James McLean, Edward Wilson, and Lindsey Groves of the Los Angeles Museum of Natural History; Roger Portell and Kurt Auffenburg of the Florida Museum of Natural History at Gainesville; Robert Van Syk of the California Academy of Sciences; Gary Rosenberg of the Academy of Natural Sciences, Philadelphia, and David Lindberg of the University of California Museum of Paleontology. I am grateful for the access to these collections and the generous hospitality and assistance provided by these individuals.

References

Anderson, L.C., 1994, Paleoenvironmental control of species distributions and intraspecific variability in Neogene Corbulidae (Bivalvia: Myacea) of the Dominican Republic, *J. Paleontol.*, 68, **3**:460–473.

Anderson, L.C., Geary, D.H., Budd, A.F., Nehm, R.H., Johnson K., and Stemann T., 1992, Paleoenvironmental control of species distributions in Neogene invertebrate taxa of the Dominican Republic, *Paleontol. Soc. Spec. Publ.*, **6**:6.

Bold, W.A. van den, 1988, Neogene paleontology of the northern Dominican Republic. 7. The subclass Ostracoda (Arthropoda: Crustacea), *Bull. Am. Paleontol.,* **94**:1–105.
Bookstein, F.L., 1996, Combining the tools of geometric morphometrics, in: *Advances in Morphometrics* (L.F. Marcus, M. Corti, A. Loy, G.J.P. Naylor, and D.E. Slice, eds.), Plenum, New York, pp. 131–151.
Budd, A.F., 1991, Neogene Paleontology in the Northern Dominican Republic, 11, The Family Faviidae (Anthozoa: Scleractinia), Part I, The Genera *Montastraea* and *Solenastrea, Bull. Am. Paleontol.*, **101**, 338:5–83.
Cheetham, A.H., Sanner, J., and Jackson, J.B.C., 2007, *Metrarabdotos* and related genera (Bryozoa: Cheilostomata) in the late Paleogene and Neogene of Tropical America, *Paleontol. Soc. Mem.,* **67**:1–96.
Eldredge, N. and Gould, S.J., 1972, Punctuated equilibria: an alternative to phyletic gradualism, in: *Models in Paleobiology* (T.J.M. Schopf, ed.), Freeman Cooper, San Francisco, pp. 82–115.
Eldredge, N., Thompson, J.N., Brakefield, P.M., Gavrilets, S., Jablonski, D., Jackson, J.B.C., Lenski, R.E., Lieberman, B.S., McPeek, M.A., and Miller, W. III, 2005, The dynamics of evolutionary stasis. *Paleobiology*, **31**:133–145.
Erwin, D.H. and Anstey, R.L., 1995, New *Approaches to Speciation in the Fossil Record.* Columbia University Press, New York.
Fortey, R.A. and Owens, R.M., 1990, Trilobites, in: *Evolutionary Trends* (K.J. McNamara, ed.), Belhaven, London, pp. 121–42.
Futuyma, D., 1987. On the role of species in anagenesis, *Am. Nat.,* **130**:465–473.
Grant, P., 1999, *Ecology and Evolution of Darwin's Finches.* Princeton University Press, Princeton, NJ, 512 pp.
Geary, D.H., Staley, A.W., Müller, P., and Magyar, I., 2002, Iterative changes in Lake Pannon Melanopsis reflect a recurrent theme in gastropod morphological evolution, *Paleobiology,* 28, **2**:208–221.
Gill, P.J. and Kelly, D.C., 2003, Iterative patterns of wall texture evolution and its significance for phylogenetic reconstruction among cenozoic planktonic foraminifera, *Geol. Soc. Am. Abs. Prog.,* 35, **6**:161.
Gould, S.J., 1969, An evolutionary microcosm: pleistocene and recent history of the land snail P. (Poecilozonites) in Bermuda, *Bull. Mus. Comp. Zool.,* **138**:407–532.
Gould, S.J., 1977, Eternal Metaphors of Paleontology, pp. 1–26 in: Patterns of Evolution as Illustrated by the Fossil Record (Hallam, A., ed.), Amsterdam: Elsevier.
Gould, S.J., 2002, *The Structure of Evolutionary Theory*. Harvard University Press, Cambridge, MA.
Gulick, J.T., 1904, The fossil land shells of Bermuda, *Proc. Acad. Natl. Sci. Philadelphia,* **56**:406–425.
Jacobs, D.K., Landman, N.H., and Chamberlain, J.A., 1994, Ammonite shell shape covaries with faces and hydrodynamics: iterative evolution as a response to changes in basinal environment, *Geology*, **22**:905–908.
Landman, N.H., 1989, Iterative progenesis in Upper Cretaceous ammonites, *Paleobiology,* **15**:95–117.
Lauridsen, B.W. and Nielsen, A.T., 2005, The Upper Cambrian trilobite Olenus at andrarum, Sweden: a case of iterative evolution?, *Palaeontology*, 48, **5**:1041–1056.
Lieberman, B.S., Brett, C.E., and Eldredge, N., 1994, A study of stasis and change in two species lineages from the middle Devonian of New York state, *Paleobiology,* **21**:15–27.
Martin, L.D. and Meehan, T.J., 2001, Does the iterative evolution of ecomorphs reflect climatic cycles?, *Geol. Soc. Am. Abs. Prog*., p. 68.
Maury, C.J., 1917a, Santo Domingo type sections and fossils. Part 1, *Bull. Am. Paleontol.,* 5, **29**:1–251.
Maury, C.J., 1917b, Santo Domingo type sections and fossils. Part 2, *Bull. Am. Paleontol.,* 5, **30**:1–43.
Nehm, R.H., 2001a, Neogene Paleontology of the Dominican Republic: The Genus *Prunum, Bull. Am. Paleontol.*, **359**:1–46.

Nehm, R.H., 2001b, The developmental basis of morphological disarmament in *Prunum* (Neogastropoda; Marginellidae), in: *Beyond Heterochrony* (M.L. Zelditch, ed.), Wiley, New York, pp. 1–26.

Nehm, R.H., 2001c, Linking macroevolutionary pattern and developmental process in marginellid gastropods, in: *Evolutionary Patterns: Growth, Form, and Tempo in the Fossil Record* (J.B.C. Jackson, S. Lidgard, and F.K. McKinney, eds.), University of Chicago Press, Chicago, pp. 159–195.

Nehm, R.H., 2005, Patterns and processes of evolutionary stasis and change in *Eratoidea* (Gastropoda: Marginellidae) from the Dominican Republic Neogene, *Carib. J. Sci.*, 41, **2**:189–214.

Nehm, R.H. and Geary, D., 1994, A gradual morphologic transition during a rapid speciation event in marginellid gastropods (Neogene; Dominican Republic), *J. Paleontol.,* 68, **4**:787–795.

Norris, R.D., 1991, Parallel evolution in the keel structure of planktonic foraminifera, *J. Foram. Res.,* **21**:319–331.

Oster, G.F., Shubin, N., Murray, J.D., and Alberch, P., 1988, Evolution and morphogenetic rules: the shape of the vertebrate limb in ontogeny and phylogeny, *Evolution*, 42, **5**:862–884.

Papadopoulos, L.N., Todd, J.A., and Michel, E., 2004, Adulthood and phylogenetic analysis in gastropods: character recognition and coding in shells of Lavigeria (Cerithioidea, Thiaridae) from Lake Tanganyika, *Zool. J. Linn. Soc.,* **140**:223–240.

Peile, A.J., 1926, The Mollusca of Bermuda, *J. Mollus. Stud.*, **17**:71–98.

Pelosi, L., Kuhn, L., Guetta, D., Garin, J., Geiselmann, J., Lenski, R.E., and Schneider, D., 2006. Parallel Changes in Global Protein Profiles During Long-Term Experimental Evolution in Escherichia coli, *Genetics*, **173**:1851–1869.

Raup, D.M. and Stanley, S.M., 1978, *Principles of Paleontology*, W.H. Freeman, New York.

Rickards, R.B. and Wright, A.J., 2003, The *Pristiograptus dubius* (Suess, 1851) species group and iterative evolution in the Mid- and Late Silurian, *Scott. J. Geol*., 39, **1**:61–69.

Rohlf, F.J., 1998, TPSRELW, version 1.18. A program to performs a relative warp analysis. Department of Ecology and Evolution, State University of New York, Stony Brook, NY.

Rohlf, F.J., 2000, TPSDIG, version 1.20. A program for digitizing landmarks and outlines for geometric morphometric analyses. Department of Ecology and Evolution, State University of New York, Stony Brook, NY.

Rohlf, F.J. and Marcus, L.F., 1993, A revolution in morphometrics, *Trends Ecol. Evol*., **8**:129–132.

Rudall, P.J. and Bateman, R.M., 2004, Evolution of zygomorphy in monocot flowers: iterative patterns and developmental constraints, *New Phytol.,* 162, **1**:25–44

Sanderson, M.J. and Hufford, L. (eds.), 1996, *Homoplasy, the Recurrence of Similarity in Evolution,* Academic, San Diego, CA.

Saunders, J.B., Jung, P., Geister, J., and Biju-Duval, B., 1982, The Neogene of the south lank of the Cibao Valley, Dominican Republic: a stratigraphic study, *Transactions of the 9th Caribbean Geological Conference, Santo Domingo*, 1980.

Saunders, J.B., Jung, P., and Biju-Duval, B., 1986. Neogene paleontology of the northern Dominican Republic 1. Field surveys, lithology, environment, and age, *Bull. Am. Paleontol.,* **89**:1–79.

Schlögl, J., Elmi, S., Rakus, M., Mangold, C., and Ouahhabi, M., 2006. Specialisation and iterative evolution of some Western Tethyan Bathonian ammonites [*Benatinites* (*B.*) *nov*., *B*. (*Lugariceras*) *nov*. and *Hemigarantia*], *Geobios*, **39**:113–124.

Schindewolf, O.H., 1950 (1993) *Basic Questions in Paleontology*. University of Chicago Press, Chicago, IL.

Shapiro, M.D., Bell, M., and Kingsley, D.M., 2006, Parallel genetic origins of pelvic reduction in vertebrates, *Proc. Natl. Acad. Sci. USA* **103**(37):13753–13758.

Shubin, N., Wake, D.B., and Crawford, A.J., 1995, Morphological variation in the limbs of *Taricha granulosa* (Caudata: Salamandridae): evolutionary and phylogenetic implications, *Evolution,* **49**:874–884.

Sucena, E., Delon, E., Jones, I., Payre, F., and Stern, D.L., 2003, Regulatory evolution of *shavenbaby/ovo* underlies multiple cases of morphological parallelism, *Nature*, **424**:935–938.

Verill, A.E., 1905, The Bermuda Islands: Part IV. Geology and paleontology, *Trans. Conn. Acad. Arts Sci.,* **12**:45–204.

Vermeij, G., 1995, *A Natural History of Shells*. Princeton University Press, Princeton, NJ.

Vermeij, G.J. and Signor, P.W., 1992, The geographic, taxonomic and temporal distribution of determinate growth in marine gastropods, *Biol. J. Linn. Soc.,* **47**:233–247.

Vokes, E.H., 1979, The age of the Baitoa Formation, Dominican Republic, using Mollusca for correlation, *Tulane Stud. Geol. Paleontol.,* **15**:105–116.

Vokes, E.H., 1989, Neogene paleontology of the northern Dominican Republic 8. The Family Muricidae (Mollusca; Gastropoda), *Bull. Am. Paleontol.,* **97**:5–94.

Wake, D.B., 1996, Introduction, pp. xvii–xxv, in: *Homoplasy, the Recurrence of Similarity in Evolution* (M.J. Sanderson and L. Hufford, eds.), Academic, San Diego, CA.

Woods, R., Schneider, D., Winkworth, C.L., Riley, M.A., and Lenski, R., 2006, Tests of parallel molecular evolution in a long-term experiment with Escherichia coli, *Proc. Natl. Acad. Sci. USA* **103**:9107–9112

Yoon, H.S. and Baum, D.A., 2004, Transgenic study of parallelism in plant morphological evolution, *Proc. Natl. Acad. Sci. USA* **101**:6524–6529.

Zelditch, M.L. and Fink, W.L., 1996, Heterochrony and heterotopy: stability and innovation in the evolution of form, *Paleobiology,* **22**:241–254.

Chapter 9
Assessing Community Change in Miocene to Pliocene Coral Assemblages of the Northern Dominican Republic

James S. Klaus[1], Donald F. McNeill[1], Ann F. Budd[2], and Kenneth G. Johnson[3]

Contents

9.1 Introduction

Records of Cenozoic scleractinian coral species from the Caribbean region document a history of pulsed origination and extinction. Periods of elevated species richness during the middle to late Eocene (40–36 Ma), late Oligocene to earliest Miocene (28–22 Ma), and early Pliocene (5–2 Ma) are interrupted by pulses of extinction that drop regional diversity between 15% and 30% of standing levels (Budd, 2000). These periods of accelerated biotic change can be correlated to episodes of significant environmental perturbation. The middle to late Eocene

[1] Department of Geological Sciences, University of Miami, 43 Cox Science Building, Coral Gables, FL, 3133. Email: jsklaus@rsmas.miami.edu

[2] Department of Geoscience, University of Iowa, 121 Trowbridge Hall, Iowa City, IA 52242. Email: ann-budd@uiowa.edu

[3] Department of Paleontology, Natural History Museum, Cromwell Road, London SW7 5BD, United Kingdom. Email: K.Johnson@nhm.ac.uk

R.H. Nehm, A.F. Budd (eds.) *Evolutionary Stasis and Change in the Dominican Republic Neogene,*

extinctions have been attributed to global drops in temperature (Prothero, 1994; Budd, 2000), the early Miocene extinctions have been linked to increased upwelling and associated turbidity and cooling of the Caribbean (Edinger and Risk, 1994; Edinger and Risk, 1995), and Plio-Pleistocene extinctions have been linked to declining productivity and drops in temperature associated with closure of the Central American Seaway, and the onset of northern hemisphere glaciation (Budd et al., 1996; Collins et al., 1996a; Budd et al., 1998; Budd and Johnson, 1999b; Jackson and Johnson, 2000; Allmon, 2001; Todd et al., 2002). Studies of extinction selectivity suggest that during these intervals of faunal turnover coral communities underwent fundamental changes beyond that expected from mere species replacement. Cold-tolerant, eurytopic reef coral species that brood larvae were more likely to survive extinction during the late Oligocene to early Miocene (Edinger and Risk, 1995), while reef coral species with large maximum colony sizes and longer generation times were more likely to survive extinction during the Plio-Pleistocene extinctions (Johnson et al., 1995).

These events of accelerated turnover have had a significant influence on the formation and ecology of modern Caribbean coral communities. This is most evident in the late Pliocene faunal turnover in which approximately 80% of Mio-Pliocene corals became extinct and more than 60% of species now living in the region originated (Budd et al., 1996; Budd and Johnson, 1997; Budd and Johnson, 1999b; Budd, 2000). Records of Caribbean coral occurrences suggest that between 7 and 4 Ma, prior to the main extinction peak (2 Ma), origination rates were high, and reef communities were dominated by *Stylophora*, *Pocillopora*, *Goniopora*, and a suite of agariciid and poritid species that more closely resemble modern Indo-Pacific species than modern Caribbean species (Budd et al., 1996; Klaus and Budd, 2003; however see Johnson et al. chapter 11, this volume). Furthermore, diverse assemblages of free-living meandroid corals (*Trachyphyllia*, *Placocyathus*, *Manicina*) inhabited nearby seagrass and soft-bottom areas (Budd et al., 1996; Budd and McNeill, 1998). In contrast, today's Caribbean reefs are dominated by *Acropora*, *Diploria*, and members of the *Montastraea annularis* species complex (Goreau and Wells, 1967; Geister, 1977).

While faunal turnover has been linked to closure of the Central American Seaway (CAS) and the onset of northern hemisphere glaciation, comparison of the origination (7–4 Ma) and extinction (2–1 Ma) peaks with the timing of regional environmental perturbations suggests that the roots of Plio-Pleistocene faunal turnover may lie as far back as 7 Ma or older (Johnson et al., Chapter 11, this volume). Ocean circulation between the Atlantic and Pacific was likely to have been affected by the late Miocene (6–10 Ma), with final closure of the CAS no later than 3 Ma (Keller and Barron, 1983; Keller et al., 1989; Coates et al., 1992; Coates and Obando, 1996; Collins et al., 1996b). In a generalized model of marine faunal turnover events of the late Pliocene, Allmon (2001) proposed that the early stages of CAS closure would cause an increase in formation of isolated populations due either to disruption and decrease in food supplies or to the instability of food and nutrient conditions, which would subsequently lead to an increase in rate of speciation.

In this chapter we characterize coral communities and examine how they vary during the period of high species richness prior to the main Plio-Pleistocene coral extinction peak. In particular, we seek to determine if a high rate of faunal persistence masks the effect of origination and extinction on the structure of coral communities through this time period, or, if changes in the available species pool greatly affect the dynamic of ecological interactions between species, thereby changing community structure. Furthermore, we examine whether changes in community structure reflect changes that might predict survivorship during the subsequent Plio-Pleistocene extinction event.

To explore these questions, in the current study we analyze the stability of reef coral assemblages in a continuous sequence of late Miocene and early Pliocene (6.2–3.2 Ma) sediments exposed in the northern Dominican Republic using three approaches. First, based on compiled records of over 4,000 coral specimens, we have found that 61% of local coral species persist from the oldest (Cercado) to the youngest formation (Mao). These values are just above the 60% cut-off proposed by Brett and Baird (1995) for determining faunal stability within a given interval. Second, quantitative community analyses of coral species presence/absence data of coral species from 21 different lithostratigraphic units collected throughout the sequence suggest that, despite background levels of origination and extinction, reef coral assemblages are relatively stable. Third, contrary to results of the first two approaches, analyses of relative abundance data based on detailed ecological surveys conducted at three localities of exceptional reef development (Arroyo Bellaco, Cañada de Zamba, Cana Gorge) suggest that coral communities were changing significantly over this interval. We examine this contradiction and the nature of these changes within the context of broader environmental trends.

9.2 Geologic Setting and Collections

The Cercado, Gurabo, and Mao Formations of the Cibao Valley, northern Dominican Republic make up a remarkably continuous northward (seaward) prograding wedge of sediments shed off the Cordillera Central during the Miocene and Pliocene (6.2 to 3.3 Ma; see McNeill et al., Chapter 2, this volume for details). The Cercado Formation is stratigraphically oldest, and predominately sandstone, but contains variations ranging from pebble stringers, conglomerate lenses, lignite beds, and reef limestone (Saunders et al., 1986; Evans, 1986). The depositional setting is interpreted to be a shallow shelf. The Gurabo Formation is predominately a siltstone, but again, lithologic variations are considerable, ranging from interbedded silts and coral limestone, to plain massive siltstone with occasional sands and gravel lenses (Saunders et al., 1986; Evans, 1986). The depositional setting is thought to include a transition from a relatively shallow shelf setting to a steep upper slope setting. Based on ostracode assemblages (Bold, 1988) as well as the abundance of menardiform

Globorotalias (Saunders et al., 1986) this transition is thought to occur at approximately 400 m in both the Río Gurabo and Río Cana sections (Fig. 2.1, McNeill et al., Chapter 2, this volume,). The presence of seagrass indicators at approximately 450 m indicates this deepening event to be fairly short lived in the Río Cana section (Cheetham and Jackson, 1996). Ostracode assemblages suggest these conditions may have persisted through the remainder of the Gurabo Formation of the Río Gurabo section. The Mao Formation is again highly variable. Evans (1986) recognized four different lithofacies: bedded siltstone, conglomerate, interbedded coral boundstone-siltstone, and a clean siltstone. The depositional setting is likewise variable, for the conglomerates and interbedded coral boundstone-siltstone an upper slope setting (Evans, 1986). The uppermost part of the formation is likely middle to deep shelf facies (McNeill et al., Chapter 2, this volume)

Here we report on an extensive collection of Miocene and Pliocene hermatypic corals from the Cercado, Gurabo, and Mao Formations. This report is based primarily on collections from the two river sections of the Río Gurabo and Río Cana with supplemental collections from exposed sections in the Río Mao, Río Amina, and Río Yaque del Norte (Fig. 2.1, McNeill et al., Chapter 2, this volume). This collection was initiated as part of a multidisciplinary project on the paleontology and stratigraphy of the Neogene mixed carbonate and siliciclastic sequence in the Cibao Valley region, lead by researchers of the Naturhistorisches Museum Basel (Saunders et al., 1986). Between 1978 and 1980, 2189 hermatypic coral specimens were collected and deposited at the Naturhistorisches Museum Basel. These initial collections have served as the basis for several systematic monographs and ecological studies (Foster, 1986; Foster, 1987; Budd, 1991; Budd et al., 1996; Budd and Johnson, 1999a). Subsequent field trips have expanded this collection to 4,236 coral specimens. Collections were made with a hammer and chisel during standardized ecological transect surveys, as well as during haphazard collections from narrowly defined outcrops. The coral specimens in this collection have been identified using a standard set of characters and character states developed on the basis of morphometric analyses of Neogene and Recent samples collected across the Caribbean region (Budd et al., 1994). Illustrations of diagnostic characters from each species are available from the Neogene Biota of Tropical America taxonomic database (NMITA: http://nmita.geology.uiowa.edu). The genus *Montastraea* consists of two or more species complexes and is currently the subject of extensive morphometric investigation (see Budd and Klaus, Chapter 5, this volume, and Schultz and Budd, Chapter 7, this volume). For the purposes of the present study, identification of species of *Montastraea* therefore follows the classification system of Budd (1991).

We have used three approaches to characterize coral communities and examine how they vary within the study sequence: (1) The extent to which coral species persist from one formation to the next; (2) quantitative community analyses of presence/absence data of coral species; and (3) quantitative community analyses of relative abundance data of coral species obtained from line transect surveys.

To perform analyses of the coral presence/absence data, collecting sites were grouped into 21 lithostratigraphic units using the stratigraphic sections of Saunders et al. (1986) in a manner similar to Budd et al. (1996). Based on current best estimates obtained from microfossils, paleomagnetics, and strontium isotopes (Fig. 2.7, McNeill et al., Chapter 2, this volume) these intervals range in median age from roughly 6.2 to 3.3 Ma, and include nine lithostratigraphic units from the Río Gurabo, eight from the Río Cana, and one each from the Río Mao, Río Amina, and Río Yaque del Norte (Table 9.1).

To assess quantitatively the distribution and relative abundance of species within more restrictively defined reef assemblages, a line transect sampling technique was applied at three regions of exceptional reef development within the sequence. Every 0.5 m, the coral colony intersecting the transect line was recorded. A total of 27 20 m line transects were collected from three localities, 10 from Arroyo Bellaco, 10 from Cañada de Zamba, and 7 from the Cana Gorge. The number of transects was determined by the amount of available outcrop at each locality.

In community analyses of both the presence/absence data obtained from the 21 lithostratigraphic units, and the relative abundance data obtained from the 27 line transects, the species-sample data matrix was analyzed using the Bray-Curtis similarity coefficient, one-way analysis of similarity (ANOSIM), and similarity percentages (SIMPER) (Clarke and Warwick, 2001). The ANOSIM analysis is based on a non-parametric permutation procedure applied to the rank similarity matrix. If samples within a group are identical, Global R = 1. These analyses were performed using the computer software package PRIMER (Clarke and Warwick, 2001). Mantel matrix comparisons (Mantel, 1970) were computed with the software package zt (Bonnet and Van de Peer, 2002), and used to further assess the correlation between coral assemblages and age. Rarefaction analyses were used to compare

Table 9.1 List of lithostratigraphic units used in coral assemblage analyses

Lithostrat. unit	Formation	River	Meters in section	Age (Ma)	Specimens	Species
G12	Mao	Gurabo	740–770	3.3	51	29
C11	Mao	Cana	960–1010	3.8	196	42
G11	Mao	Gurabo	660–710	4.2	81	29
C9	Mao	Cana	730–760	4.2	35	18
G9	Gurabo	Gurabo	387–430	4.8	53	16
C7	Gurabo	Cana	425–440	4.9	93	13
G8	Gurabo	Gurabo	360–386	5.0	316	46
C5	Gurabo	Cana	330–385	5.1	867	74
G6	Gurabo	Gurabo	250–290	5.2	633	57
C4	Gurabo	Cana	280–330	5.2	360	21
G5	Gurabo	Gurabo	210–250	5.3	118	17
A1	Gurabo	Amina	***	5.3	38	7
G3	Gurabo	Gurabo	150–185	5.5	229	19
C3	Gurabo	Cana	225–280	5.5	59	14
G2	Cercado	Gurabo	140–150	5.8	117	16
G1	Cercado	Gurabo	120–140	5.9	79	4
M1	Cercado	Mao	***	5.9	33	7
Y4	Cercado	Yaque del Norte	***	5.9	26	11
C2	Cercado	Cana	185–225	6.0	30	7
C1	Cercado	Cana	140–185	6.1	68	17
B1	Cercado	Cana	135–185	6.2	754	46

species richness values. Rarefaction calculations and 95% bootstrapped error estimates were obtained using Analytic Rarefaction 1.3 (Holland, 2003).

9.3 Assemblage Analyses

A total of 104 species belonging to 38 genera have been identified from the Cercado, Gurabo, and Mao Formations. Comparisons among the three collected formations show that many more specimens and species were collected in the Gurabo Formation than the Cercado and Mao Formations (Fig. 9.1). Of the 103 species identified, 52 have global first occurrences within the study sequence, 30 have global last occurrences, and 21 are known only from the Dominican Republic (Appendix 1). This high endemism is likely an artifact of the excellent preservation

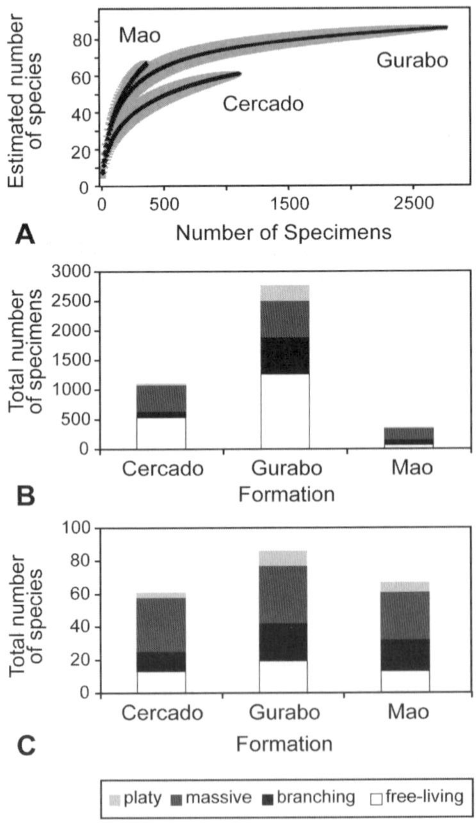

Fig. 9.1 Rarefaction curve and bar charts showing the total number of specimens and species collected in the Cercado, Gurabo, and Mao Formations. Bars are shaded according to four colony shape categories. Formations are arranged in chronological order from oldest (left) to youngest (right)

of the specimens and the large amount of monographic work based on this material. In general, the effect is a large number of first occurrences within the Cana and Gurabo Formations and a large number of last occurrences within the Gurabo and Mao Formations. Among the 52 species with first occurrences within the study interval, 27 have first occurrences within the Cercado Formation, 22 within the Gurabo Formation, and three within the Mao Formation. Two genera, *Scolymia* and *Mussismilia* also have first occurrences within the Cercado Formation. Of the 30 last occurrences found within this sequence, three are known from the Cercado Formation, nine from the Gurabo Formation, and 18 from the Mao Formation.

9.3.1 Faunal Persistence

One method for assessing the stability of the coral fauna over this roughly 3 million year interval is the percentage of taxa persisting from one formation to the next. Eighty-nine percent of species from the Cercado Formation persist to the Gurabo Formation, and 79% of species from the Gurabo Formation persist to the Mao Formation. Moreover, 61% of the species found in the Cercado Formation persist to Mao Formation. After eliminating rare taxa, these numbers are even higher, well above the 60% cut-off proposed by Brett and Baird (1995) for determining faunal stability within Middle Devonian intervals of similar duration. However, the rarefaction curves generated from each formation suggest higher diversity in the Gurabo and Mao Formations compared to the Cercado Formation. Thus faunal persistence may not be the best indicator of faunal stability through the sequence.

9.3.2 Presence/Absence Analyses

Comparisons among the 21 lithostratigraphic units shows significant variation in both the number of specimens and the number of species collected within lithostratigraphic units (Fig. 9.2A–C). In general, numbers of specimens collected per locality range from 33 to 867 (mean = 202), and numbers of species collected per locality range from 4 to 74 (mean = 24). If the amount of sampling was equal in different localities, and species richness was equal at all localities, the two histograms in Figs. 9.2A, B would be bell-shaped. Instead, they both appear to be skewed to the right. This skewed variation can be attributed in part to the accessibility of the localities, the amount of available outcrop, the nature of coral preservation, and habitat variation within and among lithostratigraphic units. Nevertheless, rarefaction curves (Fig. 9.1) suggest that a major portion of species within each formation has been sampled. The two curves presented in Fig. 9.2D represent: (1) the data randomly resampled (n = 2000) using the Analytic Rarefaction 1.3 program (Holland, 2003) to determine the average number of species as a function of the number of lithostratigraphic units sampled, and (2) the lithostratigraphic units

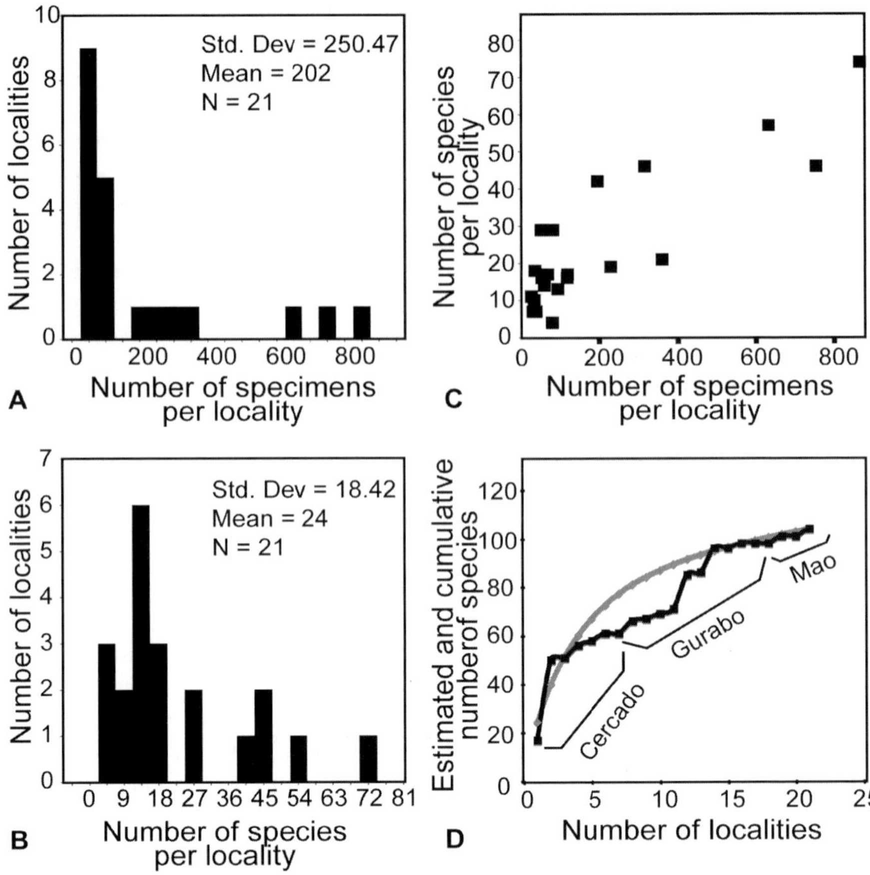

Fig. 9.2 Plots assessing sampling adequacy. (A–C) Histograms and scatterplot showing the numbers of species and specimens collected per lithostratigraphic unit. (D) Rarefaction and cumulative number of species curves. The gray curve was constructed by rarefaction analysis. The black curve was constructed by adding the lithostratigraphic units in stratigraphic order beginning with the oldest and continuing to the youngest

added in roughly temporal order beginning with the stratigraphically oldest continuing to the youngest. The second curve appears to level off in a series of plateaus. These plateaus correspond to the top of the Cercado Formation, the middle Gurabo Formation, the top of the Gurabo Formation, and the top of the Mao Formation. Comparison of these two curves suggest species occurrences are not randomly distributed through the section.

An occurrence matrix containing presence/absence data of 103 species from these 21 lithostratigraphic units was assembled. To assess the variation in community structure between the different units we used average linkage clustering of the Bray-Curtis similarity coefficient (Bray and Curtis, 1957) calculated for all sample pairs. With presence/absence data, the Bray-Curtis index is equivalent to the Sorensen similarity index. The resulting dendrogram of these analyses revealed two distinct clusters (Fig. 9.3). The distinctness of these two clusters was further verified by one-way analysis of similarity (ANOSIM) ($R = 0.674$, $p < 0.001$).

Forty-three of 103 species were detected within the lithostratographic units of cluster B (Fig. 9.3), consisting primarily of free-living taxa (62.6%) with reduced numbers of branching (21.7%), massive (12.6%) and platy (3.0%) growth forms. In contrast, cluster A consists of 99 species of primarily massive (41.0%) and branching (29.8%) corals, with fewer platy (8.9%) and free-living (20.1%) morphologies. To more precisely identify the species responsible for the recognition of these assemblages in the multivariate cluster analysis, the SIMPER procedure was performed. The average dissimilarity between these two assemblages was 81.46%. Using presence/absence data, the contribution of any one species to this overall dissimilarity is relatively small (<4.9% of the total dissimilarity) (Appendix 1). From the SIMPER analysis, we see the free-living assemblage is composed primarily of free-living members of the genus *Antillia*, *Antillophyllia*, *Manicina*, *Meandrina*, *Placocyathus*, *Thysanus* and *Trachyphyllia* (Fig. 9.4). Branching corals associated with this assemblage include species of the genus *Stylophora* (most commonly *Stylophora minor*) and *Madracis*. Rare massive corals include species

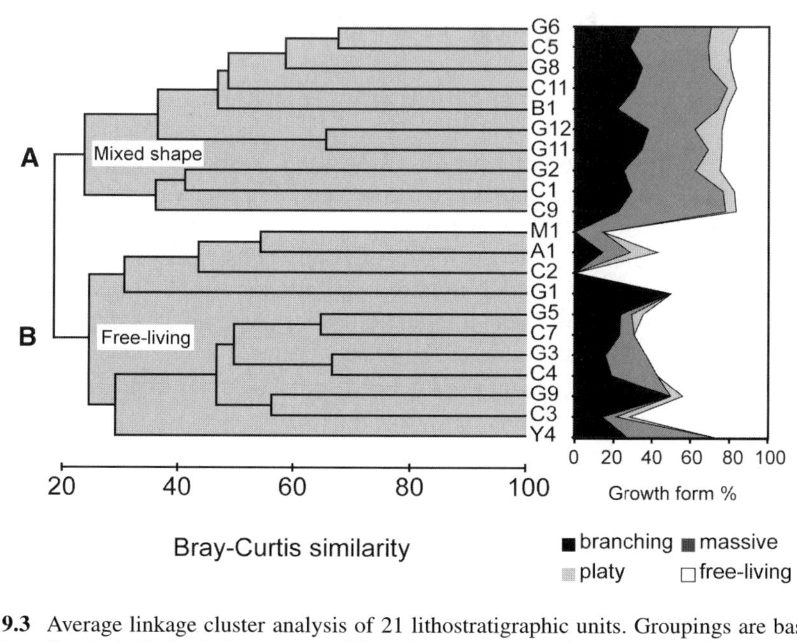

Fig. 9.3 Average linkage cluster analysis of 21 lithostratigraphic units. Groupings are based on the Bray-Curtis similarity index calculated from the presence or absence of 104 different species. Two types of assemblages were identified and characterized as having mixed shape and free-living growth morphologies

of *Montastraea* and *Goniopora*. The common occurrence of species from the genus *Porites*, *Stylophora*, *Montastraea*, *Stephanocoenia*, *Goniopora* and *Undaria* characterized the mixed assemblage (Fig. 9.4).

In comparing other independent paleoenvironmental interpretations for each lithostratigraphic unit, it was determined that the free-living assemblages were generally associated with silty, soft-bottom areas (Saunders et al., 1986). Many of these lithostratigraphic units contain evidence of seagrass associations. Identifiable patterns of seagrass cells have been found on bryozoan attachment bases of the genera *Metrarabdotos* within units G1, C3, G3, C4 and C7 (Cheetham and Jackson, 1996). Furthermore, seagrass associated limpets or the seagrass associated gastropod *Smaragdia* have been detected within units C2, G1, C3, G3, and C4 (Costa et al., 2001). *Smaragdia* was also detected within the mixed shape units G6 and C5. These units contain interbedded limestones and siltstones indicative of high-frequency sea-level cycles and significant environmental variation. It should be noted that the seagrass beds documented here may differ from shallow (less than 10 m) nearshore grass flats common today in the Caribbean in that they may have extended to depths of 20–30 m on forereef slopes (Cheetham and Jackson, 1996). The lithostratographic units of cluster B (mixed-shape) are more difficult to generalize. Based on foraminiferal assemblages these units encompass a mix of shallow near-shore environments to deeper reef slope environments.

Clustering along axis 1 in Figs. 9.6A and B appears to be related to species richness. Units G1, C2, A1 and M1 of the free-living assemblages contained fewer than seven species. Units C1, G2, and C9 of the mixed shape assemblages contained fewer than 18 species.

While there was no apparent clustering by stratigraphic position within the free-living assemblages (Figs. 9.5, 9.6A), the mixed shape assemblages did show some clustering by stratigraphic position (Figs. 9.5, 9.6B). To further assess the stability of these two assemblages types over the 3 my interval, the Bray- Curtis similarity coefficient between all locality-group pairs was plotted over the estimated age difference between locality-groups (Fig. 9.6C–D). Mantel tests comparing the similarity matrices for either free-living or mixed-shape assemblages with matrices of the estimated age differences between assemblages did not indicate there to be a strong decrease in faunal similarity between assemblages of increasingly disparate ages (free-living $r = -0.114$ $p = 0.212$; mixed-shape $r = -0.212$ $p = 0.075$).

Rarefaction curves for each of the three formations suggested increased diversity within the Gurabo and Mao Formations compared to the Cercado Formation (Fig. 9.1). Rarefaction curves generated independently for the free-living and mixed-shape assemblages suggest that most of the diversity within the Cibao Valley sequences is associated with the mixed-shape assemblages (Fig. 9.7a). Free-living assemblages level off around 40 species while the mixed shape assemblages level off near 100 species. The rarefaction curves for each lithostratigraphic unit also reflect these overall trends. Mixed-shape lithostratigraphic units from the Gurabo and Mao Formations in general level off between 50 and 70 species while the mixed-shape

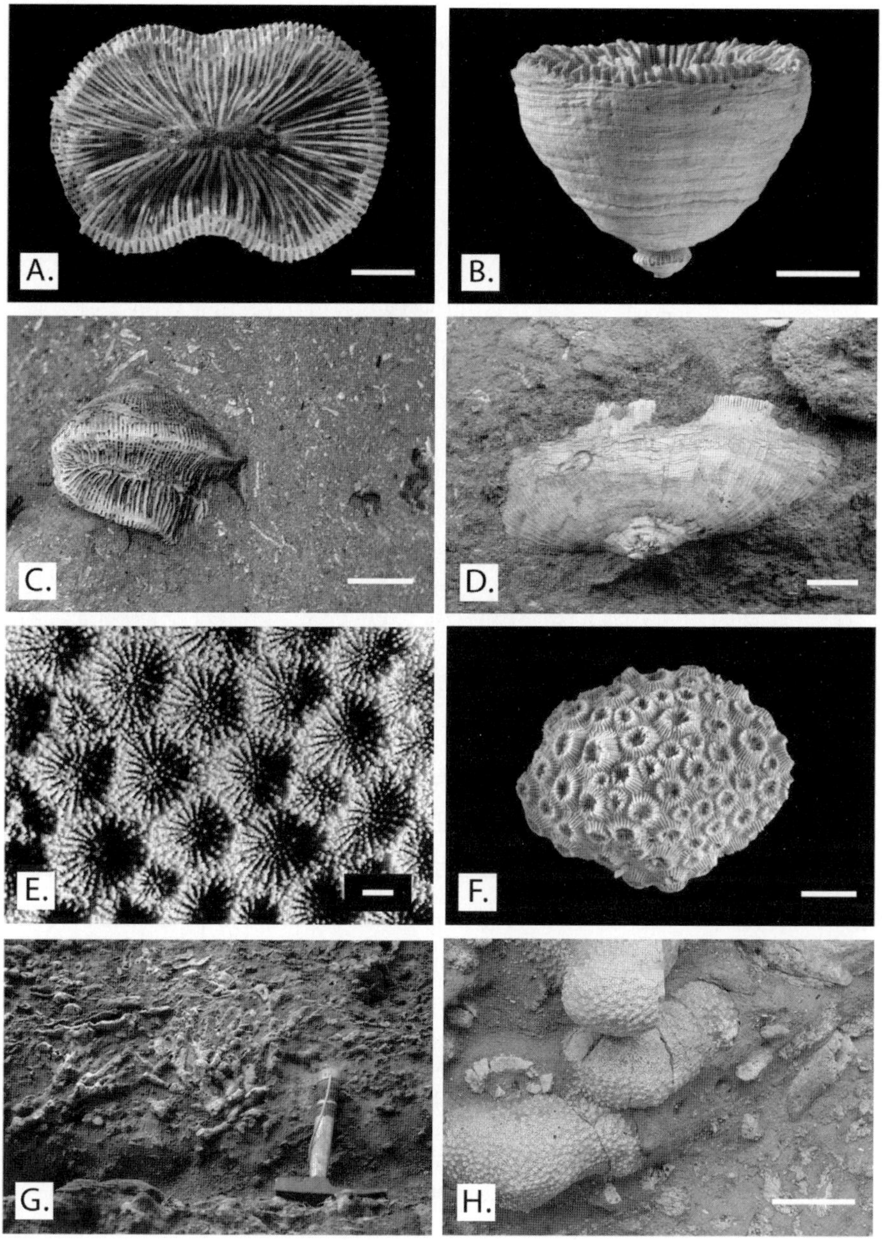

Fig. 9.4 Outcrop photos and calical surfaces of representative corals collected from the northern Dominican Republic. Free-living coral assemblages: (A) *Trachyphyllia bilobata*, NMB D5905, Cercado Fm. (NMB 15899), scale bar is 10 mm. (B) *Antillia dentata*, Gurabo Fm. (NMB 15867), scale bar is 10 mm. (C) *Manicina grandis*, Cercado Fm. (Arroyo Bellaco), scale bar 10 mm. (D) *Placocyathus variabilis*, Gurabo Fm. (NMB 15888), scale bar is 10 mm. Mixed shape coral assemblages. (E) *Goniopora hilli*, NMB D5853, Gurabo Fm. (NMB 15861), scale bar 1 mm. (F) *Montastraea brevis*, NMB D5586, Gurabo Fm. (NMB 15850), scale bar 10 mm. (G) *Stylophora affinis*, Gurabo Fm. (NMB 16822). (H) *Pocillopora crassoramosa*, Cercado Fm (Arroyo Bellaco Bel-8), scale bar 5 cm

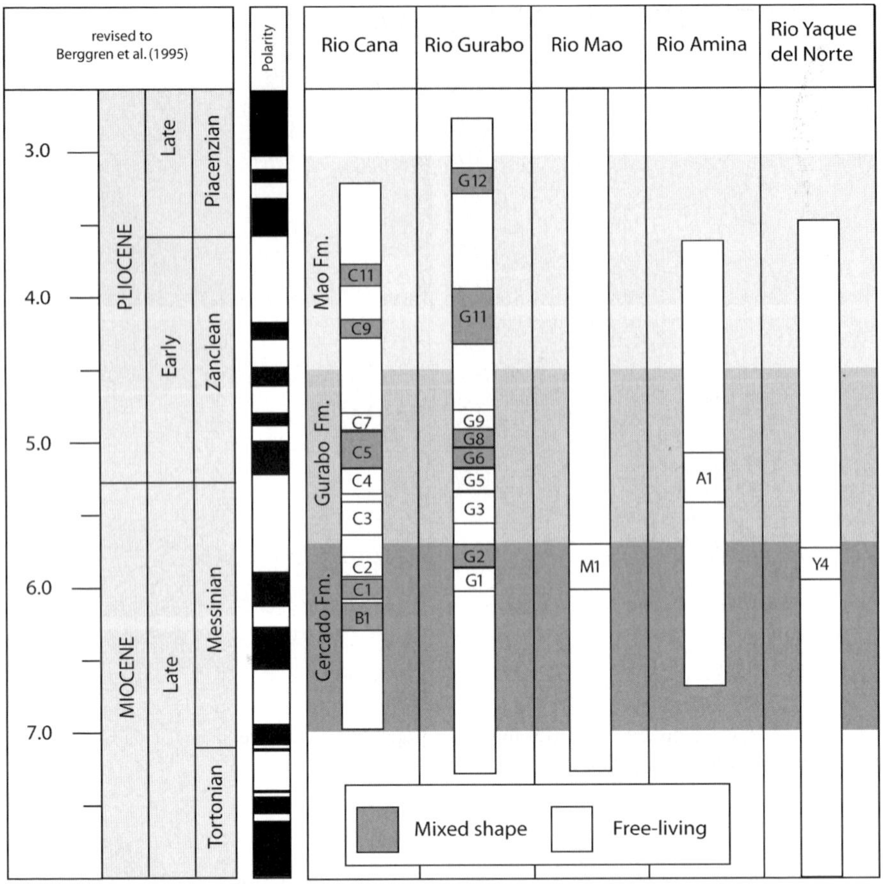

Fig. 9.5 Stratigraphic range chart showing the distribution of mixed-shape and free-living coral assemblages from Fig. 9.3

assemblages from the Cercado Formation (C1, B1) appear to level off below 40 species (Fig. 9.7b). A similar pattern was found for the free-living assemblages. Rarefaction curves of free-living assemblages C1, G1, and M1 from the Cercado Formation, all lie below the very consistent trends of free-living assemblages from the Gurabo Formation that suggest a levelling off at around 20 species (Fig. 9.7c).

9.3.3 *Transect Analyses*

A total of 27 20-m line transects were collected from three localities, 10 from Arroyo Bellaco, 10 from Cañada de Zamba, and 7 from the Cana Gorge. The

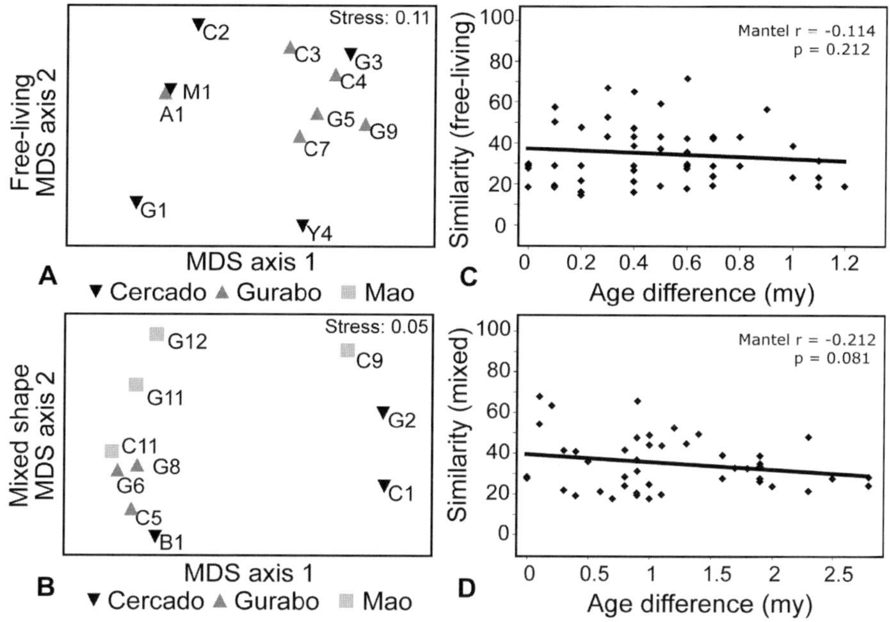

Fig. 9.6 Analyses of coral presence/absence data within 21 litostratigraphic units. (A, B) Multidimensional scaling (MDS) plots showing the relatedness of samples within the free-living (A) and mixed shape (B) clusters of Fig. 9.3 (C, D) Plots of the Bray-Curtis similarity coefficient of free-living (C) and mixed shape (D) lithostratigraphic unit pairs versus the estimated age difference between lithostatigraphic unit pairs

number of transects was determined by the amount of available outcrop at each locality. Again, the species-sample data matrix was analyzed using the Bray-Curtis similarity coefficient, one-way analysis of similarity (ANOSIM), and similarity percentages (SIMPER) (Clarke and Warwick, 2001).

9.3.3.1 Arroyo Bellaco

The reef at Arroyo Bellaco is one of the most well developed reefs of the northern Dominican Republic. Exposed approximately 2.5 km upstream of the confluence of Arroyo Bellaco with the Río Cana (Fig. 9.1 from Klaus and Budd, 2003), the reefs lie within lithostratigraphic unit B1 (6.2 Ma) of the Cercado Formation, approximately 80 m below the contact with the overlying Gurabo Formation. The excellent exposures of the approximately 19 m thick reefal section reveal a well-developed coral framework with a predominately siliciclastic matrix.

Qualitatively, four different reef zones can be recognized in the Arroyo Bellaco sequence: (1) A fine-branching thicket zone, (2) a thick-branching *Stylophora/Pocillopora* zone, (3) a massive head coral zone, and (4) a mixed free-living/small head coral zone. To characterize the community structure and zonation of the

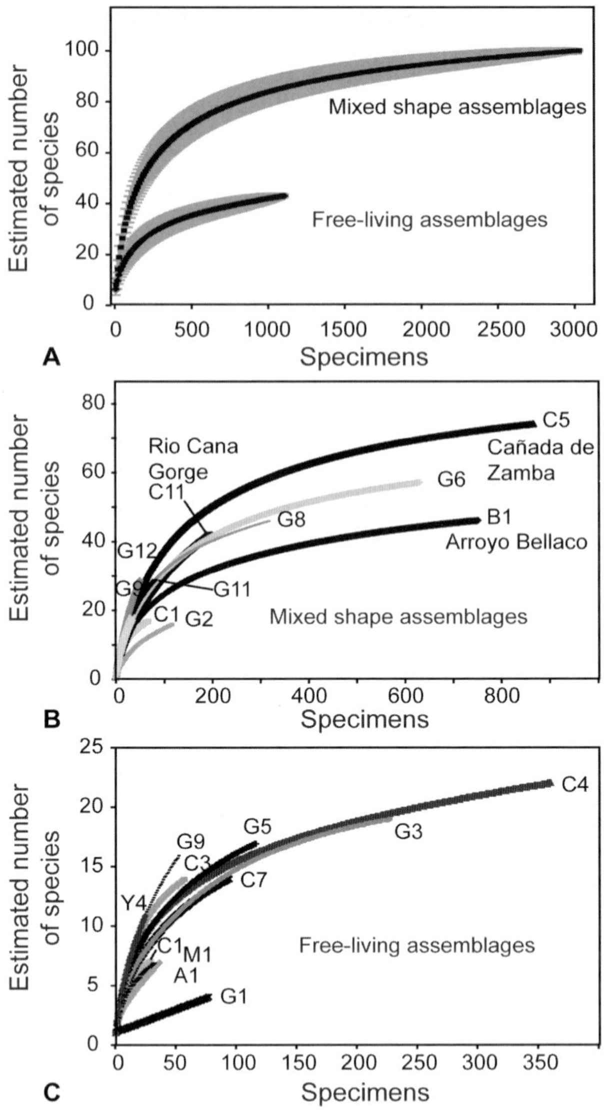

Fig. 9.7 Rarefaction curves of coral assemblage data. (A) Rarefaction curves generated independently from free-living and mixed-shape assemblages. (B) Individual rarefaction curves for each mixed-shape lithostratigraphic unit. (C) Individual rarefaction curves for each free-living lithostratigraphic unit

Arroyo Bellaco reef, 10 transect samples were collected at well-exposed outcrops displaying *in situ* coral assemblages: two within the fine-branching thicket zone, four within the thick-branching *Stylophora/Pocillopora* zone, and four within the massive head coral zone. The mixed free-living/small head coral zone is a component of ongoing investigations, and will not be discussed here.

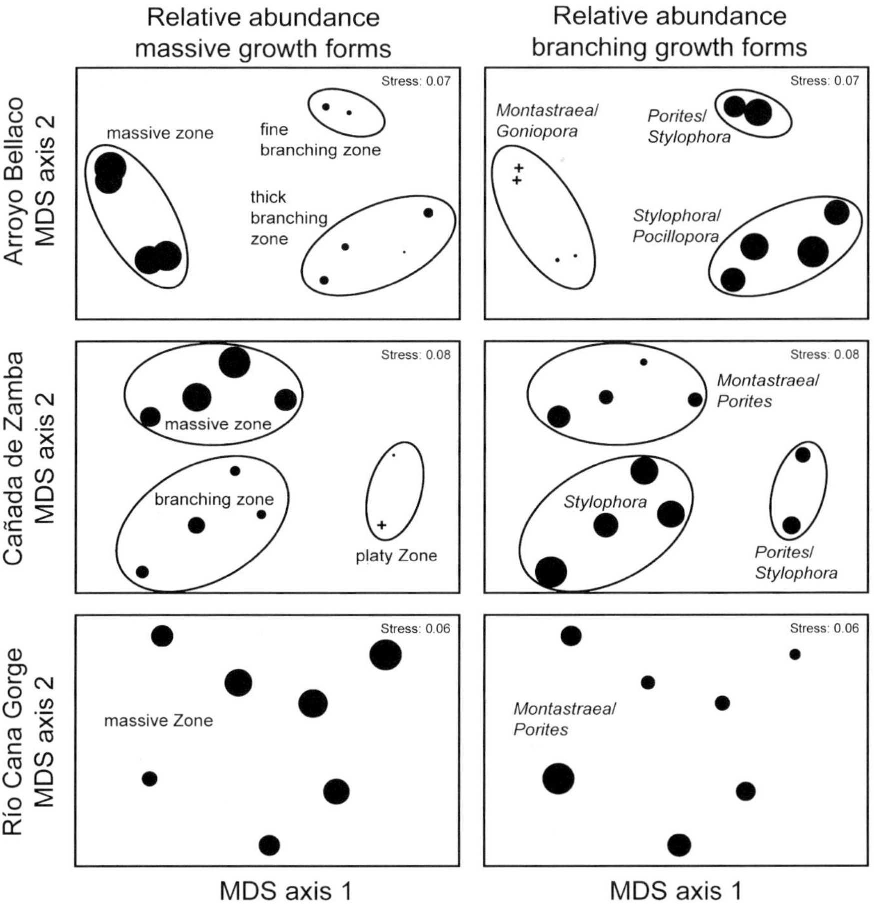

Fig. 9.8 Ordination of coral relative abundance data within 27 transect samples collected at Arroyo Bellaco, Cañada de Zamba, and Cana Gorge. The two multidimensional scaling (MDS) plots for each reef sequence are based on the same ordination. The circles in the three MDS plots on the left represent the relative abundance of massive growth morphologies within each reef sequence. The circles in the three MDS plots on the right represent the relative abundance of branching morphologies within each reef sequence. The (+) symbol marks samples with either no massive or branching corals. Dominant taxa in each cluster are listed on the right

Multidimensional scaling (MDS) ordination of the Arroyo Bellaco occurrence matrix (10 samples X 28 species) reveals three clusters of transect samples (Fig. 9.8A, B). This interpretation is supported by one-way analysis of similarity ($R = 0.952$, $p < 0.001$). SIMPER analysis reveals transect samples collected within the fine-branching zone can be characterized by small branching corals (*Porites portoricensis* 41.2%, *Stylophora minor* 10.0%, *Stylophora affinis*

10.0%) as well as relatively high numbers of the platy coral *Undaria agaricities* (11.2%) and the ubiquitous massive coral *Montastraea limbata* (5.0%). The thick branching zone was characterized by very large branching colonies of *Stylophora affinis* (27.6%) and *Pocillopora crassaramosa* (17.3%) as well as the smaller branching coral *P. portoricensis* (20.9%) and thin columns of *M. limbata* (8.7%). The massive coral zone was found to be dominated by *M. limbata* (33.9%), *Goniopora imperatoris* (12.7%), *Goniopora hilli* (9.8%), *Stephanocoenia spongiformis* (5.0%) and *Porites waylandi* (4.9%). The average abundance of all identified species within each reef zone is detailed in Appendix 2. These studies are consistent with previous surveys of the Arroyo Bellaco reef (Evans, 1986; Klaus and Budd, 2003).

9.3.3.2 Cañada de Zamba

The reefs at Cañada de Zamba are part of a topographically distinct ridge that runs parallel to and south of the larger Zamba Hills trend constructed by the Mao Adentro Limestone. The reefs lie within lithostratigraphic unit C5 (5.1 Ma) of the Gurabo Formation, approximately 150 m below the contact with the overlying Mao Formation, and 250 m below the base of the Mao Adentro Limestone Member of the Mao Formation. The coral assemblages at Cañada de Zamba show considerable variation within fairly narrow stratigraphic and geographic distances. These assemblages vary from interbedded limestones and siliciclastics, to massive limestones with fewer siliciclastics. To characterize the composition of the most prominent and accessible reef facies, ten transects were collected from the area near the confluence of the Cana and Cañada de Zamba rivers.

The results of an MDS ordination of the Cañada de Zamba occurrence matrix (10 samples X 30 species) reveals three distinct clusters of samples; a massive coral zone, a branching coral zone, and a platy coral zone (Fig. 9.8C, D). This result is further supported by one-way analysis of similarity ($R = 0.925$, $p < 0.005$). SIMPER analysis reveals that the branching coral zone is dominated by the large branching coral *S. affinis* (35.9%) with additional branching colonies of *Stylophora granulata* (7.9%), *P. portoricensis* (4.3%) and *Porites baracoaensis* (3.1%). Other important species of this assemblage include the platy corals *U. agaricites* (7.8%) and *Porites macdonaldi* (6.3%) as well as the massive corals *P. waylandi* (8.5%), *M. limbata* (4.8%), *Diploria zambensis* (4.2%), and *Siderastrea siderea* (3.3%). The platy coral zone is dominated by very delicate plates of *P. macdonaldi* (50.0%), small branching colonies of *S. minor* (24.3%) and delicately branching colonies of *P. baracoensis* (9.7%). The massive coral zone is dominated by massive growth morphologies of *M. limbata* (25.3%), *P. waylandi* (13.6%) and *D. zambensis* (6.1%) as well as the branching corals *P. baracoensis* (9.8%), and *S. granulata* (5.0%). The average abundance of all identified species within each reef zone is detailed in Appendix 2.

9.3.3.3 Río Cana Gorge

The reefs of the Río Cana Gorge are exposed where the Río Cana cuts through the Zamba Hills trend of the Mao Adentro Limestone. The apparent thickness of the Mao Adentro Limestone has been estimated at 340 m and consists of massive limestones as well as thin calcareous siltstones alternating with coral-rich, flaggy limestones dipping at angles between 7 and 20 degrees (Evans, 1986; Saunders et al., 1986). While the base of the Mao Adentro Limestone has been shown to contain a very shallow-water foraminiferal fauna, the paleoenvironment of the 340 m limestone section is less well understood. While one sample taken in a softer clay revealed a foraminiferal fauna with a high ratio of planktic to benthic species, and ostracod occurrences suggesting deeper water depths (>100 m) (Saunders et al., 1986), the presence of upright corals indicate considerably shallower conditions during accumulation of much of the limestone (<30 m). The most accessible exposures are found within lithostratigraphic unit C11 (3.5 Ma) at the very top of the Mao Adentro Limestone at the northern end of the Río Cana Gorge. The dipping nature of the limestones in this region and number of in situ corals suggest coral growth on prograding clinoforms. The bedded nature of these coral limestones reflect high-frequency sea level cycles. To characterize the composition of the most prominent and accessible reef assemblages, seven transects were collected at the northern end of the Cana Gorge.

The results of an MDS ordination of the Cana Gorge occurrence matrix (7 samples X 28 species) suggests all transect samples to be similar in nature (Fig. 9.8E, F), and dominated by the massive corals *P. waylandi* (29.1%), *M. limbata* (14.7%), and *Montastraea cavernosa* (4.9%), as well as the branching corals *P. portoricensis* (12.6%), *P. baracoensis* (7.3%), *S. granulata* (9.0%), and *Caulastraea portoricensis* (4.9%).

9.3.3.4 Community Comparison

To assess the stability of more narrowly defined assemblages in these three reef sequences, MDS ordinations were performed using: (1) the samples containing predominately massive coral morphologies, and (2) the samples containing predominately branching coral morphologies. A total of 15 transect samples containing predominantly massive morphologies were compared; four from Arroyo Bellaco, four from Cañada de Zamba, and seven from the Cana Gorge. MDS ordination of these 15 samples shows a clear separation of the samples from each of the three reefs, and a clear gradation from the oldest (Arroyo Bellaco) to the youngest (Cana Gorge) reef (Fig. 9.9A). This result is supported by one-way analysis of similarity ($R = 0.664$, $p < 0.001$). Furthermore, a plot of the Bray-Curtis similarity coefficient calculated between all transect pairs of massive coral assemblages versus the estimated age difference between localities (Fig. 9.9C) and tests of mantel matrix correlation ($r = -0.814$, $p = 0.000$) indicate a strong negative correlation. SIMPER

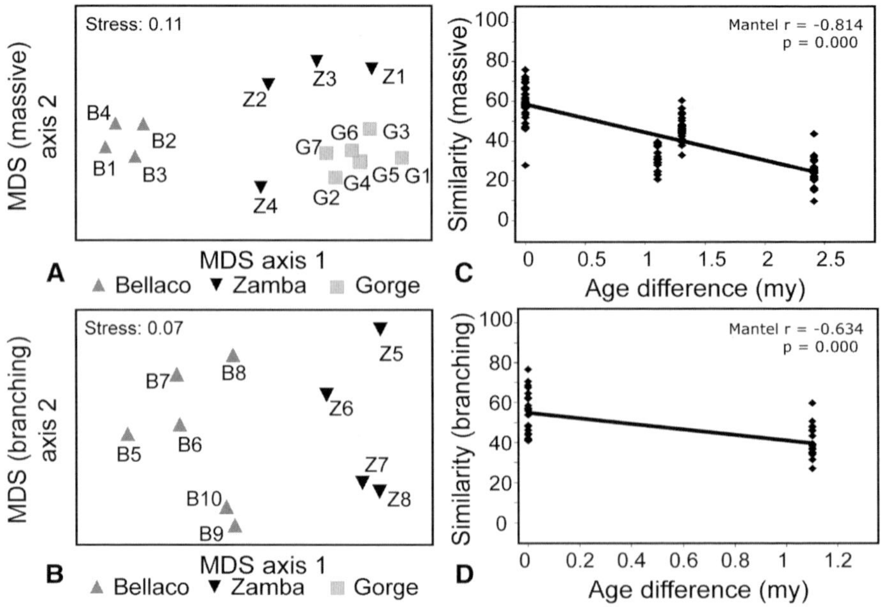

Fig. 9.9 Analyses of coral relative abundance data within 25 transect samples from massive and branching reef zones of Arroyo Bellao, Cañada de Zamba, and Cana Gorge. (A, B) Multidimensional scaling plots showing the relatedness of samples within the massive (A) and branching (B) clusters of Fig. (9.8). (C, D) Plots of the Bray-Curtis similarity coefficient of massive (C) and branching (D) transect pairs versus the estimated age difference between transect pairs

analysis reveals a decreasing trend in the relative abundance of the massive corals *G. imperatoris* (AB = 12.7%, Z = 0.6%, G = 0.0%), *G. hilli* (AB = 9.87%, Z = 0.6%, G = 0.0%), *Gardineroseris planulata* (AB = 3.1%, Z = 0.0%, G = 0.0%), and *Montastraea endothecata* (AB = 3.9%, Z = 3.0%, G = 0.7%); and an increasing trend in the relative abundance of *P. waylandi* (AB = 4.9%, Z = 13.6%, G = 29.1%) and *M. cavernosa* (AB = 0.0%, Z = 4.9%, G = 4.9%). The massive coral *D. zambensis* has greater abundance in Zamba (6.13%) compared to Arroyo Bellaco (0.0%) and the Cana Gorge (2.82%).

A total of 10 transect samples containing predominantly branching morphologies were compared; six from Arroyo Bellaco, and four from Cañada de Zamba. MDS ordination of these 10 samples shows a clear separation of the samples from each of the two reefs (Fig. 9.9B). This result is supported by one-way analysis of similarity ($R = 0.664$, $p < 0.001$). Furthermore, the mantel test of matrix correlation ($r = -0.634$, $p = 0.000$) and the plot of the Bray-Curtis similarity coefficient calculated between all transect pairs from branching coral assemblages versus the estimated age difference between localities indicate reduced similarity between versus within the two localities (Fig. 9.9D)

9.4 Discussion

9.4.1 Assemblage Analyses

Coral collections from the northern Dominican Republic have played an important role in deciphering Cenozoic coral systematics (Vaughan, 1919; Foster, 1986; Foster, 1987; Budd, 1991; Budd and Johnson, 1999a; Klaus and Budd, 2003) and evaluating trends in diversity through time (Budd et al., 1996; Budd and Johnson, 1999b; Budd, 2000). Here we provide a comprehensive assessment of the ecological context from which these collections are drawn and insights into community change leading up to the Plio-Pleistocene extinction event. In general, coral assemblages of the three formations can be subdivided into two distinct community types: (1) Free-living coral assemblages composed primarily of members of the genus *Antillia*, *Antillophyllia*, *Manicina*, *Meandrina*, *Placocyathus*, *Thysanus* and *Trachyphyllia*, and (2) mixed-shape assemblages of species from the genus *Porites*, *Stylophora*, *Montastraea*, *Stephanocoenia*, *Goniopora* and *Undaria*. Based on other environmental indicators (Saunders et al., 1986; Cheetham and Jackson, 1996; Costa et al., 2001), the free-living assemblages occur in more restricted or soft-bottom grass flat environments adjacent to the mixed-shape assemblages. Unlike modern grass flats, these Mio-Pliocene grass flats may have extended to depths as great as 20–30 m (Cheetham and Jackson, 1996; Budd et al., 1996).

Three intervals of more significant reef development occur once within each of the Cercado (Arroyo Bellaco), Gurabo (Cañada de Zamba) and Mao (Cana Gorge) Formations. The reefs of Cañada de Zamba and the Cana Gorge are exposed within two east-west trending limestone ridges easily traced for at least 91 km on the geologic map of Antonini (1979) (Fig. 2.4 McNeill et al., Chapter 2, this volume). The lateral extent of the Arroyo Bellaco reef is less well understood, but in all likelihood corresponds at the very least to a smaller reef build-up exposed at 120 m in the Río Gurabo section. Given the well-constrained ages of the Cibao Valley deposits, these reef intervals can be correlated to regional seal level and paleoceanographic events (Fig. 9.10). Reef growth at Arroyo Bellaco occurred immediately after a prolonged regression (~8–6.4 Ma), when sea level reversed during a short highstand (~6.4–6.0 Ma). Evidence also suggests that the Arroyo Bellaco reef formed during a period of decreased upwelling intensity between 6.2 and 5.8 Ma (Maier et al. submitted). This event was first recognized by Keigwin (1979) in the $\delta^{13}C$ record of deep-sea sediments of the Indo-Pacific region and later by Shackleton and Hall (1997) based on a 1‰ shift toward less positive values in $\delta^{13}C$ at Ocean Drilling Program (ODP) Site 926 on the Ceara Rise. Reef development within the Gurabo Formation, best developed near the Cañada de Zamba confluence with the Río Cana, occurs just prior to the early Pliocene warming and highstand (Zanclean flood) represented by a deepening of water depths at approximately 400 m in both the Río Gurabo and Río Cana sections (5.0–4.5 Ma). Lastly, reef development within the Mao Adentro

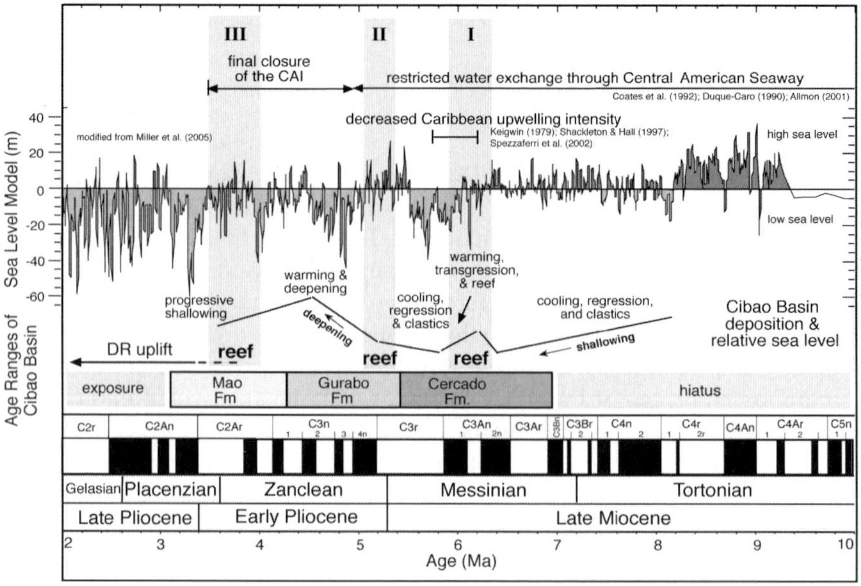

Fig. 9.10 Summary of reef development and paleoceanographic events in the Miocene and Pliocene of the Domincian Republic. A late Messinian reef is coincident with a brief highstand during an overall drop in sea level and period of reduced nutrient flux. A second reef occurs immediately preceding the early Pliocene transgression (Zanclean flood) and associated deep-water massive siltstones of the Gurabo Formation. Reef development in the Mao Formation occurs during a period of progressive shallowing, just prior to uplift and exposure of the Cibao Valley sections and final closure of the Central American Seaway

Limestone occurs during a period of progressive sea level shallowing prior to uplift and exposure of the Cibao Valley approximately 3.0 Ma.

We used three approaches to assess the extent of community change through the section: (1) the extent of faunal persistence, (2) the relative similarity of 21 lithostratigraphic units based on coral presence/absence data, and (3) the relative similarity of 27 20-m line transects collected from reef zones exposed at Arroyo Bellaco, Cañada de Zamba, and the Cana Gorge.

In their analyses of Middle Devonian invertebrate communities, Brett and Baird (1995) proposed a cut-off of 60% faunal persistence over intervals of 3–7 million years for determining community stasis. By this standard, the 61% faunal persistence from the Cercado to Mao Formation is characteristic of community stasis. Similarly, analyses of the 21 lithostratigraphic units did not detect a strong signal of community change (Fig. 9.6A–D). Despite being based on a large number of occurrences, analyses of presence/absence data within the broadly defined lithostratigraphic units is a fairly blunt tool for assessing community change. This technique only captures originations and extinctions, and gives equal weight to rare and abundant taxa. Furthermore, sampling was uneven which would contribute to significant noise in the results.

Although the line transect data reflects a stronger trend of community change, the observed differences are undoubtedly, to some degree, due to the environment. Evidence from ostracodes (Bold, 1988), foraminifera (Saunders et al., 1986) and observations of platy coral growth morphologies within the genus *Montastraea* suggest the corals of Cana Gorge inhabited slightly deeper reef slope environments. Within our transect surveys we found the dominant coral species *G. imperatoris*, *G. hilli*, and *M. endothecata* being replaced by *P. waylandi*, and *M. cavernosa*. It is difficult to determine whether the decreasing abundance of the two *Goniopora* species is due to environmental differences or to evolutionary changes in the community structure of Caribbean corals. Modern *Goniopora* in the Western Pacific are most common within shallow turbid environments (Veron, 2000), suggesting the decreasing abundance of *Goniopora* could be associated with deeper water depths at Cana Gorge. On the other hand, while *Goniopora* can be found within late Pliocene deposits of the Bahamas (1.9–1.8 Ma) (Budd and Manfrino, 1999) and Cuba (3.5–1.6 Ma) (Vaughan, 1919), they are rarely common after about 3 Ma. The decreasing abundance in the transect surveys could be mirroring this regional trend. The decreasing abundance of *M. endothecata* is less easily explained by a shift in depositional environment because it is thought to be common within deeper reef environments (Budd et al., 1996). Furthermore, the global last occurrence of *M. endothecata* within the Mao Formation suggests it could have been declining on Caribbean reefs at this time. The increasing abundance of *M. cavernosa* and *P. waylandi* is interesting in that they both have occurrences as far back as the middle Miocene and co-occurred with the two *Goniopora* species and *M. endothecata* until the late Pliocene. *Montastraea cavernosa* is a common species on Recent Caribbean reefs. *Porites waylandi* is last found within the Lomas del Mar Formation of Costa Rica (1.9–1.5 Ma) (Budd et al., 1999).

The three approaches used here reflect trade offs in scale and data resolution. The degree of faunal persistence records the timing and extent of species originations and extinctions in the regional species pool through the studied sequence. However, this approach provides no information on the co-occurrence of taxa and potential ecological or environmental interactions. The analysis of the 21 lithostratigraphic units overcomes this shortcoming by comparing assemblages of broadly defined co-occurring taxa. This scale of analysis is important in recognizing broad ecological patterns (free-living vs. mixed shape), and characterizing changes in the regional paleoecological setting. An adequate documentation of community change through time however, requires relative abundance data of species occurrences from narrowly defined environmental settings. This is a considerable challenge in fossil reef studies given the high variability of reef coral assemblages between adjacent reef zones. This obstacle is overcome in the Pleistocene to Recent fossil record through the use of modern indicator species such as *Acropora palmata*, with narrow ecological ranges (Pandolfi and Jackson, 2001). While we apply a similar approach in the transect analyses presented here by comparing coral growth morphologies, the ability to constrain reef zones in the Miocene and Pliocene is less certain than in the Pleistocene to Recent, and undoubtedly contributes to increased variability between reef assemblages.

An additional shortcoming in our characterization of community change was taxonomic resolution. For example, Budd and Klaus (this volume) identified nine different species of *Montastraea "annularis"* –like corals within the Dominican Republic sequences. In the current study these species were all identified as *M. limbata* (Duncan, 1863). Given the short durations of the nine species identified by Budd and Klaus (Chapter 5, this volume), 100 kyr to 2.73 myr, this level of taxonomic scrutiny would likely increase the signal of community change in each of the three methods we used to assess community change. This is also the case in *M. "cavernosa"* – like corals (Schultz and Budd, Chapter 7, this volume). Thus, while we found the impact of originations and extinctions to have only a moderate impact on reef communities through the sequence, this does not take into account originations and extinctions within large and ecologically dominant species complexes. A significant driver of community change may be the changing ecological characteristics within these complexes. Additional studies are needed to integrate detailed taxonomic studies such as Budd and Klaus (Chapter 5, this volume) and Schultz and Budd (Chapter 7, this volume) into studies of community change.

9.4.2 Future Work

The trend from *G. imperatoris*, *G. hilli*, and *M. endothecata* dominance to *P. waylandi* and *M. cavernosa* dominance is similar to the slightly younger transition from *Stylophora* to *Acropora* dominated reefs. The two modern dominants *Acropora cervicornis* and *Acropora palmata* first appear at 5.9–4.6 Ma and 3.2–2.9 Ma, respectively (McNeill et al., 1997; Budd and McNeill, 1998; Budd et al., 1998; Budd et al., 1999). These species co-occur with as many as seven different species of *Stylophora* for as much as 4 my until the last Caribbean occurrence of *Stylophora* in the early Pleistocene (Budd et al., 1998; Budd and Manfrino, 1999).

The coral collections from the Cercado, Gurabo, and Mao Formations contain only a single occurrence of the genus *Acropora*; a specimen of *Acropora saludensis* was collected within lithostratigraphic unit A1 (Table 9.1). This is of considerable contrast to the over 290 specimens of the genus *Stylophora* collected. Based primarily on the taxonomic interpretations of Vaughan (1919), there are six species of *Stylophora* found within these collections.

One limitation in understanding this important transition is our ecological understanding of Mio-Pliocene *Stylophora* species. The wide range of environments and excellent preservation in the Cibao Valley sections makes it an ideal setting for future studies of this genus. By revising the systematics of this group using extensive collections made within the context of narrowly defined transect samples, we hope to better understand the environmental constraints on this group, and the changing ecologic role of the genus in the preliminary stages of faunal turnover. Comparisons with contemporaneous and younger deposits in Curaçao, the Bahamas and Panama should provide a regional perspective to the timing and ecologic factors associated with the extinction of *Stylophora* and increased dominance of *Acropora*.

In addition to the transition from *Stylophora* to *Acropora* dominance, Plio-Pleistocene faunal turnover is characterized by high extinction rates among free-living coral species. Twenty free-living coral species were identified in the present study. In contrast, only two free-living coral species are present in the Caribbean and western Atlantic today, *Manicina areolata*, and *Meandrina brasiliensis*. The free-living mode of existence allows corals to thrive in soft-substrate environments typically unsuitable for most reef corals. Furthermore, while free-living corals are not by definition solitary, they typically encompass very few individual polyps and have population dynamics more like nonclonal animals than typical colonial reef corals (Johnson, 1992). Colonial species, especially the best space competitors, are typically more limited in range of depth and substrate compared to more solitary species (Jackson, 1977). Thus it is somewhat surprising that free-living corals would be among the hardest hit during Plio-Pleistocene environmental perturbation. Much remains to be learned about the ecology of pre-turnover free-living coral communities and environmental impacts to their primary grass flat habitats during the Plio-Pleistocene. The extensive free-living coral assemblages of the Dominican Republic provide the best setting in which to study these communities. By integrating a modified transecting technique to obtain free-living species abundance data from narrowly defined assemblages with independent environmental data obtained from benthic foraminifera, we are currently working to better our understanding of the environmental constraints on these communities. We hope to use this information to determine how these communities responded to the protracted period of environmental perturbation during which faunal turnover took place and after which most free-living species were extinct.

Acknowledgments We thank Jörn Geister, Peter Jung, and John Saunders for providing the initial collections from the Dominican Republic. Furthermore, we acknowledge the collections and taxonomic work of Tom Stemann as well as additional help from Troy Fadiga, Jessica Jacobs, and Katie Maier. Tiffany Adrain provided assistance with specimen curation. This research was supported by funds from the National Science Foundation (EAR 0445789 to A.F. Budd, and EAR 0446768 to D.F. McNeill).

Appendix 1

Species Occurrence Data from Mixed and Free-living Coral Assemblages. Colony shape codes are B = branching, P = platy, M = massive, and F = freeliving. Avg. Occ. is the probability of finding a given species within a free-living or mixed shape assemblage. Avg. Diss. is the average contribution of a species to the overall dissimilarity between the free-living and mixed shape communities (81.46%). #L-S Units is the number of free-living, mixed or total lithostratigraphic units a coral species is found within. Species with first occurrences in the study sequence are marked[F], species with last occurrences are marked[L]

Species	Colony shape	Avg. occ. free-living	Avg. occ. mixed	Avg. diss	# L-S units free-living	# L-S units mixed	# L-S units total
Acropora saludensis	B	0.09	0	0.65	1	0	1
Agaricia fragilis[F]	P	0	0.1	0.07	0	1	1
Agaricia lamarcki	P	0	0.5	0.49	0	5	5
Agaricia sp.A[F,L]	P	0	0.3	0.43	0	3	3
Agaricia undata	P	0	0.3	0.33	0	3	3
Antillia dentata[F]	F	0.64	0.4	2.79	7	4	11
Antillophyllia sawkinsi	M	0.45	0.4	2.17	5	4	9
Caulastraea portoricensis	B	0	0.4	0.53	0	4	4
Cladocora sp.A[F,L]	B	0	0.1	0.17	0	1	1
Colpohyllia natans	M	0	0.3	0.49	0	3	3
Dichocoenia caloosahatcheensis	M	0	0.1	0.17	0	1	1
Dichocoenia sp.A[F,L]	M	0	0.1	0.11	0	1	1
Dichocoenia sp.B[F,L]	M	0	0.1	0.11	0	1	1
Dichocoenia sp.C[F,L]	M	0	0.1	0.12	0	1	1
Dichocoenia tuberosa	M	0	0.5	0.62	0	5	5
Diploria clivosa[F]	M	0	0.1	0.11	0	1	1
Diploria strigosa[F]	M	0	0.3	0.3	0	3	3
Diploria zambensis[F]	M	0	0.5	0.75	0	5	5
Eusmilia sp.A[F]	B	0	0.4	0.38	0	4	4
Eusmilia fastigiata	B	0	0.1	0.12	0	1	1
Favia dominicensis[F,L]	M	0	0.3	0.3	0	3	3
Favia maoadentrensis[F,L]	M	0	0.2	0.41	0	2	2
Favia sp.A[F,L]	M	0	0.1	0.11	0	1	1
Favia vokesae[F]	M	0	0.1	0.29	0	1	1
Galaxea excelsa[L]	M	0	0.2	0.38	0	2	2
Gardineroseris planulata	M	0	0.4	0.54	0	4	4
Goniopora calhounensis[L]	M	0	0.2	0.39	0	2	2
Goniopora hilli	M	0.09	0.4	0.54	1	4	5
Goniopora imperatoris	M	0	0.4	0.75	0	4	4
Hadrophyllia saundersi[F]	M	0.09	0.4	0.53	1	4	5
Isophyllia sp.A[L]	M	0	0.5	0.78	0	5	5

(continued)

Appendix 1 (continued)

Species	Colony shape	Avg. occ. free-living	Avg. occ. mixed	Avg. diss	# L-S units free-living	# L-S units mixed	# L-S units total
Leptoseris cailleti[F]	B	0.09	0.3	0.5	1	3	4
Leptoseris gardineri[L]	B	0.09	0.3	0.5	1	3	4
Leptoseris glabra	P	0.18	0.4	1.25	2	4	6
Leptoseris sp.A[F,L]	P	0.09	0.3	0.48	1	3	4
Leptoseris sp.B[F,L]	B	0	0.2	0.18	0	2	2
Leptoseris sp.C[F,L]	B	0	0.1	0.07	0	1	1
Madracis cf.herricki[F,L]	B	0	0.5	0.61	0	5	5
Madracis decactis	M	0	0.4	0.38	0	4	4
Madracis decaseptata[L]	B	0	0.4	0.44	0	4	4
Madracis mirabilis	B	0.09	0.4	0.63	1	4	5
Madracis sp.A[F,L]	B	0	0.4	0.47	0	4	4
Manicina aff.mayori[F,L]	F	0.09	0.3	0.63	1	3	4
Manicina geisteri[F]	F	0.55	0.5	2.29	6	5	11
Manicina grandis[F]	F	0.91	0.8	3.65	10	8	18
Manicina jungi[F]	F	0.55	0.3	1.66	6	3	9
Manicina pliocenica	F	0	0.2	0.19	0	2	2
Manicina puntagordensis[F]	F	0.27	0.2	1.19	3	2	5
Meandrina brasiliensis[F]	F	0.36	0.5	1.71	4	5	9
Meandrina meandrites	M	0	0.1	0.07	0	1	1
Montastraea-I brevis[F]	M	0	0.4	0.37	0	4	4
Montastraea-I limbata	M	0.09	1	1.66	1	10	11
Montastraea-I trinitatis	M	0.09	0	0.41	1	0	1
Montastraea-II canalis	M	0	0.5	0.71	0	5	5
Montastraea-II cavernosa[F]	M	0	0.1	0.07	0	1	1
Montastraea-II cavernosa-2	M	0	0.5	0.7	0	5	5
Montastraea-II cylindrical	M	0	0.6	1.01	0	6	6
Montastraea-II endothecata	M	0	0.3	0.26	0	3	3
Mussa angulosa[F]	B	0	0.4	0.38	0	4	4
Mussismilia aff.hartti[F]	M	0	0.4	0.4	0	4	4
Mussismilia sp.A[F,L]	M	0	0.4	0.38	0	4	4
Mussismilia sp.B[F,L]	M	0	0.1	0.12	0	1	1
Mycetophyllia bullbrooki[L]	M	0	0.1	0.17	0	1	1
Pavona sp.A[F]	B	0	0.5	0.62	0	5	5

(continued)

Appendix 1 (continued)

Species	Colony shape	Avg. occ. free-living	Avg. occ. mixed	Avg. diss	# L-S units free-living	# L-S units mixed	# L-S units total
Pavona sp.B[F,L]	B	0.18	0.5	1.01	2	5	7
Pavona trinitatis	B	0.45	0.2	1.43	0	1	1
Placocyathus alveolus[F]	F	0.27	0.4	1.07	3	4	7
Placocyathus barretti[F]	F	0.27	0.1	1.26	3	1	4
Placocyathus costatus[F]	F	0.55	0.6	1.62	6	6	12
Placocyathus trinitatis	F	0.09	0	0.41	5	2	7
Placocyathus variabilis	F	0.91	0.9	2.91	10	9	19
Pocillopora baracoaensis[F]	B	0.27	0.8	1.39	0	2	2
Pocillopora crassoramosa	B	0.18	0.5	1.79	2	5	7
Porites-I macdonaldi[L]	P	0	0.6	1.03	0	6	6
Porites-I portoricensis	B	0	1	1.72	0	10	10
Porites-I sp.A[F,L]	M	0	0.1	0.07	0	1	1
Porites-I waylandi	M	0.27	0.7	1.32	3	7	10
Porites-II baracoaensis	B	0	0.2	0.4	3	8	11
Porites-II convivatoris[F,L]	M	0.09	0.2	0.45	1	2	3
Psammocora trinitatis	M	0	0.1	0.11	1	0	1
Scolymia cubensis	F	0.45	0.6	2.4	5	6	11
Siderastrea mendenhalli	M	0.09	0.3	0.68	1	3	4
Siderastrea radians[F]	M	0	0.1	0.11	0	1	1
Siderastrea siderea	M	0.18	0.4	0.87	2	4	6
Siderastrea silencensis	M	0	0.2	0.16	0	2	2
Solenastrea bournoni	M	0.18	0.9	1.38	2	9	11
Solenastrea hyades	M	0	0.2	0.61	0	2	2
Stephanocoenia duncani	M	0	0.4	0.54	0	4	4
Stephanocoenia spongiformis	M	0	0.5	0.66	0	5	5
Stylophora affinis	B	0.18	0.5	0.81	2	5	7
Stylophora affinis-2[F]	B	0	0.3	0.28	0	3	3
Stylophora granulata	B	0.27	0.7	1.57	3	7	0
Stylophora imperatoris	B	0	0.2	0.23	0	2	2
Stylophora minor	B	0.55	0.8	1.68	6	8	14
Stylophora monticulosa	B	0.09	0.8	1.43	1	8	9

(continued)

Appendix 1 (continued)

Species	Colony shape	Avg. occ. free-living	Avg. occ. mixed	Avg. diss	# L-S units free-living	# L-S units mixed	# L-S units total
Thysanus corbicula	F	0.09	0	0.65	1	0	1
Thysanus excentricus[F]	F	0.27	0.4	1.03	3	4	7
Thysanus navicula[F]	F	0.36	0.2	1.21	4	2	6
Trachyphyllia bilobata	F	1	0.8	4.06	1	8	19
Trachyphyllia sp.A[F]	F	0.18	0.3	0.71	2	3	5
Undaria agaricites[F]	P	0.09	0.8	1.32	1	8	9
Undaria crassa[F]	M	0	0.6	0.77	0	6	6
Undaria sp.A[F]	P	0	0.2	0.16	0	2	2

Appendix 2

Average % abundance of coral species identified in transect sample groups

	Arroyo Bellaco			Cañada de Zamba			Cana Gorge
Species	Massive	Thick branching	Thin branching	Massive	Branching	Platy	Massive
Agaricia fragilis	0	0	0	0.61	0	0	0
Agaricia lamarcki	*0.61*	*2.75*	*1.25*	*3.06*	*0*	*1.22*	*0*
Agaricia undata	*0*	*0*	*0*	*0*	*0*	*1.22*	*0*
Caulastraea portoricensis	*0*	*0*	*0*	*1.22*	*0.63*	*0*	*4.9*
Colophyllia natans	*1.29*	*0*	*0*	*0*	*0*	*0*	*0*
Dichocoenia sp.A	*0.61*	*0*	*1.25*	*0*	*0*	*0*	*0*
Dichocoenia sp.B	*0.68*	*0*	*0*	*0*	*0*	*0*	*0*
Diploria zambensis	*0*	*1.88*	*0*	*6.13*	*4.26*	*0*	*2.82*
Favia dominicensis	*0.61*	*2.08*	*1.25*	*0.63*	*0.78*	*0*	*0*
Favia maoadentrensis	*0*	*00*	*0*	*0*	*0.63*	*0*	*1.42*
Gardineroseris planulata	*3.09*	*0*	*0*	*0*	*0*	*0*	*0*
Goniopora hilli	*9.87*	*0*	*0*	*0.63*	*0*	*0*	*0*
Goniopora imperatoris	*12.7*	*3.43*	*2.5*	*0.61*	*0*	*0*	*0*
Hadrophyllia saundersi	*0*	*0*	*0*	*0*	*0*	*0*	*0*
Isophyllia sp.A	*0*	*0*	*0*	*1.88*	*0.78*	*0*	*1.05*
Leptoseris gardineri	*0*	*0*	*0*	*0*	*0.63*	*0*	*0*
Madracis decactis	*0*	*0*	*0*	*0*	*1.39*	*0*	*0*
Madracis decaseptata	*0*	*0*	*0*	*1.88*	*0*	*0*	*0*
Madracis mirabilis	*0*	*3.96*	*1.25*	*0*	*0.78*	*0*	*0*
Madracis sp.A	*0*	*0.63*	*1.25*	*0*	*0.78*	*0*	*0*
Manicina grandis	*0*	*0*	*0*	*1.25*	*0*	*0*	*0*
Montastraea-I limbata	*33.96*	*8.71*	*5.00*	*25.38*	*4.88*	*0*	*14.77*

(continued)

Appendix 2 (continued)

	Arroyo Bellaco			Cañada de Zamba			Cana Gorge
Species	Massive	Thick branching	Thin branching	Massive	Branching	Platy	Massive
Montastraea-I trinitatis	*0*	*0*	*0*	*0*	*0*	*0*	*0*
Montastraea-II canalis	*1.88*	*0*	*0*	*1.22*	*0*	*0*	*0.70*
Montastraea-II cavernosa	*0*	*0*	*0*	*4.91*	*1.39*	*0*	*4.93*
Montastraea-II cylindrica	*0*	*0*	*0*	*0*	*0*	*0*	*1.75*
Montastraea-II endothecata	*3.87*	*1.35*	*2.50*	*3.05*	*0*	*0*	*0*
Mussa angulosa	*0*	*0*	*0*	*1.22*	*0*	*0*	*0*
Mussismilia aff.hartti	*0*	*0*	*0*	*1.83*	*0*	*0*	*1.06*
Mussismilia sp.A	*0*	*0.63*	*1.25*	*0*	*0*	*0*	*0*
Pavona sp.A	*0*	*3.16*	*5*	*0*	*0*	*0*	*0*
Pavona trinitatis	*0*	*0*	*0*	*0.61*	*0*	*0*	*0*
Placocyathus variabilis	*0*	*0*	*1.25*	*1.25*	*0*	*0*	*0.70*
Pocillopora crassoramosa	*3.66*	*17.36*	*0*	*0.61*	*1.48*	*0*	*0*
Porites-I macdonaldi	*0*	*0*	*0*	*1.88*	*6.35*	*50*	*0.71*
Porites-I portoricensis	*0*	*20.94*	*41.25*	*3.05*	*4.31*	*0*	*12.61*
Porites-I waylandi	*4.97*	*0*	*3.75*	*13.63*	*8.5*	*3.66*	*29.11*
Porites-II baracoaensis	*0*	*1.47*	*0*	*9.88*	*3.09*	*9.76*	*7.34*
Siderastrea siderea	*3.05*	*0.74*	*1.25*	*0*	*3.34*	*0*	*0*
Solenastrea bournoni	*2.57*	*2.11*	*0*	*0*	*0*	*0*	*0*
Stephanocoenia duncani	*3.75*	*0*	*0*	*0*	*2.80*	*1.22*	*1.41*
Stephanocoenia spongiformis	*5.03*	*0*	*0*	*0*	*0*	*0*	*0*
Stylophora affinis	*3.13*	*27.61*	*10*	*4.92*	*35.95*	*0*	*0*
Stylophora granulata	*0*	*0*	*0*	*5*	*7.97*	*0*	*9.09*
Stylophora minor	*0*	*0.63*	*10*	*2.47*	*1.14*	*24.39*	*0*
Undaria agaricites	*4.68*	*0.61*	*11.25*	*0*	*7.89*	*8.54*	*1.4*
Undaria crassa	*0*	*0.63*	*0*	*1.22*	*0*	*0*	*0*

References

Allmon, W.D., 2001, Nutrients, temperature, disturbance, and evolution: a model for the late Cenozoic marine record of the western Atlantic, *Palaeogeogr. Palaeoclim. Palaeoecol.*, **166**:9–26.

Antonini, G.A., 1979, Physical geography of the northwest Dominican Republic, in: *Hispaniola: Tectonic Focal Point of the Caribbean – Three geologic studies in the Dominican Republic* (B. Lidz, and F. Nagle, eds.), Miami Geological Society, Miami, 96 pp.

Bold, W.A. van den, 1988, Neogene paleontology of the northern Dominican Republic. 7. The subclass Ostracoda (Arthropoda: Crustacea), *Bull. Am. Paleontol.*, **94**:1–105.

Bonnet, E. and Van de Peer, Y., 2002, zt: a software tool for simple and partial mantel tests, *J. Stat. Soft.*, **7**:1–12

Bray, J.R. and Curtis, J.T., 1957, An ordination of the upland forest communities of southern Wisconsin, *Ecol. Monogr.*, **34**:77–87.

Brett, C.E. and Baird, G.C., 1995, 'Cordinated stasis' and evolutionary ecology of Silurian and Middle Devonian faunas in the Appalachian Basin, in: new Approaches to Speciation in the Fossil Record (D. H. Erwin, R. L. Anstey, eds.), Columbia University Press, New York, pp. 285–315.

Budd, A.F., 1991, Neogene paleontology in the Northern Dominican Republic. 11. The Family Faviidae (Anthozoa: Scleractinia), *Bull. Am. Paleontol.*, **101**:5–83.

Budd, A.F., 2000, Diversity and extinction in the Cenozoic history of Caribbean reefs, *Coral Reefs*, **19**:25–35.

Budd, A.F., Stemann, T.A., and Johnson, K.G., 1994, Stratigraphic distributions of genera and species of Neogene to Recent Caribbean reef corals, *J. Paleontol.*, **68**:951–977.

Budd, A.F., Johnson, K.G., and Stemann, T.A., 1996, Plio-Pleistocene turnover and extinctions in the Caribbean reef coral fauna, in: *Evolution and Environment in Tropical America* (J.B.C. Jackson, A.F. Budd, and A.G. Coates, eds.), University of Chicago Press, Chicago, IL, pp. 168–204.

Budd, A.F. and Johnson, K.G., 1997, Coral reef community dynamics of 8 myr of evolutionary time: stasis and turnover, *Proc. 8th Int. Coral Reef Symp.*, **1**:423–428.

Budd, A.F. and McNeill, D.F., 1998, Zooxanthellate Scleractinian corals from the Bowden Shell Bed, SE Jamaica, *Contrib. Tert. Quat. Geol.*, **35**:49–65.

Budd, A.F., Petersen, R.A., and McNeill, D.F., 1998, Stepwise faunal change during evolutionary turnover: a case study from the Neogene of Curacao, Netherlands Antilles, *Palaios.*, **13**:170–188.

Budd, A.F. and Johnson, K.G., 1999a, Neogene paleontology in the northern Dominican Republic 19. The family Faviidae (Anthozoa: Scleractinia) Part II. The genera *Caulastrea*, *Favia*, *Diploria*, *Thysanus*, *Hadrophyllia*, *Manicina*, and *Colpophyllia*, *Bulls. Am. Paleontol.*, **109**:5–83.

Budd, A.F. and Johnson, K.G., 1999b, Origination preceding extinction during late Cenozoic turnover of Caribbean reefs, *Paleobiology*, **25**:188–200.

Budd, A.F., Johnson, K.G., Stemann, T.A., and Tompkins, B.H., 1999, Pliocene to Pleistocene reef coral assemblages in the Limon Group of Costa Rica, in: *The Neogene of the Isthmus of Panama: A Paleobiotic Survey of the Caribbean Coast* (L.S. Collins and A.G. Coates, eds.), Bulletins of the American Paleontological Society, pp. 119–158.

Budd, A.F. and Manfrino, C., 1999, Coral assemblages and reef environments in the Bahamas Drilling Project cores, in: *Subsurface Geology of a Prograding Carbonate Platform Margin, Great Bahama Bank: Results of the Bahamas Drilling Project* (R. N. Ginsburg, ed.), SEPM (Society for Sedimentary Geology), Tulsa, pp. 41–59.

Budd, A.F. and Klaus, J.S., This volume, Early evolution of the *Montastraea "annularis"* species complex (Anthozoa: Scleractinia): evidence from the Mio-Pliocene of the Dominican Republic, in *Evolutionary Stasis and Change in the Dominican Republic Neogene* (R.H. Nehm and A.F. Budd, eds.), Kluwer /Plenum, New York.

Cheetham, A.H. and Jackson, J.B.C., 1996, Speciation, extinction, and the decline of arborescent growth in Neogene and Quaternary cheilostome bryozoa of tropical America, in: *Evolution and Environment in Tropical America* (J.B.C. Jackson, A.F. Budd, and A.G. Coates, eds.), University of Chicago Press, Chicago, IL, pp. 205–233.

Clarke, K.R. and Warwick, R.M., 2001, *Change in Marine Communities: An Approach to Statistical Analysis and Interpretation*, Primer-E Ltd, Plymouth.

Coates, A.G., Jackson, J.B.C., Collins, L.S., Cronin, T.M., Dowsett, H.J., Bybell, L.M., Jung, P., and Obando, J.A., 1992, Closure of the Isthmus of Panama: the near-shore marine record of Costa Rica and western Panama, *Geol. Soc. Am. Bull.*, **104**:814–828.

Coates, A.G. and Obando, J.A., 1996, The geologic evolution of the Central American Isthmus, in: *Evolution and Environment in Tropical America* (J.B.C. Jackson, A.F. Budd, and A. G. Coates, eds.), University of Chicago Press, Chicago, IL, pp. 21–56.

Costa, F.H.A., Nehm, R.H., and Hickman, C.S., 2001, Neogene paleontology in the northern Dominican Republic 22. The family Neritidae (Mollusca: Gastropoda), *Bull. Am. Paleontol.*, **359**:47–69.

Collins, L.S., Budd, A.F., and Coates, A.G., 1996a, Earliest evolution associated with closure of the tropical American seaway, *Proc. Natl. Acad. Sci. USA*, **93**:6069–6072.

Collins, L.S., Coates, A.G., Berggren, W.A., Aubry, M.-P., and Zhang, J., 1996b, The Late Miocene Panama isthmian strait, *Geology.*, **24**:687–690.

Duncan, P.M., 1863, On the fossil corals of the West Indian Islands. Part 1, *Quart. J. Geol. Soc. Lond.*, **19**:406–458, pls. 13–16.

Duque-Caro, H., 1990, Neogene stratigraphy, paleoceanography and paleobiogeography in northwest South America and the evolution of the Panama seaway, *Palaeogeogr. Palaeoclim. Palaeoecol.*, **77**:203–234.

Edinger, E.N. and Risk, M.J., 1994, Oligocene-Miocene extinction and geographic restriction of Caribbean corals: roles of turbidity, temperature, and nutrients, *Palaios.*, **9**:576–598.

Edinger, E.N. and Risk, M.J., 1995, Preferential survivorship of brooding corals in a regional extinction, *Paleobiology*, **21**:200–219.

Evans, C.C., 1986, *Facies Evolution in a Neogene Transpressional Basin: Cibao Valley, Dominican Republic*, Unpublished Ph.D. dissertation, University of Miami, Coral Gables.

Foster, A.B., 1986, Neogene paleontology in the Northern Dominican Republic 3. The Family Poritidae (Anthozoa: Scleractinia), *Bull. Am. Paleontol.*, **90**:47–123.

Foster, A.B., 1987, Neogene paleontology in the northern Dominican Republic 4. The Genus *Stephanocoenia* (Anthozoa: Scleractinia: Astrocoeniidae), *Bull. Am. Paleontol.*, **93**:5–22.

Geister, J., 1977, The influence of wave exposure on the ecological zonation of Caribbean coral reefs, *Proc 3rd Int Coral Reef Symp.*, 23–29.

Goreau, T.F. and Wells, J.W., 1967, The shallow-water Scleractinia of Jamaica: revised list of species and their vertical distribution range, *B. Mar. Sci.*, **17**:442–453.

Holland, S.M., 2003, Analytic Rarefaction 1.3., Available online at: http://www.uga.edu/strata/software/AnRareReadme.html.

Jackson, J.B.C., 1977, Competition on marine hard substrata: the adaptive significance of solitary and colonial strategies, *Am. Nat.*, **111**:743–767.

Jackson, J.B.C. and Johnson, K.G., 2000, Life in the last few million years, *Paleobiology* (*Suppl.*), **26**:221–235.

Johnson, K.G., 1992, Population dynamics of a free-living coral: recruitment, growth and survivorship of *Manicina areolata* (Linnaeus) on the Caribbean coast of Panama, *J. Exp. Mar. Biol. Ecol.*, **164**:171–191.

Johnson, K.G., Budd, A.F., and Stemann, T.A., 1995, Extinction selectivity and ecology of Neogene Caribbean reef corals, *Paleobiology*, **21**:52–73.

Johnson, K.G., Budd, A.F., Klaus, J.S. and McNeill, D.F. This volume, The impact of fossils from the northern Dominican Republic on origination estimates for Miocene and Pliocene Caribbean reef corals, in: *Evolutionary Stasis and Change in the Dominican Republic Neogene* (R. H. Nehm and A.F. Budd, eds.), Kluwer/Plenum, New York.

Keigwin, L.D., 1979, Late Cenozoic stable isotopic stratigraphy and paleoceanography of DSDP sites from the east equatorial and central Pacific Ocean, *Earth Planet. Sci. Lett.*, **45**: 361–382.

Keller, G. and Barron, J.A., 1983, Paleoceanographic implications of Miocene deep-sea hiatuses, *Geol. Soc. Am. Bull.*, **94**:590–613.

Keller, G., Zenker, C.E., and Stone, S.M., 1989, Late Neogene history of the Pacific-Caribbean gateway, *J. S. Am. Earth Sci.*, **2**:73–108.

Klaus, J.S. and Budd, A.F., 2003, Comparison of Caribbean coral reef communities before and after Plio-Pleistocene faunal turnover: analyses of two Dominican Republic reef sequences, *Palaios.*, **18**:3–21.

Maier, K.L., Klaus, J.S., McNeill D.F., and Budd, A.F., 2007, A late Miocene low-nutrient window for Caribbean reef formation?, *Coral Reefs,* **26**:635–639.

Mantel, N., 1970, A technique of nonparametric multivariate analysis, *Biometrics.*, **26**, 547–558.

McNeill, D.F., Klaus, J.S., Evans, C.C., Budd, A.F., and Maier, K. L., This volume, An overview of the regional geology and stratigraphy of the Neogene deposits of the Cibao Valley, Dominican Republic, in: *Evolutionary Stasis: species and communities through geologic time* (R.H. Nehm and A.F. Budd, eds.), Kluwer/Plenum, New York.

McNeill, D.F., Budd, A.F., and Borne, P.F., 1997, An earlier (Late Pliocene) first appearance of the reef-building coral *Acropora palmata*: stratigraphic and evolutionary implications, *Geology.*, **25**:891–894.

Miller, K.G., Kominz, M.A., Browning, J.V., Wright, J.D., Mountain, G.S., Katz, M.E., Sugarman, P.J., Cramer, B.S., Christie-Blick, N., and Pekar, S.F., 2005, The Phanerozoic record of global sea-level change, *Science*, **310**:1293–1298.

Pandolfi, J.M. and Jackson, J.B.C., 2001 Community structure of Pleistocene coral reefs of Curaçao, Netherlands Antilles, *Ecol. Monogr.*, **71**:49–67.

Prothero, D.R., 1994, The Late Eocene-Oligocene extinctions, *Ann. Rev. Earth Planet. Sci.*, **22**:145–165.

Saunders, J.B., Jung, P., and Biju-Duval, B., 1986, Neogene paleontology of the northern Dominican Republic 1. Field surveys, lithology, environment, and age, *Bull. Am. Paleontol.*, **89**:1–79.

Shackleton, N.J. and Hall, M.A., 1997, The late Miocene stable isotope record, Site 926, *Proc. Ocean Drill. Prog.*, **154**:367–373.

Spezzaferri, S., McKenzie, J.A., and Isern, A., 2002, Linking the oxygen isotope record of late Neogene eustasy to sequence stratigraphic patterns along the Bahamas margin: results from a paleoceanographic study of ODP Leg 166, Site 1006 sediments, *Mar. Geol.*, **185**:95–120.

Todd, J.A., Jackson, J.B.C., Johnson, K.G., Fortunato, H.M., Heitz, A., Alvarez, M., and Jung, P., 2002, The ecology of extinction: molluscan feeding an faunal turnover in the Caribbean Neogene, *Proc. Roy. Soc. Lond. B Biol.*, **269**:571–577.

Vaughan, T.W., 1919, Fossil corals from Central America, Cuba, and Porto Rico with an account of the American Tertiary, Pleistocene, and Recent coral reefs, *US Nat. Hist. Mus. Bull.*, **130**:189–524.

Veron, J.E.N., 2000, *Corals of the world*, Australian Institute of Marine Science, Townsville, Australia.

Chapter 10
Mollusc Assemblage Variability in the Río Gurabo Section (Dominican Republic Neogene): Implications for Species-Level Stasis

Rysanek Rivera, Jermaine Lawson, Maria Harvey, and Ross H. Nehm

Contents

10.1 Introduction

A central question in evolutionary biology is whether community-level processes facilitate, maintain, or constrain the tempo and mode of species-level evolutionary change (Jackson, 1994; Brett and Baird, 1995; Morris et al., 1995; Ivany, 1996; Gould, 2002). The fossil record of abundant and well-preserved invertebrates from the Neogene of the Dominican Republic has been central to investigations of the tempo and mode of speciation (e.g., Cheetham, 1986, 1987; Foster, 1986, 1987; Nehm and Geary, 1994; Nehm, 2005). Surprisingly, the community ecological context of these speciation events remains unstudied despite more than 20 years of systematic and taxonomic investigations of the Dominican Republic Neogene invertebrate fauna (e.g., Saunders et al., 1982; Costa et al., 2001). The question of how community composition, stability, and change may have influenced the timing and cause of these well-known and extensively studied

---

Department of Biology, The City College C.U.N.Y. Convent Avenue at 138th Street, New York, NY 10031, USA. Email: nehm.1@osu.edu

R.H. Nehm, A.F. Budd (eds.) *Evolutionary Stasis and Change in the Dominican Republic Neogene,*

speciation events remains in need of investigation (Nehm and Budd, 2001). The Dominican Republic Neogene is an untapped resource for empirical and theoretical studies of how community-level change may promote or limit species-level change (Baird and Brett, 1983; Brett and Baird, 1995; Morris et al., 1995; Bennington and Bambach, 1996; Ivany, 1996; Holland and Patzkowsky, 2004; Bonelli et al., 2006).

The overarching goal of our study is to document mollusc assemblages from the same stratigraphic section as several well-documented cases of gastropod and bryozoan stasis and change in order to determine if species-level stasis is associated with assemblage-level stasis (e.g., Cheetham, 1986, 1987; Nehm and Geary, 1994; Nehm and Budd, 2001). We employ approximately 16,000 specimens from more than 300 mollusc species in order to quantify the magnitude of change in mollusc assemblages in the Rio Gurabo section, which remains the best-studied and most complete stratigraphic unit in the Dominican Republic Neogene (Saunders et al., 1982, 1986). We find that although several well-documented cases of morphological stasis within species have been found to occur in the Rio Gurabo section, mollusc assemblages lack stability in composition, relative abundance, species richness, and trophic distributions.

10.2 Geologic Setting

The seminal systematic work of Maury (1917a, b) and the many years of field research in the Dominican Republic Neogene (DRN) by Saunders, Jung, Biju-Duval, and E. and H. Vokes served as the foundation for our study. Saunders, Jung, and Biju-Duval's (1986) monograph contains a broad overview of the DRN research system as well as the maps and locality descriptions for all of the samples that were used in our study. Additional information may be found in the Neogene Marine biota of Tropical America online database (http://nmita.geology.uiowa.edu) as well as the Dominican Republic Project website (www.dominicanrepublicproject.org).

In the Neogene, the Cibao Valley of the northern Dominican Republic was a sea-filled graben, bordered on the south by the Cordillera Central and on the north by the Cordillera Septentrionale (Saunders et al., 1986). Several north-south trending tributaries of the Rio Yaque del Norte exposed approximately 1,100 m of the siliciclastic sediments along the south side of the Cibao Valley. Exposed river sections range in age from the Middle Miocene to early Middle Pliocene, including the Baitoa Formation (Upper Lower to Lower Middle Miocene). Cercado Formation (Upper Miocene), Gurabo Formation (uppermost Miocene to Lower Pliocene), and Mao Formation (Lower to Middle Pliocene). Our study focused on the mollusc-rich assemblages of the Rio Gurabo section (Fig. 10.1).

Previous paleoenvironmental studies suggest that the Río Gurabo section contained brackish, very shallow marine, marine, and deep marine sediments deposited in progressively offshore and more open marine conditions upsection (Saunders et al., 1986). The Cercado Formation (0–~150 m in the section) contains brackish and very shallow marine deposits, whereas the Gurabo Formation contains shallow and deep marine deposits. The brackish water *Larkinia-Mytilus-Melongena* mollusc assemblage, which occurs in the Cercado Formation of the Río Cana, also occurs

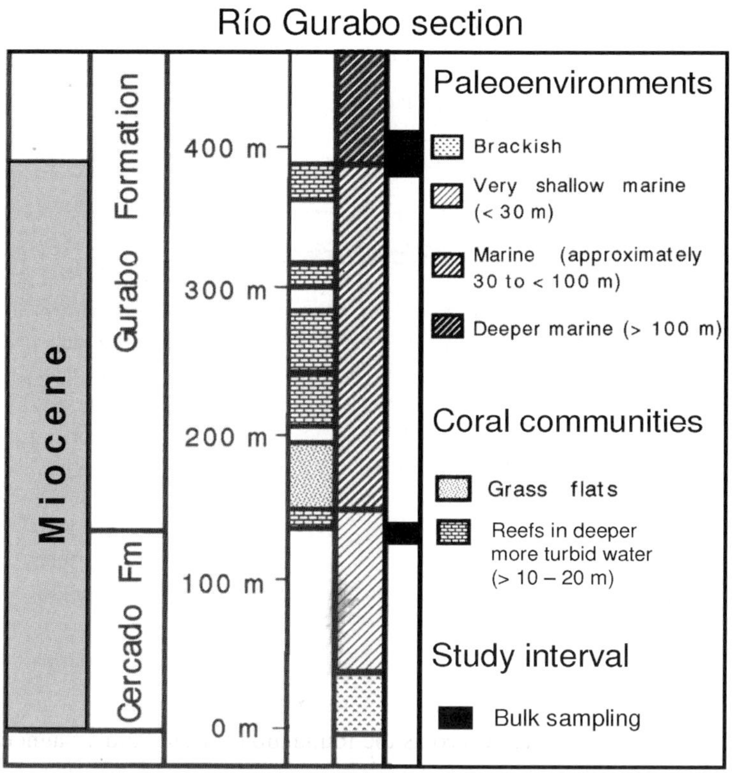

Fig. 10.1 Temporal, paleoenvironmental, and paleoecological overview of the Rio Gurabo section in the Cibao Valley of the northern Dominican Republic (Modified from Anderson, 1994)

in the lower Cercado Formation of the Río Gurabo from approximately 20–50 m in the section. Brackish water ostracod species also occur near 50 m in the middle Cercado Formation (Bold, 1988; Saunders et al., 1986). Very shallow marine conditions (<30 m paleodepth) occur from 60–150 m in the section, whereas deeper marine conditions (30–100 m paleodepth) occur from 150 to 380 m in the section. Gradual deepening (from 40 to 200 m paleodepth) is thought to have occurred from 100 to 380 m in the section (Saunders et al., 1986). Reef forming, branching, and unstable substrate corals occur throughout 120–400 m in the section.

Klaus et al. (this volume) partitioned the Neogene sections of the northern Dominican Republic into 21 lithostratigraphic units. Our study focuses on Klaus et al.'s units G1 and G9. Unit G1 occurs within the Cercado Formation of the Rio Gurabo from 120–140 m. The estimated median age of unit G1 is 5.9 Ma (McNeill et al. this volume). Unit G9 occurs in the Gurabo Formation from 387–430 m and has an estimated median age of 4.8 Ma. In their analysis of coral assemblages, Klaus et al. (this volume) found that units G1 and G9 both contain a large percentage of free-living corals, including *Antillia, Antillophyllia, Manicina, Meandrina, Placocyathus, Thysanus*, and *Trachyphyllia*. Free-living corals are typically associated with silty and soft-bottom substrates.

10.3 Materials and Methods

For our study, 20 bulk samples were taken from units G1 and G9 of the Cercado and Gurabo Formations of the Rio Gurabo section (Fig. 10.2, Table 10.1). Along each shell bed we measured a 25 foot zone and selected locations for bulk sampling from along this

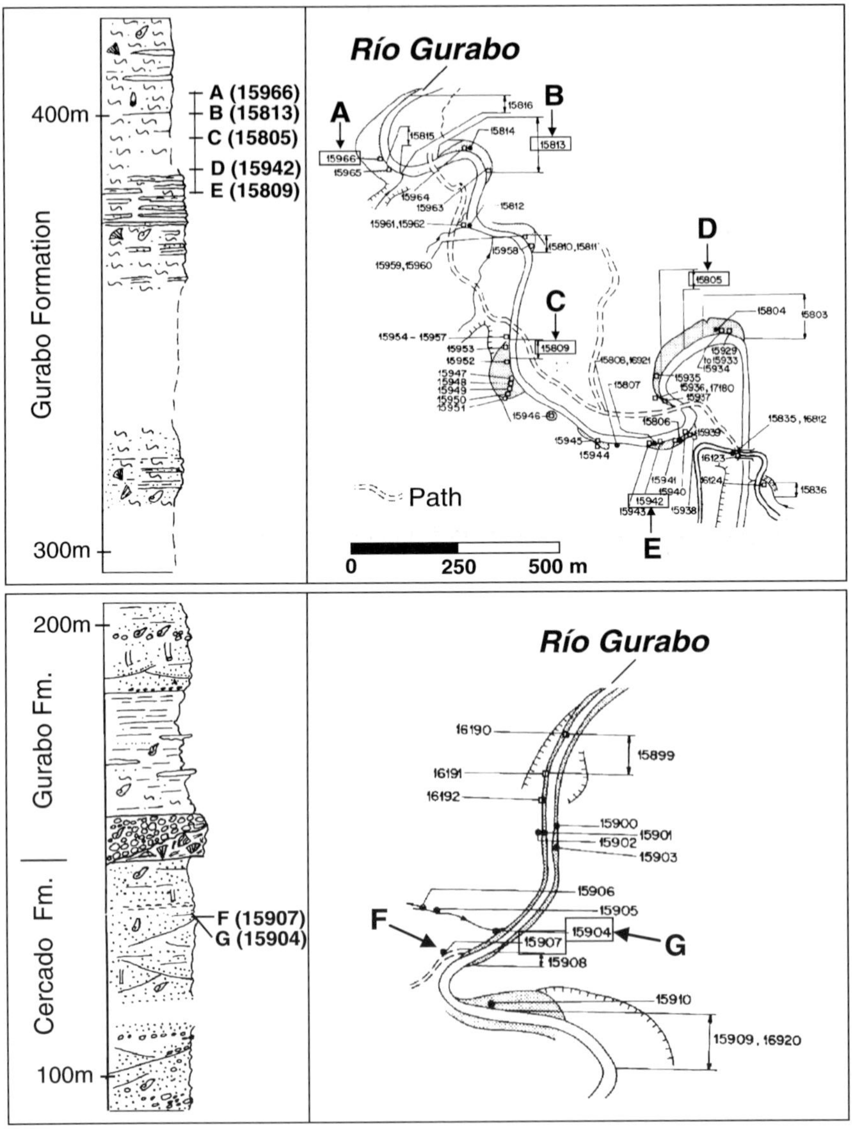

Fig. 10.2 Diagrams indicating the stratigraphic and geographic location of samples from the Rio Gurabo section. Upper panel: location of NMB samples 15966–15809 in the Gurabo Formation, along with the geographic location of each sample along the banks of the Rio Gurabo (see Saunders et al., 1986 for a complete map). Lower panel: location of NMB samples 15907 and 15904. See also Table 10.1 for more detailed information

Table 10.1 Summary of sample data. Locality numbers refer to the NMB sites from Saunders et al. (1986). 'Lateral' and 'vertical' refer to the sampling areas for each bulk. See Figs. 10.1 and 10.2 for additional data about these samples

Locality	Bulk	m	Kg	Lateral (cm)	Vertical (cm)	Formation	Individuals	Families	Genera	Species	% Bivalves	% Gastropods
15966	A	405	3.6	35.6	38.1	Gurabo Fm.	806	31	44	65	82.13	17.87
15966	B	405	3.2	35.6	25.4	Gurabo Fm.	613	42	55	72	65.09	34.91
15966	C	405	2.9	43.2	15.2	Gurabo Fm.	660	36	55	64	49.55	50.45
15813	A	400	2.8	40.6	25.4	Gurabo Fm.	579	34	31	46	86.01	13.99
15813	B	400	3.5	66.0	43.2	Gurabo Fm.	542	34	44	71	54.06	45.94
15813	C	400	2.8	40.6	48.3	Gurabo Fm.	339	28	31	50	68.44	31.56
15805	1	395	~3	33.0	23.0	Gurabo Fm.	452	28	36	45	69.91	30.09
15805	2	395	~3	33.0	23.0	Gurabo Fm.	372	32	42	50	70.43	29.57
15805	3	395	~3	38.1	20.3	Gurabo Fm.	291	25	32	43	63.57	36.43
15942	A	387	3.0	28.0	25.4	Gurabo Fm.	282	23	32	37	47.16	52.84
15942	B	387	3.1	35.6	23.0	Gurabo Fm.	282	25	37	44	52.84	47.16
15942	C	387	2.6	23.0	25.4	Gurabo Fm.	128	21	30	34	70.31	29.69
15809	A	382	3.2	33.0	43.2	Gurabo Fm.	54	13	15	21	61.11	38.89
15809	B	382	3.0	33.0	43.2	Gurabo Fm.	92	23	22	28	48.91	51.09
15809	C	382	2.8	38.1	28.0	Gurabo Fm.	100	24	24	30	55.00	45.00
15904	1	127	~3	103.0	20.0	Cercado Fm.	2,559	24	34	41	54.83	45.17
15904	2	127	~3	103.0	20.0	Cercado Fm.	1,503	20	24	31	88.52	11.48
15904	3	127	~3	103.0	20.0	Cercado Fm.	3,057	24	27	37	51.49	48.51
15907	1	127	~3	46.0	18.0	Cercado Fm.	1,574	19	21	26	60.17	39.83
15907	3	127	~3	46.0	18.0	Cercado Fm.	1,876	28	30	42	54.32	45.68

horizon by randomly picking three numbers from a hat. Each bulk was taken from a region approximately 30 cm vertical by 50 cm horizontal (Table 10.1). Each bag of sediment that we collected weighed approximately 3 Kg and had a volume of approximately 2 L (Table 10.1). Lithological descriptions were made in the field at each location.

All bulk samples were shipped to The City College, soaked in a detergent solution for several weeks, disaggregated, washed through stacked sediment sieves, and the fossil material was separated from sediment and sorted into taxonomic groups. Identification of fossil material to the species level was made using Maury (1917a, b) as well as the many systematic monographs in the "Neogene Paleontology in the Northern Dominican Republic" series in the journal *Bulletins of American Paleontology*. All gastropod and bivalve specimens greater than 2.5 mm were included in our analyses. The maximum count method was used to calculate species abundances (each bivalve shell was counted as one individual). Photographs and detailed descriptions of most of the mollusc taxa discussed herein may be found in Maury (1917a, b) and Pflug (1969).

The trophic modes and life habits of mollusc genera were identified using the database of Todd (2001). Specifically, we assumed that if a genus from our samples was extant and all living species in the genus shared the same life habit (e.g., mobile carnivore, sessile filter feeder, etc.), then the species in our sample also shared this trophic strategy. Although this method is prone to error, it serves as a general proxy for the life habits of the species in our samples.

It is well-established in the ecological and paleontological literature that different numbers of individuals and samples influence the number of species uncovered. If diversity comparisons are made between temporal units with different numbers of individuals or samples, then diversity differences are not necessarily meaningful (Miller and Foote, 1996). Rarefaction analyses were therefore used to compare species richness values. Rarefaction calculations and 95% bootstrapped error estimates were obtained using Analytic Rarefaction 1.3 (Holland, 2003).

Paleocommunity analyses began with the raw data in our sample-taxon matrix. Prior to analysis, we removed rare taxa (comprising <3% of all individuals) and transformed species abundances to percentages in order to compensate for the differences in sample sizes apparent among bulks. Bray-Curtis similarity coefficients (also known as Sorenson coefficients) were calculated in PAST 2.0 (Hammer and Harper, 2006) in order to compare bulk samples. The Bray-Curtis coefficient is well-suited to paleocommunity studies because the absence of a species does not contribute to the overall measure.

Non-metric Multidimensional Scaling (NMS) (also referred to as MDS) was used to ordinate and visualize the differences among bulk samples from the Gurabo section. In the NMS ordination, differences among graphed data points were quantified as a 'stress value'. Low stress values (generally <5) were considered to represent relationships among samples accurately and minimize misinterpretation risks (McCune and Grace, 2002:132). NMS analyses were performed in PAST 2.0 (Hammer and Harper, 2006) and PRIMER (Clarke and Warwick, 2001).

ANOSIM (ANalysis Of SIMilarities) is a non-parametric method used to test for significant differences among multivariate data points (Hammer and Harper, 2006).

ANOSIM has been used recently to test for faunal stability in the fossil record (Bonelli et al., 2006). Similarly to Bonelli et al. (2006), we used ANOSIM to test for assemblage differences using rank-ordered Bray-Curtis values derived from bulk samples in the Rio Gurabo section. We performed tests comparing the upper and lower regions of the Gurabo section (lithostratigraphic units G1 and G9) and also performed pair-wise comparisons among all bulk samples. The null hypothesis for these tests was that no significant differences occurred among mollusc samples in the Rio Gurabo section. In order to interpret significant ANOSIM results, we performed SIMPER analyses (Clarke and Warwick, 2001).

10.4 Results

10.4.1 Sample Abundance and Species Richness

Our 20 bulk samples from Rio Gurabo units G1 and G9 contained 16,261 individuals from 329 species. Of these 329 species, approximately 34% (111 species) were bivalves and 66% (218 species) were gastropods. Despite collecting similar volumes of material, there were drastically different numbers of individuals among bulks (Fig. 10.3). Bulk 15809A for example had only 54 individuals whereas Bulk 15904-3 had more than 3,000 individuals (Fig. 10.3). Similarly, Bulk 15904-3 contained 3,057 individuals (37 species) whereas replicate bulk 15904-2 contained only 1,603 individuals (31 species).

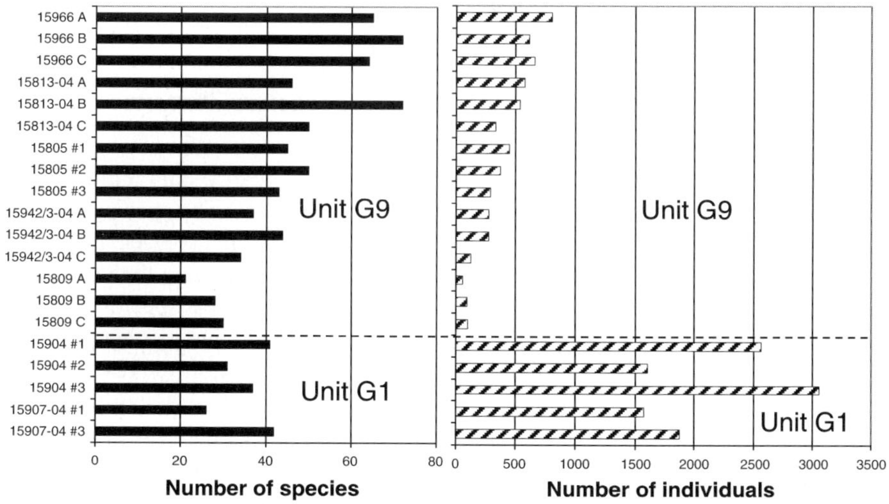

Fig. 10.3 Plot of the raw number of individuals and raw number of species from each bulk sample

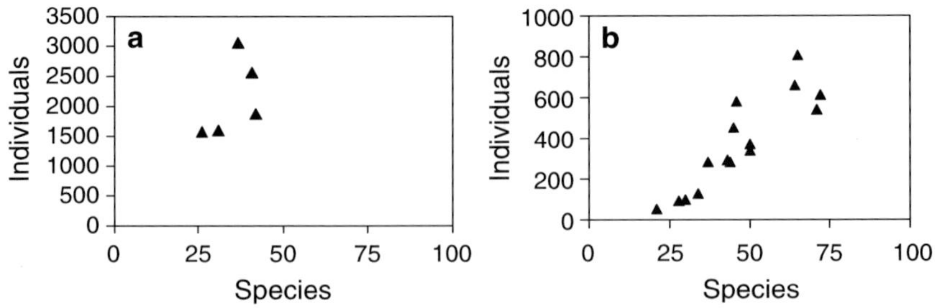

Fig. 10.4 (**a**) Plot of the relationship between the number of individuals sampled and the number of species uncovered in lithostratigraphic unit G1. (**b**) Number of individuals sampled and the number of species uncovered in lithostratigraphic unit G9

Significant differences in abundance occurred in different parts of the section. The samples from unit G9 in the Gurabo Formation had dramatically lower abundances than samples from unit G1 in the Cercado Formation. On average, bulks from unit G1 in the Cercado Formation contained 1,820 individuals whereas bulks from the Gurabo Formation contained 384 individuals. Since each of the twenty bulks had a similar volume, one can infer that the bulks from unit G1 contained a higher density of individuals per volume. Samples from unit G9 had the highest species richness even though they had the lowest abundance (Figs. 10.3 and 10.4).

There were also significant differences when we compared the replicate bulk samples within units. In lithostratigraphic unit G9, for example, bulks 15966-A, 15966-B, 15966-C, and 15813-B had high species richness values whereas bulks 15809-A, 15809-B, 15809-C had low species richness values. The distribution of species in replicate bulk samples varied significantly within localities, which suggests that patchiness characterizes the distribution of species in these units.

10.4.2 Rarefaction Analysis of Species Richness Patterns

Rarefaction curves were used to standardize sample abundance data and compare species richness patterns in the Rio Gurabo section (Fig. 10.5). The species richness values among bulk samples ranged from 20 to 72 species (Fig. 10.5A). The upper Gurabo unit G9 had greater species richness values (42–72 species) than the lower Gurabo unit G1 (20–35 species). Although unit G1 had the lowest species richness, it had the highest number of individuals per bulk sample, with an average of 2,128 specimens. When analyzed individually, most of the rarefaction curves from G9 samples did not level off, indicating that more sampling of these localities would likely capture more species; one bulk sample is clearly insufficient for measuring species richness in this unit. In contrast, the rarefaction curves for the lower Gurabo unit G1 appear to be levelling off, indicating the majority of species have probably been sampled with few bulk samples.

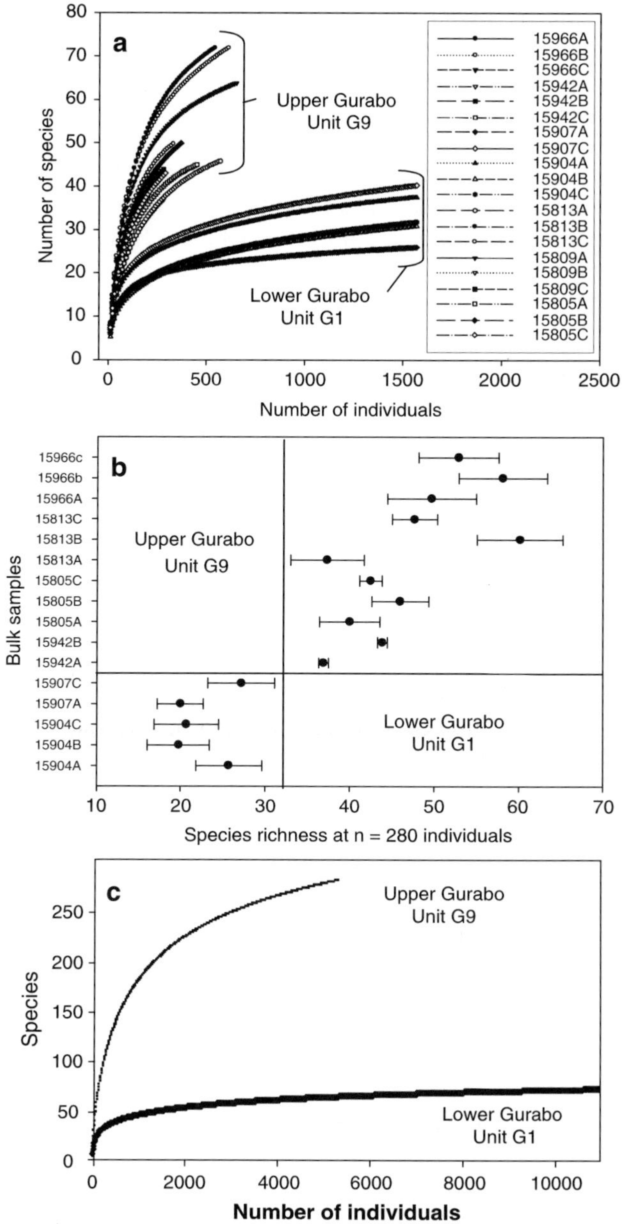

Fig. 10.5 **a.** Individual rarefaction curves for each bulk sample. **b.** Rarefaction estimates of species richness (at $n = 280$ individuals) for each bulk with 95% confidence intervals. **c.** Composite rarefaction curves for lithostratigraphic unit G9 and lithostratigraphic unit G1

In order to compare species richness values among bulks, our rarefaction study compared species richness estimates at n = 280 individuals (Fig. 10.5B). We excluded four bulk samples from unit G9 that had fewer than 280 individuals in this analysis. The species richness estimates that we obtained using rarefaction methods ranged from 16 to 65 species. The upper Gurabo unit G9 had a range of 33–65 species, whereas the lower Gurabo unit G1 had a range of 16 to 31 species. We calculated 95% confidence intervals for each species richness estimate, and none of the values from units G1 and G9 overlapped, indicating that these two regions of the Gurabo section had significantly different species richness values.

Significant differences in estimated species richness values also occurred within lithostratigraphic units G1 and G9 (Fig. 10.5B). In the upper Gurabo unit G9, for example, species richness values from localities 15813 and 15942 did not overlap. Differences also occurred within the lower Gurabo unit G1. Estimated species richness values from all three lower Gurabo bulk samples from 15904 overlapped, whereas replicate bulk samples from 15907 did not.

In addition to bulk-by-bulk comparisons of estimated species richness, we also generated rarefaction curves for each lithostratigraphic unit. This served two purposes: (1) To compare species richness values between the two units based on larger abundances, which likely provides greater accuracy; and (2) to determine how many bulks are likely needed to sample all the species within each unit. In our comparison of the two units, estimated species richness values (at n = 1,790) produced values of 48 species for unit G1 and 146 species for unit G9. Thus, similar to our bulk-level rarefaction results, we found no overlapping species richness estimates in the unit-level comparisons. The fact that unit G9 had significantly greater species richness at all levels of analysis suggests that this pattern is robust.

Using rarefaction curves, we estimated the number of bulk samples needed to estimate species richness values within lithostratigraphic units. As Figure 5C illustrates, in the lower Gurabo unit G1, 5 bulk samples captured most species, as indicated by the levelling off of the sampling curve. In contrast, in the upper Gurabo lithostratigraphic unit G9, 15 bulk samples were not sufficient to obtain all the species that initially inhabited this unit as indicated by the lack of a plateau in the sampling curve (Fig. 10.5C).

10.4.3 Paleocommunity Analysis

The Non-metric Multidimensional Scaling (NMS) analysis of mollusc abundance data produced clear differences between the lithostratigraphic units G1 and G9 from the Rio Gurabo section. The two-dimensional NMS ordination of the 20 Gurabo bulks shown in Fig. 10.6 forms two non-overlapping clusters: one cluster includes the upper Gurabo samples from unit G9 and the other cluster includes the lower Gurabo samples from unit G1. Within the G9 cluster, almost all of the replicate bulk samples from a single locality cluster together, indicating that these replicate bulks are similar in terms of faunal composition. Of notable exception are the

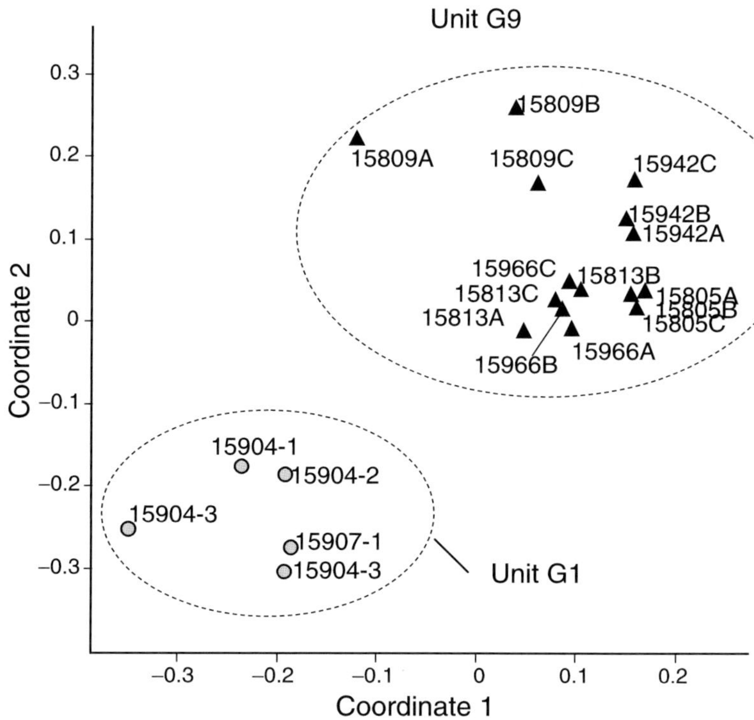

Fig. 10.6 NMS plot of Bray-Curtis distance values for all 20 bulk samples

bulks from locality 15809; these bulks are dispersed widely. Localities 15966 and 15813 display close proximity in the G9 cluster (see also Fig. 10.7) and were the most similar in composition among the G9 localities. The NMS analysis also illustrates the close proximity of the G1 bulks, from localities 15904 and 15907, with the exception of bulk 15904-3 which is separated from the others. This bulk sample had the greatest number of individuals.

The Analysis of Similarities (ANOSIM) was used to test the null hypothesis that there were no significant faunal similarities between lithostratigraphic units G1 and G9. The ANOSIM produced R values that ranged from 0 to 1; a value of 1 indicated no similarity between units. We used a cut-off value of $p < 0.05$ (5%) to test our null hypothesis. The results indicated that the null hypothesis should be rejected (Global $R = 0.865$, $p = 0.02$). We used pair-wise tests to explore which localities accounted for the between-unit differences (Table 10.2). These results indicated that most samples had very high R values, indicating considerable dissimilarity.

The pair-wise comparisons did reveal a few similarities between localities, however, such as those between localities 15966 and 15813 ($R = 0.11$, $p = 0.4$). In addition, there was significant similarity between the two localities of the lower Gurabo unit G1 (15904 and 15907; $R = 0.33$ $p = 0.3$). Finally, we found that localities 15942 and 15809 displayed moderate similarity ($R = 0.48$, $p = 0.2$). It is worth noting that these two localities are not proximal in time. Because the ANOSIM

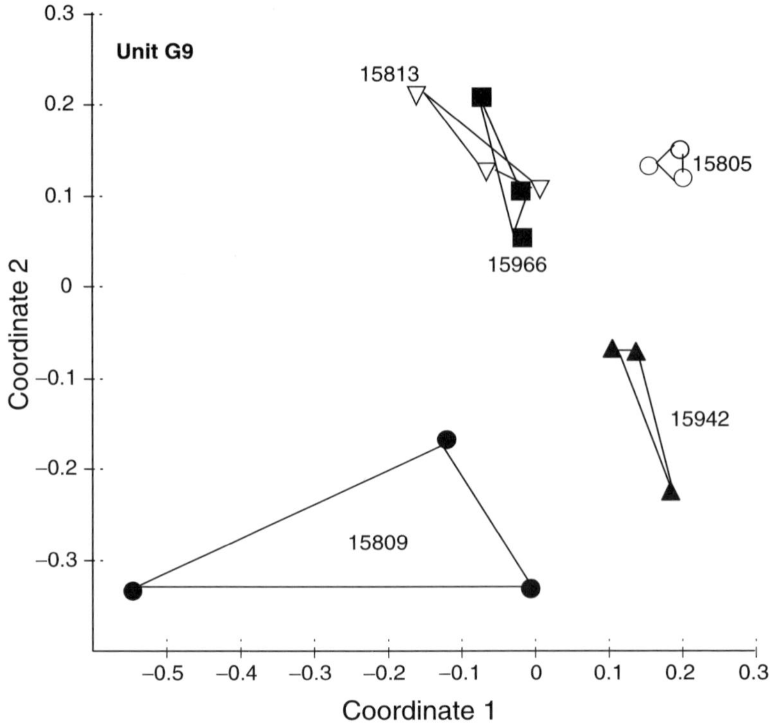

Fig. 10.7 NMS plot of Bray-Curtis distance values for bulk samples from upper Gurabo lithostratigraphic unit G9

pair-wise tests produced low similarity values between most of the localities, we used SIMPER to explore which taxa contributed to these dissimilarities.

Because of the faunal similarities documented in the NMS and ANOSIM between localities 15966 and 15813, we partitioned lithostratigraphic unit G9 into two zones (A and B). Localities 15966 and 15813 comprised the first zone, which we will subsequently refer to as G9 Zone A, and the localities 15942, 15805, and 15809 comprised the second zone, which we will refer to as G9 Zone B. Using this faunal framework, we performed three SIMPER analyses: G9 Zone A vs. G9 Zone B; G9 Zone A vs. G1; and G9 Zone B vs. G1. Bray Curtis values were used in the SIMPER and the cut off for low contributions was set at 90 percent.

SIMPER comparisons between G9 Zone A and G9 Zone B produced an average dissimilarity value of 65.95, indicating that there are some faunal differences between these two units. *Pecten* species accounted for a large percentage of the dissimilarities that we documented. *Pecten thetidis* was the primary discriminating species between zones, and it accounted for 7.6% of the faunal differences. *Pecten cercadicus* accounted for 6.17%, and *Pecten eugrammatus* accounted for 3.72% of faunal differences. The suspension feeding gastropod *Petaloconchus laddfranklinae* accounted for 3.54% of faunal differences and the bivalves *Barbatia reticulata*

Table 10.2 Results from the ANOSIM analysis of all bulk samples

Null hypothesis: Significant similarities between the Upper and Lower Gurabo Zones	
Sample statistic (global R):	0.865
Significance level of sample statistic:	0.02%
Number of permutations (random sample from a large number):	5000
Number of permuted statistics greater than or equal to global R:	0

Outcome: Reject Null hypothesis

Pairwise tests

Groups (sample comparisons)	R-value	Significance level % (p%)	Possible permutations	Actual permutations
15966, 15813	0.111	40	10	10
15966, 15942	0.926	10	10	10
15966, 15805	1.000	10	10	10
15966, 15809	1.000	10	10	10
15966, 15904	1.000	10	10	10
15966, 15907	1.000	10	10	10
15813, 15942	1.000	10	10	10
15813, 15805	0.963	10	10	10
15813, 15809	0.963	10	10	10
15813, 15904	1.000	10	10	10
15813, 15907	1.000	10	10	10
15942, 15805	1.000	10	10	10
15942, 15809	0.481	20	10	10
15942, 15904	1.000	10	10	10
15942, 15907	1.000	10	10	10
15805, 15809	1.000	10	10	10
15805, 15904	1.000	10	10	10
15805, 15907	1.000	10	10	10
15809, 15904	1.000	10	10	10
15809, 15907	1.000	10	10	10
15904, 15907	0.333	30	10	10

and *Corbula caimitica* each accounted for 5% of the faunal differences between G9 Zone A and G9 Zone B.

SIMPER analyses indicated that the faunal differences between unit G9 and unit G1 were significantly greater than the differences between G9-A and G9-B (89.31 vs. 65.95 respectively). Discriminating species included the epifaunal bivalves *Scapharca corcupidonis* (contribution of 8.52%) and *Scapharca riocanensis* (7.47%); the epifaunal recliner *Pecten thompsoni* (7.28%); the infaunal siphonate bivalves *Pitaria acuticostata* (7.03%), *Phacoides yaquensis* (6.89%), and *Strigilla pisformis* (6.68%).

The greatest differences documented by the SIMPER analysis were between G9 Zone B and unit G1 (average dissimilarity 93.9). *Scapharca* was most responsible for the dissimilarity that we documented: *Scapharca corcupidonis* accounted for 9.12% and *Scapharca riocanensis* accounted for 7.94% of faunal differences. The epifaunal recliner *Pecten thompsoni* accounted for 7.85% of faunal differences. The infaunal siphonate bivalves *Phacoides yaquensis* (7.54%), *Pitaria acuticostata* (7.53%) and *Strigilla psiformis* (7.16%) were also discriminating species between the two units. In summary, all three of the stratigraphic units that we compared displayed large faunal dissimilarities.

10.4.4 Faunal Carryover and Turnover

In addition to NMS, ANOSIM, and SIMPER analyses, we also calculated faunal carryover and holdover values among units and zones because these simple values have been used as proxies for community stasis (Brett and Baird, 1995) (Table 10.3). Carryover was defined as the ratio of the number of taxa that continue into the next time period relative to the total number of taxa in the time period being considered. Holdover was defined as the number of taxa persisting from the previous time period.

Carryover and holdover measures took into account time succession, beginning from the oldest locality and ending at the youngest locality (i.e., 15907–15966, respectively). The highest carryover percent was from localities 15907–15904 (83%: 90% of the bivalves and 81% of gastropods found in 15907). The lowest carryover (23%) occurred between localities 15904 and 15809 (33% of bivalves and 16% gastropods found in 15904 persisted). The carryover and holdover results are concordant with our NMS and ANOSIM results.

We also compared turnover and holdover values between the three larger faunal units: G9 Zone A *vs.* G9 Zone B; G9 Zone A *vs.* G1; and G9 Zone B *vs.* G1. From the lower Gurabo unit G1, only 33% of taxa carried over to G9 Zone B, and 15% of these taxa were holdovers. From G9 Zone B, 47% of taxa carried over to the G9 Zone A. None of these results support the hypothesis of faunal stability between zones.

10.4.5 Bivalve and Gastropod Life Habit Distributions

Mollusc life habits are fundamental ecological attributes that may be used to better understand the assemblage differences identified using NMS and ANOSIM. Our

Table 10.3 Holdover and Carryover values by NMB locality. Temporal and geographic information on each locality is shown in Figs. 10.1 and 10.2

	Holdover/Carryover indices					
	Total Taxa		Bivalves		Gastropods	
Locality	Carryover	Holdover	Carryover	Holdover	Carryover	Holdover
15966	n/a	53	n/a	66	n/a	47
15813	60	37	68	43	55	33
15942	59	52	71	67	52	43
15805	45	30	46	29	44	30
15809	40	28	39	35	42	23
15904	23	48	33	33	16	57
15907	83	n/a	90	n/a	81	n/a

paleocommunity analyses delineated three distinct faunal zones within the Rio Gurabo, and we therefore explored whether these zones differed in mollusc life habit distributions. For bivalves we studied two variables: trophic habits and substrate relationships. For gastropods, we studied trophic habits.

Within the Bivalvia, Todd (2001) defined nine modes of substrate relationships: (1) Epifaunal (EP, e.g., *Ostrea*); (2) Infaunal Siphonate (IS, e.g., *Cardium*); (3) Infaunal Asiphonate (IA, e.g., *Nucula*); (4) Epifaunal and Infaunal Siphonate (EP/IS, e.g., *Glycimeris*); (5) Epifaunal Recliner (ER, e.g., *Pecten*); (6) Infaunal Siphonate and Semi-Infaunal (IS/SI, e.g., *Chione*); (7) Epifanual and Infaunal Asiphonate (EP/IA, e.g., *Crassinella*); (8) Borer, nesting on hard substrate (WB, e.g., *Petricola*); and (9) Nestler within a burrow of another organism (WU, e.g., *Montacuta*). We tallied and compared the distributions of these trophic and substrate relationship habits in bulk samples from faunal units G1, G9 Zone A, and G9 Zone B.

The trophic habits of bivalves from faunal units G1, G9 Zone A, and G9 Zone B did not differ significantly through time. This result is not surprising considering that the vast majority (~95%) of bivalves in our samples were suspension feeders.

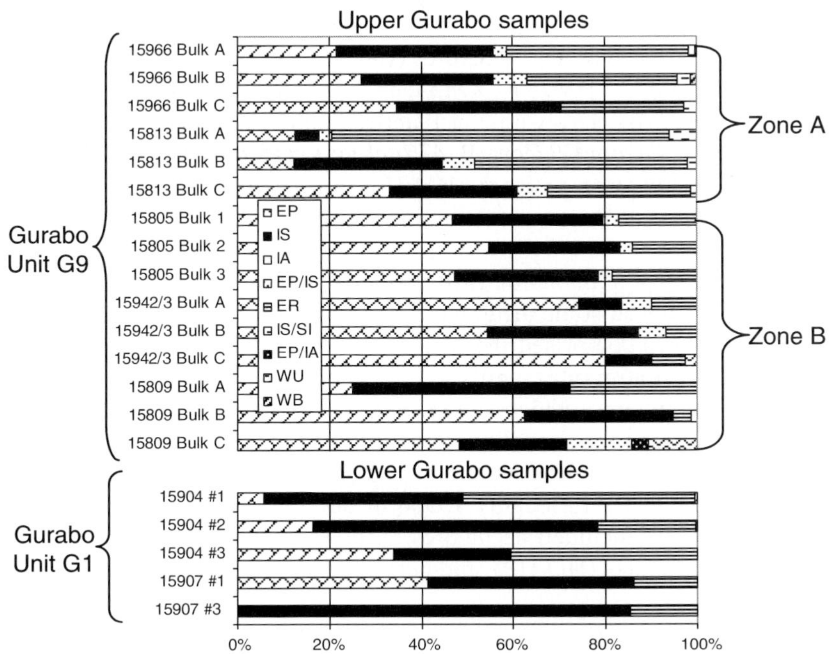

Fig. 10.8 Substrate relationship distributions of bivalves through time in the Rio Gurabo section, lithostratigraphic units G1 and G9 The shell fixation modes are: EP: Epifaunal; IS: Infaunal Siphonate; IA: Infaunal Asiphonate; EP/IS: Epifaunal/Infaunal Siphonate; ER: Epifaunal Recliner; IS/SI: Infaunal Siphonate/Semi-Infaunal; EP/IA: Epifanual/Infaunal Asiphonate; WB: Borer, nesting on hard substrate; and WU: Nestler within burrow of another organism. This classification was taken from Todd (2001)

We also studied the substrate relationships of bivalves. The bulk samples from lithostratigraphic units G1 and G9 were dominated by modes EP (Epifaunal), IS (Infaunal Siphonate), and ER (Epifaunal Recliner). There were some substrate relationship modes other than EP, IS, and ER, however, that appeared in small percentages in one zone but not in the others (Fig. 10.8). In unit G1, for example, only modes EP, IS, and ER were present. Notably, however, EP was not present in bulk sample 15907-3.

In unit G9 Zone B, there were some modes that were not present in unit G1. Throughout G9, there were bivalves that were both Epifaunal and Infaunal Siphonate (EP/IS); these modes were uncommon or absent in Unit G1. Additional differences included borers nesting on hard substrates (WB); nestlers within the burrow of another organism (WU); and bivalves that were both Epifaunal and Infaunal Asiphonate (EP/IA). The last mode, however, only occurred in bulk sample 15809-C. Unit G9 Zone B contained the greatest percentage of epifaunal bivalve species. G9 Zone A is unique in that modes WB and EP/IA were not present, modes IS/SI were only present in this zone, and ER were more abundant relative to other samples. Overall, however, the life habits and trophic modes of bivalves did overlap among zones and units.

Within the gastropoda, Todd (2001) defined eight trophic modes: (1) Predatory Carnivores (CP, e.g., *Phos* and *Conus*); (2) Browsing Carnivores (CB, e.g., *Cancellaria* and *Prunum*); (3) Herbivorous Omnivores (HO, e.g., *Strombus*); (4) Herbivores on Fine-Grained substrates (HM, e.g., *Xenophora*); (5) Herbivores on Rock (HR, e.g., *Arene* and *Astrea*); (6) Herbivores on Plant (HP, e.g., *Rissoina*); (6) Suspension Feeders (SU, e.g., *Petalochoncus* and *Turritella*); (7) Herbivores on Fine-Grained substrates and Herbivores on Rock (HM/HR, e.g., *Phasianella* and *Cerithium*); and (8) Herbivores on Rock and Browsing Carnivores (HR/CB, *Fissuridea* and *Diodora*).

Although the trophic habits of gastropods from faunal units G1, G9 Zone A, and G9 Zone B showed some differences through time, browsing carnivores and predatory carnivores comprised the majority of all bulk samples from all zones (Fig. 10.9). Lithostratigraphic unit G1 was dominated by predatory carnivores and browsing carnivores, which collectively accounted for about 95% of the trophic modes in this zone. Locality 15907 also contained 4% herbivores on rock and browsing carnivores (HR/CB) and 1% of herbivorous omnivores and herbivores on plants (HO/HP). In locality 15904 of unit G1, trophic modes other than predatory carnivores included herbivores on plant (HP), herbivorous omnivores and herbivores on plants (HO/HP), and herbivorous omnivores and herbivores on rock (HO/HR).

In unit G9, there were significantly greater percentages of suspension feeders than in unit G1. In G9 Zone B, there were herbivores on rock (HR), and herbivores on fine-grained substrates that appeared in minor quantities and were absent altogether from unit G1. Unit G9 Zone B had fewer predatory carnivores relative to most other bulk samples and many more suspension feeders (SU), browsing carnivores (CB), and herbivores on rock. In unit G9 Zone A, suspension feeders were present, as well as comparatively more herbivorous omnivores (HO). In summary,

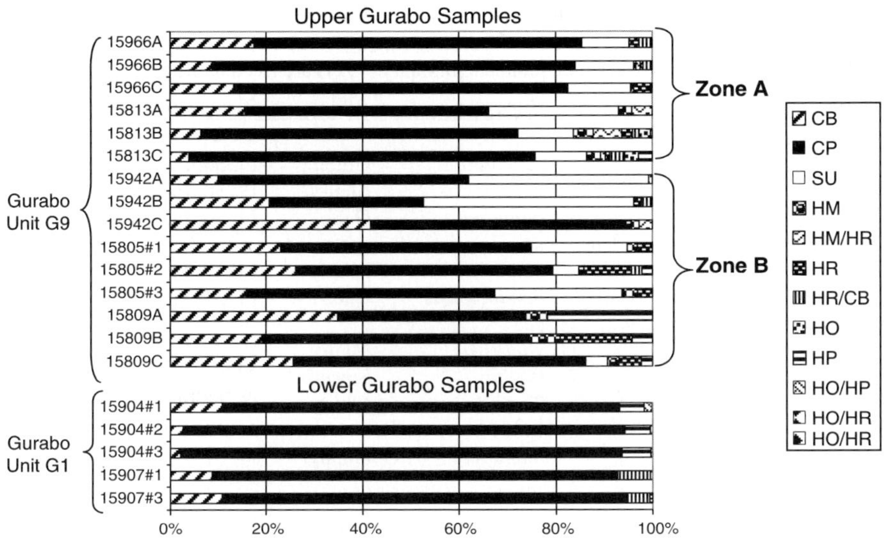

Fig. 10.9 Trophic distributions of gastropods through time in the Rio Gurabo section, lithostratigraphic units G1 and G9. The trophic modes are: CB: browsing carnivores; CP: predatory carnivores; SU: suspension feeders; HM: herbivores on fine-grained substrates; HM/HR: herbivores on fine-grained substrates and herbivores on rock; HR: herbivores on rock; HR/CB: herbivores on rock and browsing carnivores; HO: herbivorous omnivores; and HP: herbivores on plant. This classification was taken from Todd (2001)

although lithostratigraphic units G1 and G9 demonstrated a great similarity in having large percentages of suspension feeding bivalves and browsing and predatory carnivorous gastropods, unit G9 contained a greater diversity of gastropod trophic modes as well as a greater percentage of suspension feeding gastropods. These results are suggestive of ecological differences between zones.

10.4.6 Taxon Distribution and Abundance Patterns

Previous analyses revealed clear faunal differences between mollusc assemblages from lithostratigraphic units G1 and G9 in the Rio Gurabo section. We now explore how G9 Zone A, G9 Zone B, and unit G1 differed in terms of the presence and abundance of particular species as well as which species accounted for the life habit differences among zones that we documented above. All these patterns allow us to better understand mollusc assemblage changes in the Rio Gurabo section. We analyzed the taxon distribution and abundance patterns at three levels: (1) the most abundant species in each zone; (2) the species that persisted throughout the zones; and (3) characteristic genera of exhibiting life habits characteristic of particular tropic modes.

Table 10.4 List of bivalve and gastropod species that persist from lithostratigraphic units G1–G9 (the lower and upper Gurabo section, respectively). Maury (1917) contains illustrations and descriptions of these species. Plates, pages, and figure numbers in Maury (1917) are listed in the right column

Family	Genus	Species	Author	Maury (1917)
Bivalvia				
Arcidae	*Barbatia*	*reticulata*	Gmelin	Pl. 30; Pg. 166; Fig. 16
Arcidae	*Scapharca*	*riocanensis*	Maury	Pl. 30; Pg. 176; Fig. 4
Arcidae	*Barbatia*	*bonaczyi*	Gabb	Pl. 30; Pg. 165; Fig. 15
Cardiidae	*Trigoniocardium*	sp.		
Corbulldae	*Corbula*	*caimitica*	Maury	Pl. 39; Pg. 233; Figs. 18, 19
Corbulidae	*Corbula*	*dominicensis*	Gabb	Pl. 39; Pg. 232; Figs. 14, 15
Ostreidae	*Ostrea*	sp.		
Lucinidae	*Phacoides*	*yaquensis*	Maury	Pl. 35; Pg. 206; Fig. 8
Pectinidae	*Pecten*	*thetidis*	Sowerby	Pl. 34; Pg. 185; Fig. 6
Pectinidae	*Pecten*	*Thompsoni*	Maury	Pl. 34; Pg. 188; Figs. 9, 10
Pectinidae	*Pecten*	sp.		
Veneridae	*Antigona*	*blandiana*	Guppy	Pl. 37; Pg. 217; Fig. 5
Gastropoda				
Buccinidae	*Phos*	*elegans*	Guppy	Pl. 14; Pg. 86; Fig. 10
Buccinidae	*Phos*	*fasciolatus*	Dall	Pl. 14; Pg. 88; Figs. 15, 16
Conidae	*Conus*	*sp.*		
Conidae	*Conus*	*symmetricus*	Sowerby	Pl. 7; Pg. 36; Fig. 7
Mitridae	*Mitra*	*henekeni*	Sowerby	Pl. 12; Pg. 74; Figs. 5, 5a
Muricidae	*Murex*	*messorius*	Sowerby	Pl. 16; Pg. 101; Figs. 1, 2
Muricidae	*Murex*	*sp.*		
Naticidae	*Natica*	*youngi*	Maury	Pl. 23; Pg. 135; Fig. 12
Olividae	*Oliva*	*sp.*		
Turridae	*Mangilia*	*lalonis*	Maury	Pl. 9; Pg. 58; Fig. 10

Of the more than 300 mollusc species that were identified in lithostratigraphic units G1 and G9, only 21 species (12 bivalves and 9 gastropods) occurred in both units. Table 10.4 provides a list of these species along with their corresponding descriptions in Maury (1917a,b), and Fig. 10.10 illustrates the five most abundant gastropod and bivalve species from each zone. There were twelve species of bivalves that persisted from unit G1 to G9. These species belonged to the genera *Barbatia, Scapharca,* and *Ostrea,* which have an epifaunal substrate relationship; *Chama*, *Corbula*, *Phacoides*, and *Antigona*, which have an infaunal siphonate substrate relationship; and *Pecten,* which is an epifaunal recliner. The seven gastropod genera that persisted from G1 to G9 were *Phos, Conus, Mitra, Murex, Natica, Oliva,* and *Mangilia* (see Table 10.4); all of these genera are predatory carnivores.

The epifaunal bivalve genera that persisted from units G1 to G9 show interesting distribution patterns (Fig. 10.11). The genus *Barbatia*, for example, was most abundant

Fig. 10.10 (continued) *caribaea*, height 6.1 mm, width 2.2 mm. (C) *Terebra berlinerae*, height 13 mm, width 3.4 mm. (D) *Rissoina cumingic*, height 5.3 mm, width 2.3 mm. (E) *Strombina gurabensis*, height 6 mm, width 2.3 mm. (F) *Scapharca riocanensis*, length 6.2 mm, width 5.1 mm. (G) *Scapharca corcupidonis*, length 5.8 mm, width 4.8 mm. (H) *Pecten thompsoni*, length 11.2 mm, width 10.2 mm. (I) *Pitaria acuticostata*, length 10.7 mm, width 12.4 mm. (J) *Strigilla pisformis,* length 4 mm, width 4.5 mm

Fig. 10.10 The five most abundant gastropod and bivalve species from each Gurabo unit and zone. Upper panel: Unit G9, Zone A. Left to right: (A) *Phos costatus*, height 6.5 mm, width 3.8 mm; (B) *Clava plebia*, height 15 mm, width 5.9 mm.(C) *Turritella submortoni*, height 4.3 mm, width 2.2 mm. (D) *Conus marginatus*, height 7.8 mm, width 3.7 mm. (E) *Natica youngi*, height 8.3 mm, width 6.8 mm. (F) *Pecten eugrammatus*, length 13.6 mm, width 14 mm. (G) *Corbula caimitica*, length 9.4 mm, width 7.8 mm. (H) *Glycimeris acuticostata*, length 20.8 mm, width 22.5 mm. (I) *Venericardia cerrogordensis*, length 2.8 mm, width 2.8 mm. (J) *Pecten Thetidis*, length 10.4 mm, width 10.3 mm. Middle panel: Unit G9, Zone B. (A) *Petaloconchus laddfranklinae* (Fragment) 4.3 mm. (B) *Clava plebia*, height 15 mm, width 5.9 mm. (C) *Prunum coniforme*, height 11.7 mm, width 5.4 mm. (D) *Phos costatus*, height 6.5 mm, width 3.8 mm. (E) *Oliva cylindrica*, height 5.3 mm, width 2.3 mm. (F) *Chama involuta*, length 8.1 mm, width 7.9 mm. (G) *Pecten thetidis*, length 10.4 mm, width 10.3 mm. (H) *Corbula dominicensis*, length 4 mm, width 5.5 mm. (I) *Barbatia reticulata*, length 4.5 mm, width 5.5 mm. (J) *Ostrea sp*, length 5.4 mm, width 4.7 mm. Lower panel: Unit G1. (A) *Cythara caimitica*, height 6.7 mm, width 2.4 mm. (B) *Strombina*

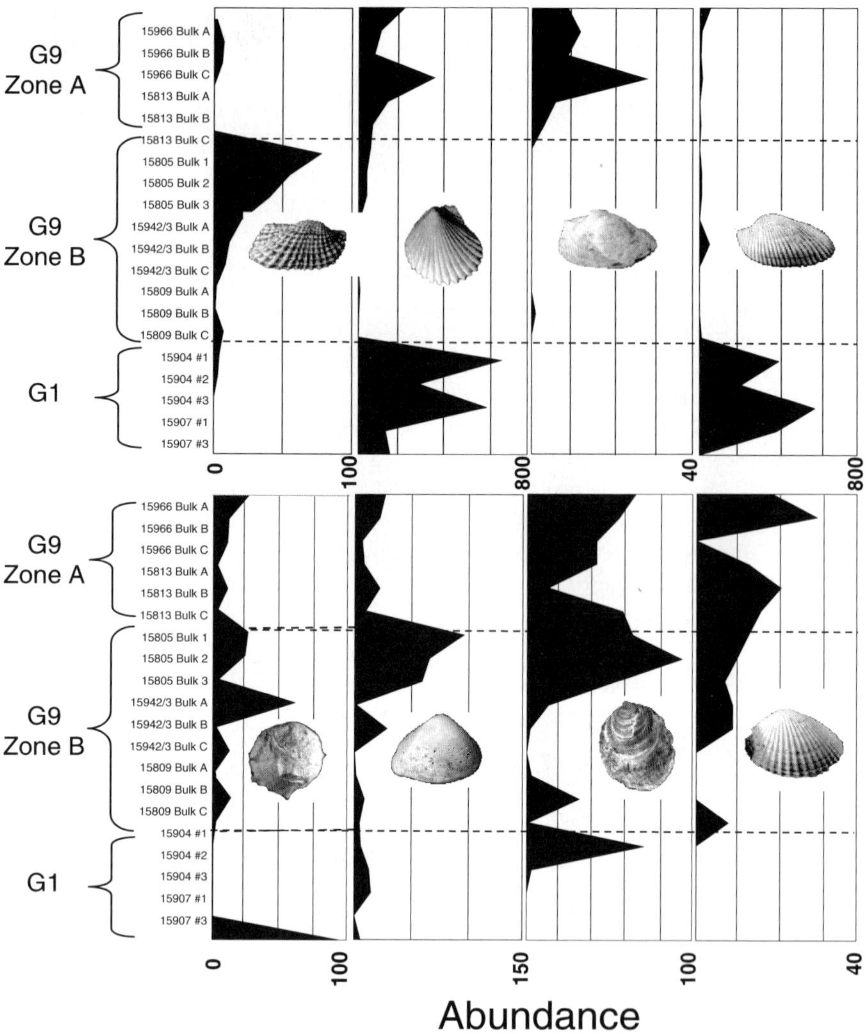

Fig. 10.11 Abundance graphs of exemplar bivalve genera through the Rio Gurabo section. Dashed lines delineate faunal zones identified herein. Top row, left to right: *Barbatia*, length 4.5 mm, *Pecten*, length 10.4 mm, *Chione*, length 7.5 mm, *Scapharca*, length 6.2 mm. Bottom row, left to right: *Chama*, length 8.1 mm, *Corbula*, length 9.4 mm, *Ostrea*, length 5.4 mm, *Glycimeris*, length 20.8 mm

in G9 Zone B, representing 18% of all bivalves. In G9 Zone A and unit G1, however, it represented less than 1% of bivalves. *Scapharca* exhibited the greatest relative abundance in unit G1, comprising 29% of the bivalves in this unit, whereas in G9 Zone A and B it represented 4% and 6% of bivalves, respectively. *Ostrea* displayed its greatest relative abundance in unit G9 Zone A, comprising 8% of the bivalves (Fig. 10.11).

The infaunal siphonate genus *Corbula* comprised less than 1% of bivalves in unit G1, whereas it comprised 19% in unit G9 Zone B and 6% of the bivalves in unit G9 Zone A. The infaunal siphonate genus *Chama* represents 1.6% of total bivalve abundance and comprised 0.8% of the bivalves in G9 zone A and 7.4% of the bivalves in G9 zone B. *Chama* was apparently absent from unit G1. The epifaunal and infaunal siphonate bivalve *Glycimeris* persisted throughout units G1 and G9 but in very low percentages: 4% in unit G1 and 6% in unit G9 (Fig. 10.11).

The predatory carnivores *Strombina, Terebra,* and *Conus* persisted throughout the Rio Gurabo units. *Strombina* was mostly abundant in unit G1 comprising 7% of the gastropods (see Fig. 10.12). The genus *Terebra* comprised 4% of the gastropods in G1 and 2% in G9 (Fig. 10.12). *Conus* displayed a similar distribution in units G1 and G9. However, *Conus* was more abundant in G9 Zone A, with 4% and 5% in G1 and G9, respectively (see Table 10.4 and Fig. 10.12).

The genera *Petalochoncus, Vermicularia*, and *Turritella* were the most abundant suspension feeders in the Rio Gurabo samples. *Petalochoncus* dominated unit G9 with a maximum abundance of over 50 individuals, while showing no evidence of existing in unit G1 (Fig. 10.13). The genera *Turritella* and *Vermicularia* were most abundant in zone G9 Zone A. *Turritella* had a total abundance of 21 individuals (comprising 2% of the gastropods in G9 Zone A); *Vermicularia* had a total abundance 54 individuals in G9 Zone A, comprising 5% of the gastropods in this zone (Fig. 10.13).

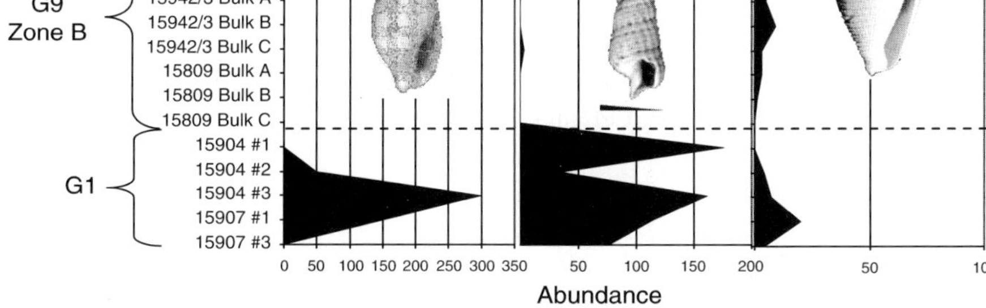

Fig. 10.12 Abundance and percentage graphs of exemplar carnivorous gastropod genera through the Rio Gurabo section. Dashed lines delineate faunal zones identified herein. Left: *Strombina*, height 6.1 mm. Middle: *Terebra*, height 13mm. Right: *Conus*, Height 12.9 mm

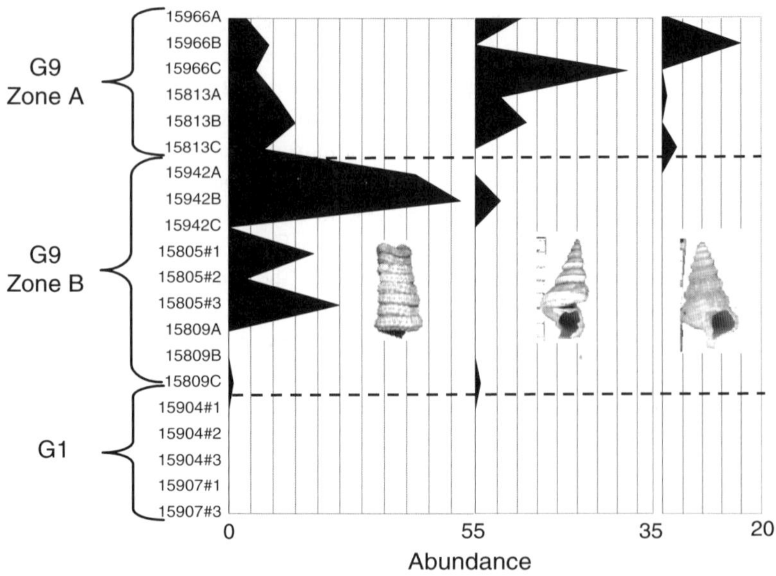

Fig. 10.13 Abundance and percentage graphs of exemplar suspension feeding gastropod genera through time in the Rio Gurabo section. Dashed lines delineate faunal zones identified herein. Left: *Petaloconchus* (fragment) 4.3 mm; Middle: *Vermicularia* (fragment) 5.2 mm; Right: *Turritella* (fragment) height 4.5 mm

10.5 Discussion

For the past 20 years the Dominican Republic Neogene has served as an important research system for investigating evolutionary patterns within and among species over macroevolutionary timescales. Detailed investigations of intraspecific change and speciation have been carried out in many invertebrate groups, including bryozoans (Cheetham, 1986, 1987; Cheetham and Jackson, 1995), corals (Foster, 1986, 1987; Klaus et al., this volume; Scultz and Budd, this volume; Beck and Budd, this volume), gastropods (Nehm and Geary, 1994; Nehm, 2005; Nehm, this volume) and bivalves (Anderson, 1994). In general, these studies have found that oscillatory stasis typifies the evolutionary history of most Dominican Neogene invertebrate lineages. Long-term directional change has not been identified in any lineages to date, with the exception of brief episodes of gradualism in *Prunum* gastropods (Nehm and Geary, 1994).

Despite considerable interest in patterns and processes of faunal evolutionary stability (Brett and Baird, 1995; Morris et al., 1995; Bennington and Bambach, 1996; Ivany, 1996; Bennington and Rutherford, 1999; Holland and Patzkowsky, 2004; Bonelli et al., 2006), and the concerted efforts of many workers to explore the processes responsible for evolutionary stasis within Dominican species (Cheetham and Jackson, 1995) very little empirical work has attempted to link these two research areas (although see Morris, 1996). Although the Dominican Neogene is well-suited to empirical explorations that couple investigations of both species and faunas, no studies have built upon the early outline of mollusc communities by Saunders et al. (1982). This is undoubtedly because of the methodology adopted by the early Dominican Republic Project. After initial processing of field samples, the fossil material was disaggregated and dispersed to representative taxonomic specialists for study. The lack of study of this material by many workers, and the failure to re-aggregate it for composite analyses, has only exacerbated the delayed investigation of the molluscan fauna. The more recent demise of the Naturhistorishes Museum Basel as a center of Neogene mollusc research has further limited paleocommunity research in the Dominican Republic. As a result of this diverse set of factors, re-sampling the Rio Gurabo section for this study emerged as the only realistic approach for moving mollusc community research forward. Unlike Dominican molluscs, coral communities have been much better studied, due in part to their lower diversity and abundance and the fact that the original material was not dispersed among many different workers (Klaus et al., this volume).

The present chapter represents one of the first attempts to explore faunal change in mollusc assemblages from the Dominican Republic Neogene (Saunders et al., 1986). Given the considerable diversity and abundance of molluscs within this section, this study began with a detailed investigation of the two endpoints of the most fossiliferous and well-studied portions of the section. The 20 bulk samples extracted from lithostratigraphic units G1 and G9 included 16,261 individuals from 329 species. This seemingly large sample nevertheless represents only a small portion of the exposed section. Future efforts will expand our current work and explore mollusc diversity and paleoecology in the intervening lithostratigraphic units. When this work is complete, it will be possible to more precisely explore the concordance and discordance of species-level and community-level evolution. The two units (G1 and G9) that served as the focus of our present study are separated by approximately 1 million years (McNeill et al., this volume), which is much less time than originally thought (Saunders et al., 1986).

Although prior work hypothesized that the Cercado and Gurabo Formations contained different biofacies (Saunders et al., 1986), this hypothesis was not tested using quantitative paleoecological analyses of appropriately collected samples. Prior work relied on a very small subset of species and did not include comparisons of samples collected in a standardized manner. Additionally, considering that morphological stasis had been documented within and among invertebrate species from both formations, it remained to be seen whether the communities from these different environments were in fact significantly different. Thus, although it was likely that the faunas differed between the two formations, the aforementioned reasons were considered justification for testing this hypothesis more rigorously (Fig. 10.14).

In order to investigate the stability of mollusc assemblages in the Rio Gurabo section, we gathered data on species richness (standardized using rarefaction analyses), faunal carryover/holdover, assemblage similarity (quantified using MDS, ANOSIM, and SIMPER) and trophic mode distributions. Only a few localities that

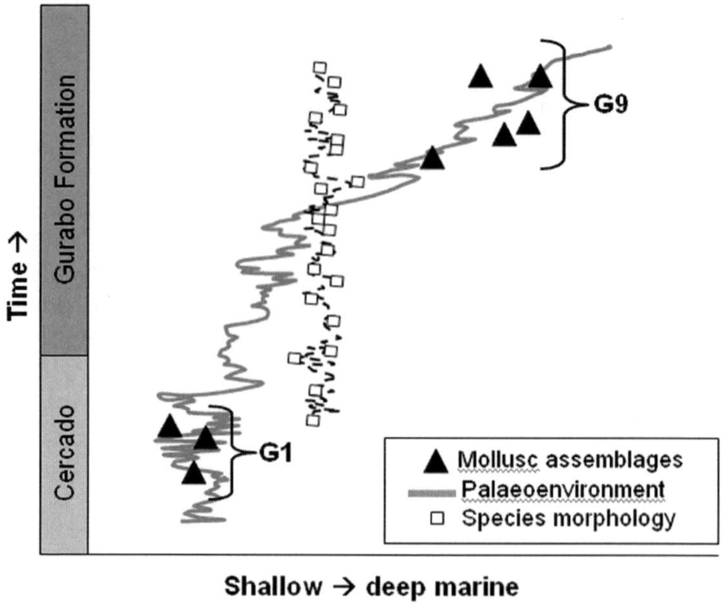

Fig. 10.14 Simplified model of the Rio Gurabo section through time. The main variables of interest in this section are mollusc assemblages, paleoenvironmental change, and species morphology

we studied were found to display high levels of faunal similarity. Localities 15966 and 15813, for example, overlapped in an MDS plot, had an ANOSIM R-value of 0.11, and were characterized by a carryover value of 60% (the highest value among localities in lithostratigraphic unit G9). The Lower Gurabo localities 15904 and 15907 were clustered together in an MDS plot, had an ANOSIM R-value of 0.33, and were characterized by a carryover value of 83% (the highest value among localities in lithostratigraphic unit G1). No other comparisons produced high levels of faunal similarity.

Our comparisons of lithostratigraphic units G1 and G9, which occur in the Cercado Formation of the lower Rio Gurabo section and the Gurabo Formation of the upper Rio Gurabo section respectively, produced very few indications of faunal similarity. Raw estimates and standardized comparisons of species richness indicated that unit G9 is characterized by significantly greater species richness than unit G1. The ANOSIM Global R value (between G1 and G9) was 0.94, which indicates low faunal similarity. A 33% carryover value between units corroborates this conclusion. SIMPER analyses revealed that six bivalve species contributed to approximately 50% of the dissimilarity between the upper and lower Gurabo units and that most of these species only occurred in lower Gurabo unit G1 (*Scapharca corcupidonis, S. riocanensis, Pitaria acuticostata* and *Strigilla psiformis*). The two remaining species (*Pecten Thompsoni* and *Phacoides yaquensis*) existed in both units but displayed different abundances.

The mollusc assemblages of lithostratigraphic units G1 and G9 differed in their faunal composition. The bivalve species that dominated G1 were *Scapharca, Pecten,* and *Phacoides*. Collectively, they accounted for 53% of the fauna. *Scapharca riocanensis* and *S. corcupidonis* represented 28% of the total individuals in G1. Unit G1 trophic mode distributions (Figs. 10.6–10.8) illustrated the dominance of infaunal siphonate genera. The bivalves that dominated unit G9 were *Pecten, Barbatia, Ostrea,* and *Glycimeris*.

The presence and distribution of gastropod species also differed between units. The most abundant gastropods in the lower Gurabo unit G1 were *Oliva* and *Terebra*. These predatory carnivores accounted for about 50% of the gastropods in this zone. Overall, predatory carnivores comprised 87% of the individuals in lower Gurabo unit G1. The most abundant gastropods in the upper Gurabo unit G9 were the suspension feeder *Petalochoncus* and several predatory carnivore species from the genera *Oliva* and *Conus*. Predatory carnivores dominated upper Gurabo unit G9, but fluctuations occurred in their abundance through time (see Fig. 10.7, notably localities 15809–15942).

Using Brett and Baird's (1995) benchmark of 60% carryover as indicative of faunal stability, both coral and mollusc data suggest that little faunal stability characterizes lithostratigraphic units G1 and G9 in the Rio Gurabo. Between these two units, mollusc carryover was very low (23%). Although Klaus and Budd (this volume) found that coral persistence was 89% from the Cercado to the Gurabo Formation, coral data from units G1 and G9 alone indicate a faunal persistence of only 6%. NMS analyses of both coral and mollusc data also reveal faunal differences between units G1 and G9. Similar to our NMS analyses of mollusc species and abundance (Fig. 10.11), Klaus et al. (this volume) illustrated the spatial separation of the coral faunas from units G1 and G9 in two-dimensional NMS plots. The corals from both lithostratigraphic units did, however, fall within Klaus and Budd's "free-living" faunal group. Although Klaus and Budd did not perform an ANOSIM analysis on units G1 and G9, the ANOSIM global R value for mollusc communities was 0.87 (Table 10.3), indicating large faunal differences. Comparisons of species richness between units also revealed similar patterns in corals and molluscs: lithostratigraphic unit G9 has more species than unit G1 (Klaus et al. this volume). Molluscs are significantly more diverse than corals, however, in Gurabo lithostratigraphic units G1 and G9: twenty coral species have been identified whereas 329 species of molluscs were identified in the present study. In summary, both mollusc and coral assemblages in Gurabo units G1 and G9 display significant faunal differences.

10.6 Conclusions

Although several well-documented cases of morphological stasis have been found to occur in the Rio Gurabo section, an analysis of 20 bulk samples containing more than 16,000 specimens from 329 mollusc species revealed that mollusc assemblages lack stability in composition, relative abundance, species richness, and some aspects of trophic distribution. Considerable patchiness also characterized individual localities, as indicated by marked variation in species richness, trophic distributions, and faunal patterns among replicate bulk samples. More extensive lateral and temporal sampling will be necessary in order to accurately characterize mollusc paleocommunities in the Rio Gurabo section. Overall, however, faunal stability does not appear to characterize this stratigraphic section.

Acknowledgments We thank Brian Beck, Novia Jarrett, Jonathan Baez, Victor Matos, and Moziah Saad for assistance in the field, James Klaus and Emily Vokes for reviews of the manuscript, and the National Science Foundation for financial support.

References

Anderson, L.C., 1994, Paleoenvironmental control of species distributions and intraspecific variability in Neogene Corbulidae (Bivalvia: Myacea) of the Dominican Republic, *J. Paleontol.,* **68**:460–473.

Baird, G.C. and Brett, C.E., 1983, Regional variation and paleontology of two coral beds in the Middle Devonian Hamilton Group of western New York, *J. Paleontol.,* **57**:417–446.

Bennington, J.B. and Bambach, R.K., 1996, Statistical testing for palaeocommunity recurrence: are similar fossil assemblages ever the same? *Palaeogr. Palaeoclim. Palaeoecol.*, **127**:107–133.

Bennington, J.B. and Rutherford, S.D., 1999, Precision and reliability in paleocommunity comparisons based on cluster confidence intervals: how to get more statistical bang for your sampling buck, *Palaios,* **14**:506–515.

Brett, C.E. and Baird, G.C., 1995, Coordinated stasis and evolutionary ecology of Silurian to Middle Devonian faunas in the Appalachian Basin, in: *New Approaches to Speciation in the Fossil Record* (Erwin, D.H. and R.L. Anstey, eds.), Columbia University Press, New York, pp. 285–315.

Bold, W.A. van den, 1988, Neogene palaeontology of the northern Dominican Republic, 7, The subclass Ostracoda (Arthropoda: Crustacea), *Bull. Am. Paleontol.,* **94**:1–105.

Bonelli, J.R., Brett, C., Miller, A.I., and Bennington, J.B., 2006, Testing for faunal stability across a regional biotic transition: quantifying stasis and variation among recurring coral-rich biofacies in the Middle Devonian Appalachian Basin, *Paleobiology*, **32**:20–37

Cheetham, A.H., 1986, Tempo of evolution in a Neogene Bryozoan: rates of morphologic change within and across species boundaries, *Palaeobiology,* **12**:190–202.

Cheetham, A.H., 1987, Tempo of evolution in a Neogene Bryozoan: are trends in single morphologic characters misleading?, *Palaeobiology*, **13**:286–296.

Cheetham, A.H. and Jackson, J.B.C., 1995, Process from pattern: tests for selection versus random change in punctuated bryozoan speciation, in: *New Approaches to Speciation in the Fossil Record* (Erwin, D. and Anstey, R., eds.), Columbia University Press, New York.

Clarke, K.R. and Warwick, R.M., 2001, *Change in Marine Communities: An Approach to Statistical Analysis and Interpretation*, Primer-E Ltd, Plymouth.

Costa, F.H.A., Nehm, R.H., and Hickman, C., 2001, Neogene Palaeontology in the northern Dominican Republic, 22, The Family Neritidae, *Bull. Am. Paleontol.,* **359**:47–71.

Foster, A.B., 1986, Neogene palaeontology in the Northern Dominican Republic, 3, The Family Poritidae (Anthozoa: Scleractinia), *Bull. Am. Paleontol.,* **90**:47–123.

Foster, A.B., 1987, Neogene palaeontology in the northern Dominican Republic, 4, The Genus *Stephanocoenia* (Anthozoa: Scleractinia: Astrocoeniidae), *Bull. Am. Paleontol.,* **93**:5–22.

Gould, S.J., 2002, *The Structure of Evolutionary Theory*, Harvard University Press, Cambridge.

Hammer, Ø. and Harper, D.A.T., 2006, *Paleontological Data Analysis*, Blackwell, Oxford.

Holland, S.M., 2003, Analytic Rarefaction 1.3., Available online at: http://www.uga.edu/strata/software/AnRareReadme.html

Holland, S.M., and Patzkowsky, M.E., 2004, Ecosystem structure and stability: middle Upper Ordovician of central Kentucky, *Palaios,* **19**:316–331.

Ivany, L.C., 1996, Coordinated stasis or coordinated turnover? Exploring intrinsic vs. extrinsic controls on pattern, *Palaeogr. Palaeoclim. Palaeoecol.*, **127**:239–256.

Jackson, J.B.C., 1994, Community unity?, *Science,* **264**:1412–1413.

Maury, C.J., 1917a, Santo Domingo type sections and fossils, Part 1, *Bull. Am. Palaeontol.,* **5**:29, 1–251.

Maury, C.J., 1917b, Santo Domingo type sections and fossils, Part 2, *Bull. Am. Palaeontol.,* **5**:30, 1–43.

McCune, B. and Grace, J.B., 2002, Analysis of ecological communities, MjM Software, Gleneden Beach, Oregon, USA

Miller, A.I. and Foote, M., 1996, Calibrating the Ordovician Radiation of Marine Life: implications for Phanerozoic Diversity Trends, *Palaeobiology*, **22**:304–309

Morris, P.J., 1996, Testing patterns and causes of faunal stability in the fossil record, with an example from the Pliocene Lusso Beds of Zaire, *Palaeogr. Palaeoclim. Palaeoecol.*, 127.

Morris, P.J., Ivany, L.C., Schopf, K.M., and Brett, C.E., 1995, The challenge of palaeoecological stasis: reassessing sources of evolutionary stability, *Proc. Natl. Acad. Sci.* USA, **92**(11):269–273.

Nehm, R.H. and Budd, A.F., 2001, Species-level and Community-level Stability: Case Studies from the Dominican Republic Neogene, (Session 11), In: Program & abstracts; NAPC 2001, North American paleontological convention 2001, Paleontology in the new millennium, *PaleoBios*, **21**, 2.

Nehm, R.H., 2005, Patterns and processes of evolutionary stasis and change in *Eratoidea* (Gastropoda: Marginellidae) from the Dominican Republic Neogene, *Carib. J. Sci.,* **41**:189–214.

Nehm, R.H. and Geary, D., 1994, A gradual morphologic transition during a rapid speciation event in marginellid gastropods (Neogene; Dominican Republic), *J. Paleontol.*, **68**:787–795.

Pflug, H.D., 1969, Molluscen aus dem Tertiar von St. Domingo, *Acta Humboltiana*, Series Geologica et Palaeontologica, NR **1**:1–107.

Saunders, J.B., Jung, P., Geister, J., and Biju-Duval, B., 1982, The Neogene of the south bank of the Cibao Valley, Dominican Republic: a stratigraphic study, *Trans. 9th Carib. Geol. Conf., Santo Domingo*, 1980.

Saunders, J.B., Jung, P., and Biju-Duval, B., 1986, Neogene palaeontology of the northern Dominican Republic, 1, Field surveys, lithology, environment, and age, *Bull. Am. Paleontol.,* **89**:1–79.

Todd, J.A., 2001, Introduction to molluscan life habits databases, accessed online at http://eusmilia.geology.uiowa.edu/database/mollusc/mollusclifestyles.htm

Chapter 11
The Impact of Fossils from the Northern Dominican Republic on Origination Estimates for Miocene and Pliocene Caribbean Reef Corals

Kenneth G. Johnson[1], Ann F. Budd[2], James S. Klaus[3], and Donald F. McNeill[3]

Contents

11.1 Introduction

Documenting the pattern of biodiversity change on global, regional, or local scales is an important use of paleontological data. Long-term trends of taxonomic richness and the rate of first and last occurrences of taxa can be used to study effects of environmental change on ecosystems and is of significant relevance during this time of increased concern over the effects of anthropogenic environmental change. Compilations of taxon occurrences provide the basic data used to describe the

[1]Department of Palaeontology, Natural History Museum, Cromwell Road, London SW7 5BD, United Kingdom. Email: K.Johnson@nhm.ac.uk

[2]Department of Geoscience, University of Iowa, 121 Trowbridge Hall, Iowa City, IA 52242, USA. Email: ann-budd@uiowa.edu

[3]Department of Geological Sciences, University of Miami, 43 Cox Science Building, Coral Gables, FL, 3133. Email: jsklaus@rsmas.miami.edu

R.H. Nehm, A.F. Budd (eds.) *Evolutionary Stasis and Change in the Dominican Republic Neogene*,

distributions of taxa in time and space. Paleontologists have been producing hard-copy and digital compilations for many decades (for example, Valentine, 1969; Raup, 1972; Sepkoski, 1982, 2002), and a large suite of sophisticated methodologies have been developed for the statistical analysis of these data (for example, Gilinsky and Signor, 1991). Most of the existing compilations are limited to particular taxonomic groups within restricted geographic and stratigraphic scope, but some intend to be global in scope (for example, Alroy, 2007). However, no compilation can ever be comprehensive or truly authoritative because ongoing research based on new field sampling as well as study of existing museum collections will continue to generate new information requiring the enlargement or modification of data compilations. Indeed, one of the most difficult tasks for compilation managers is determining how best to increase the utility of synthetic compilations. Is it better to prioritize the addition of new data over increasing the accuracy of existing information? Similar compromises must be met to determine which data to include and which to exclude in the absence of unlimited resources. For any given question, it is likely that some subsets of data will be more influential than other subsets, and a rational choice would be to prioritise increasing the accuracy of the more influential data rather than the existing practice of concentrating effort based on particular research interests of data contributors, the ready availability of existing data sets, or the search for 'even' coverage. The aim of this chapter is to explore a basic methodological approach to estimate the relative influence of subsets of data. We will apply these new techniques to ongoing study of long-term patterns of taxonomic turnover dynamics in Cenozoic Caribbean Scleractinian reef-coral occurrences to determine the importance of collections from the Dominican Republic in our understanding of patterns of species origination and extinction in response to regional environmental change.

11.1.1 Maintaining Paleontological Occurrence Data

The general approach common to many paleobiodiversity studies is to compile occurrences of taxa from a set of samples. The occurrences can be based on primary data such as specimens in collections of fossils, or secondary data such as lists of taxa extracted from the published literature. Two independent kinds of interpretations are compiled within these occurrence data sets. The first set includes information about the samples including stratigraphic position as age, geography, and paleoenvironmental interpretations. Taxonomic determinations are the second kind of data included in typical occurrence compilations. Both classes of information include inferences that are likely to change with increasing research. Specimen identifications will change as new material is collected and new systematic studies necessitate taxonomic revisions, but sample-related interpretations will also change with further research. New stratigraphic results may require changes to the stratigraphic range assigned to samples, but even in the absence of new data, the continuous refinement of geologic time scales require periodic reassessment of age assignments of individual samples. For example, the inferred age of the Chattian/Aquitanian boundary has changed from 23.7 to 23.8 to 23.03 Ma in the Berggren et al. (1985), Berggren et al. (1995), and Gradstein et al. (2004) geologic time scales. Therefore, any sample with an upper or lower limit assigned to the Chattian/Aquitanian boundary will require a new age interpretation for each update of the time scale. To date, much effort has been expended compiling existing information into occurrence data, but relatively little attention has been devoted to maintaining data sets by incorporating updated interpretations about existing occurrences (but see Adrain and Westrop, 2000; Alroy, 2000). Maintenance may include a variety of activities including both new and existing information. Examples include incorporating new information that results from resampling of key localities or key stratigraphic intervals, checking existing taxonomic determinations or age assignments by examining collections, applying new time scales or taxonomic frameworks (this may also require re-determining existing collections), or incorporating interpretations from new or newly encountered publications.

The need to maintain paleontological occurrence data sets presents potential pitfalls for compilation owners with respect to database design and prioritisation of effort. We have been compiling and maintaining an occurrence database of Cenozoic Caribbean reef-corals for the past 20 years (Budd et al., 1993, 1994) and have learned some interesting lessons regarding the challenge of data maintenance. A full discussion of database design is beyond the scope of this contribution, but future maintenance can be simplified greatly if this need is accommodated during the database design process. For example, if secondary sources are used for taxonomic occurrences, the recording of voucher specimen details will ensure that that determinations can be re-verified as taxonomic concepts are refined through ongoing study. Without this information, it is not possible to update taxonomic determinations because changing taxonomic practice can lead to splitting of taxa or partial synonymy. These changes cannot be incorporated into existing data sets if the chain of evidence between specimen and recorded name has been broken by exclusion of specimen data from the compilation. Similar design considerations apply to maintaining stratigraphic information about samples, in that tracking basic information facilitates maintenance as interpretations change. In practice, this requires tracking the source of an age pick as a set of zones that can be calibrated against various geologic time scales rather than recording the resulting chronostratigraphic pick. This means that increasingly detailed information needs to be incorporated, a process that inevitably slows the process of assimilating new occurrence records into a compilation. However, our experience shows that making short cuts at the initial stages of pulling together occurrence data sets will result in headaches and high costs to maintaining systems in the future. A more serious consequence of not tracking primary data is the introduction of artifacts into the system by uneven or incomplete revision of data because some records are amenable to update and others are not. Probably the most basic design error is to not record the source of each interpretation along with a time stamp. This allows multiple versions of the data set to be compared and provides an archive of previous interpretations that may be critical to reproduce past analyses and check the effects of new interpretations on previous studies.

The problem of maintenance grows as occurrence databases increase in size and number, and owners of data sets of even modest size eventually need to develop a set of priorities for keeping information up-to-date. Probably not every piece of information in a database carries the same influence for solving particular problems, and priority for maintenance should be placed on the more influential set of records. For example, when documenting patterns of extinction, the last occurrence of each taxon is the critical record that could be checked to ensure correct identification. Occurrence records that fall within the stratigraphic range of a taxon are less influential in determining the taxon range. When documenting extinction rates, highest priority might be assigned to checking last occurrences that are separated from 'second-to-last' occurrences by a large gap, or on verifying the last occurrences with long inferred confidence intervals (Marshall, 1994). But assigning priorities is not straightforward if the database is used to address more complex hypotheses or to test multiple hypotheses. For an example, we will consider our database of Cenozoic Caribbean reef-coral occurrences.

11.1.2 The Cenozoic Caribbean Coral Database

Previous analyses of a specimen-based compilation of Neogene occurrences suggested that two intervals were characterized by accelerated faunal change (Budd et al., 1994; Johnson et al., 1995; Budd and Johnson, 1999), including an extinction pulse and an interval of accelerated origination. A strong peak of extinction occurred during the interval from 2 to 1 Ma. This interval contains the Plio-Pleistocene boundary coinciding with intensification of northern hemisphere glaciation as well as the 'Mid Pleistocene Revolution' (MPR; Berger and Jansen, 1994; Maslin and Ridgwell, 2005). The MPR was a rapid switch in the mode of global climate cycles from a relatively low-amplitude obliquity-dominated cycle of 41 Ka to a high amplitude precession-linked cycle of 100 Ka that occurred between 1.2 and 0.5 Ma. Besides resulting in large amplitude, low frequency cycles, post-MPR cycles are characterized by a distinctive sawtooth waveform resulting from the slow buildup of ice followed by rapid melting. Sea level would have risen rapidly during these relatively brief melting phases, including intervals of increases on the order of 1–2 m/century (Overpeck et al., 2006). Such rapid rates would have drowned coral communities that were not able to keep up and triggered extinctions of species, so the MPR is likely to have played a role in the Pleistocene reef-coral extinction pulse. The interval of accelerated origination is less well constrained, but occurred either in the Late Miocene or Early Pliocene. Differences in estimates of the timing and intensity of this event result from analyses of different subsets of species occurrences and variation in taxonomic practice. Early versions of the Cenozoic Caribbean Coral Database (Budd et al., 1994; Johnson et al., 1995; Budd and Johnson, 1999) show a sharp pulse of origination during the interval from 4–3 Ma, but either no strong pulse or a less intense increase in origination during the Late Miocene is apparent in later analyses (Jackson and Johnson, 2000; Johnson, 2001). The main difference between these results is the inclusion of more samples from Early and Middle Miocene rocks including the lower part of the Seroe Domi Formation in Curacao (Budd et al., 1998) and the Tamana Formation of Trinidad (Johnson, 2001). Adding new information clearly can change estimates of taxonomic turnover, in this case, newly discovered species occurrences in the Early and Middle Miocene required extension of ranges back into these older intervals and diminished the previously observed Early Pliocene origination pulse. In this chapter, we will look at the influence of samples of fossil corals from the Dominican Republic in estimates of taxonomic turnover in Caribbean reef-coral fauna of the Caribbean. We will especially consider the effects of a new age model for Late Miocene to Pliocene sections in the Northern Dominican Republic.

11.1.3 Old and New Data from the Cibao Valley, Dominican Republic

Collections of fossil reef-corals from the Cibao Valley of the Dominican Republic are a major part of the Cenozoic Caribbean coral compilation. Of the 11,032 fossil specimens included in the Late Miocene to Late Pleistocene samples in the Caribbean coral database, 4,499 specimens (41%) were collected from sections of the Cercado, Gurabo, and Mao Formations exposed along the Cana and Gurabo rivers on the southern flank of the Cibao Valley. These collections include material obtained during the 1980 Swiss field expeditions organised by Peter Jung and John Saunders (Saunders et al., 1986) and new collections from the past 15 years that are described in Klaus et al. (this volume). Stratigraphic study of these sections during the 1980 field excursions resulted in a set of age assignments based on zonation of nannofossils and planktonic foraminifera calibrated by the Berggren et al. (1985) time scale. These dates have been used in all previous analyses of coral turnover that included records from the Cibao Valley. However, new integrated age models have been produced for sections exposed in the valley walls of the Rio Cana and Rio Gurabo (McNeill et al., this volume) suggests that the sections are considerably younger than was previously interpreted.

In this chapter, we will consider the effects of these new age dates on our understanding of diversity dynamics in Neogene Caribbean reef-corals. We also take the opportunity to assess the influence of the samples from the Cibao Valley as a whole, in an attempt to better understand difficulties in documenting Late Miocene and Early Pliocene record of species origination in the region. The general approach will be to compare sets of stratigraphic ranges of taxa and estimates of taxonomic turnover calculated from different combinations of new and old data. Analysis of the resulting differences and similarities will then be used to develop priorities for future sampling and maintenance of our existing compilation.

11.2 Materials and Methods

11.2.1 Data Sets

The data sets compiled for this study were based on the 2007 version of the Caribbean Coral Database that includes specimen-based occurrences of fossil corals from across the region supplemented with selected lists of taxa taken from the published literature. For the current study, only samples from the Late Miocene to Late Pleistocene were included (Fig. 11.1). This set includes specimen-based compilations from the Bahamas Drilling Program (Budd and Manfrino, 2001), the Limon Basin of Costa Rica (Budd et al., 1999), Curacao (Budd et al., 1998), the Cibao Valley of the Dominican Republic (Budd et al., 1994), Jamaica (Budd and McNeill, 1998), and the Bocas del Toro Basin of Panama (Budd, A.F., 2007, unpublished data). All specimens were identified to the species level using characters presented in the NMITA database (Budd et al., 2006). Collections from sections exposed along the Rio Yaque del Norte, Dominican Republic were excluded from the data set as new age models are not available for these sections. The Late Miocene to Late Pleistocene specimen-based data were supplemented with published species lists from the Late Pleistocene of the southern coast of the Dominican Republic (Klaus and Budd, 2003) and the Late Pleistocene Ironshore Formation of Grand Cayman (Hunter and Jones, 1996).

A total of 11,078 specimens were examined and when combined with Late Pleistocene records, the data set includes 2,324 occurrence records of 171 species

Fig. 11.1 Locality map showing the six regions from which collections of Late Miocene to Late Pleistocene reef-corals were recovered. The regions include the Cibao Valley of the northern Dominican Republic, the Limon Basin of Costa Rica, the Bocas del Toro Basin of Panama, coastal sections on the islands of Curacao and Jamaica, the Pleistocene carbonates of Grand Cayman and material from bore holes into the Great Bahama Bank platform west of Andros Island

(Table 11.1). The stratigraphic ranges of some of these taxa extend beyond the study interval (Late Miocene to Late Pleistocene), so a Late Oligocene - Middle Miocene list of species and a list of extant species were added to the compilation to act as range endpoints. Late Oligocene species lists were obtained from collections of fossils from the Antigua Formation of Antigua (Johnson, 2007), Lares and 'Juana Diaz Formations' of Puerto Rico (Frost et al., 1983; Johnson, K. G., 2007, unpublished data), and the White Limestone of Jamaica (Stemann, 2004; Johnson, K. G., 2007, unpublished data). Early and Middle Miocene occurrences were compiled from collections from the Anguilla Formation of Anguilla (Budd et al., 1995), Culebra and Valiente Formations of Panama (Johnson and Kirby, 2006; Budd, A. F., 2007, unpublished data), Seroe Domi Formation of Curacao (Budd et al., 1998), and Tamana Formation of Trinidad (Johnson, 2001). Species lists were also included from the Lower Miocene Tampa Limestone Member of the Arcadia Formation (Weisbord, 1973) and the Middle Miocene Chipola Formation (Weisbord, 1971). Recent records were compiled from various sources in the published literature (Wells and Lang, 1973; Dustan and Halas, 1987; Tomascik and Sander, 1987; Holst and Guzmán, 1993; Guzmán and Holst, 1994; Chiappone et al., 1996; Guzmán and Guevara, 1998; Fenner, 1999).

Age ranges are based on a variety of sources, in most cases including magnetostratigraphy. Some strontium isotope dates were also available, and a full description of age assignments can be found in the publications cited for each part of the data set. In each case, age ranges were assigned to samples. These age ranges should not

Table 11.1 Numbers of species, samples, and occurrences from each geographic region. Age dates for the Cibao Valley collections from McNeill et al. (this volume). Samples from the Cibao Valley and Late Pleistocene Terraces from the southern Dominican Republic (Dom. Rep.) are distinguished, because only those samples from the Cibao Valley were excluded from the NODR data set. Numbers of specimens are not available for Late Pleistocene samples from Grand Cayman and the Dominican Republic because presence/absence data was extracted from the published literature

	Bahamas	Cayman Islands	Costa Rica	Curacao	Dom. Rep.	Dom. Rep. Terrace	Jamaica	Panama	Total
Number of species, endemic species, occurrences, and specimens									
Species	47	31	83	89	102	20	75	55	171
Endemic Sp	0	4	5	8	21	0	1	1	–
Occurrences	67	162	649	522	499	43	298	84	2,324
Specimens	132	–	2,487	2,230	4,499	–	1,510	220	11,078
Number of faunules									
L. Pleistocene	0	10	0	2	0	3	2	0	17
E. Pleistocene	1	0	15	1	0	0	12	1	30
Late Pliocene	1	0	19	6	2	0	6	4	38
E. Pliocene	1	0	1	16	11	0	0	0	29
L. Miocene	0	0	0	2	6	0	0	0	0
Total	3	10	35	27	19	3	20	5	124

be interpreted as representing actual depositional rates for each section, but instead indicate the level of uncertainty associated with each age assignment. To facilitate analyses, collections were merged into faunules that include all samples from a limited stratigraphic range and geographic scope. This procedure has no effect on estimates of stratigraphic ranges of species, because collections within each faunule have been assigned identical age ranges (Fig. 11.2). A summary of faunules including assigned age ranges, and the number of collections, specimens, and species richness is provided in the Appendix.

Four data sets were compiled for this study. The DROLD set includes the all Late Miocene to Late Pleistocene samples with age assignments for collections from the Cibao Valley derived from the stratigraphy in Saunders et al. (1986) as compiled in Budd et al. (1994). The DRNEW data set includes the same faunules, but with age dates for the Cibao Valley obtained from the new age models of McNeill et al. (this volume). The NODR data set excludes all collections from the Cibao Valley, but includes Late Pleistocene records from the southern coast of the Dominican Republic. The NOCR data set includes all data from the Dominican Republic with new age assignments from McNeill et al. (this volume), but excludes all collections from the Pliocene and Early Pleistocene of the Limon Group of Costa Rica (Budd et al., 1999). We use the NOCR data set to determine the influence of collections from Costa Rica on the observed pattern of taxonomic turnover as compared to the impact of samples from the Cibao Valley. Collections from Costa Rica were selected because they represent the second largest regional subset of the total data set (2,487 specimens or 22.5%), and include collections that constrain the timing of the Plio-Pleistocene extinction pulse.

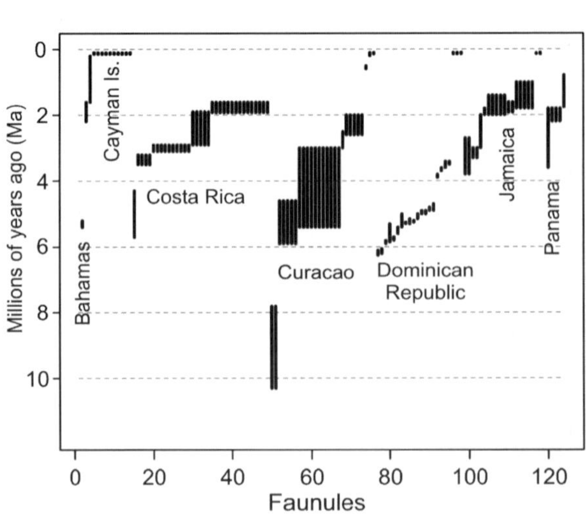

Fig. 11.2 Stratigraphic intervals assigned to each faunule in the analysis. The range of each faunule is indicated by a vertical line, and the ranges are sorted in by geographical region and then in stratigraphic order

11.2.2 Calculating and Comparing Stratigraphic Ranges

In general, the stratigraphic range of a taxon is estimated as extending from the age of the sample in which the taxon first occurs (SFO) to the age of the sample in which the taxon last occurs (SLO). However in the current case, the age of each faunule is expressed as an interval, so there is no unique 'age' for each faunule. At least three simple methods could be applied following the approach of Johnson and McCormick (1999), including 'short ranges', 'long ranges', and 'midpoint ranges' (Fig. 11.3A). Short ranges extend from the upper limit of the age interval assigned to the SFO to the lower limit of the age assigned to the SLO, and long ranges extend from the lower limit of the age interval assigned to the SFO to the upper limit of the age range assigned to the SLO. As the name suggests, midpoint ranges extend from the median of the age interval assigned to the SFO to the median of the age interval assigned to the SLO. We will calculate long ranges in this study, because (1) short ranges can have negative duration if the stratigraphic ranges of the SFO and SLO overlap such that the upper limit of the SFO is greater than the lower limit of the SLO, and (2) using median ages assumes that the uncertainty assigned to samples is symmetrical about a median value. This assumption is likely to be invalid in cases where upper and lower stratigraphic limits are based on different data sources with variable precision, for example a biostratigraphic datum and a strontium date. Long ranges are also a conservative choice, and it is appropriate to bias estimates of stratigraphic ranges towards greater extension because observed range endpoints will always underestimate the true stratigraphic ranges of taxa as it is extremely unlikely that the true first or last occurrence of a taxon will be captured in the fossil record.

Both changing the ages assigned to samples and including different subsets of samples can alter estimates of stratigraphic range for a taxon, either directly by changing the inferred age of the SFO or SLO or indirectly if the changed sample is no longer the stratigraphically lowest or highest sample in which a taxon occurs. In general, nine types of changes can occur for each range compared in two sets. These correspond to combinations of the status of range limits - so that both the upper limit and the lower limit can either not change or can shift up- or down- section. The nine modes of range change can be shown schematically in a matrix with three rows and three columns (Fig. 11.3B). Using a subset of samples to estimate stratigraphic ranges will result in fewer modes of range change because removing samples can never result in range extensions. Excluding the SFO for a taxon will result in a contraction of the base of the range to the age of the next sample in which the taxon occurs, and excluding a SLO will result in a contraction of the upper limit of the observed range. In addition to causing range contraction, excluding subsets of samples can also result in the disappearance of taxa from a range set if the taxa only occur in samples that have been excluded. To compare sets of stratigraphic ranges estimated using different data sets we will first exclude all taxa that do not occur in both range sets, and then measure the additive amount of change within each of the nine modes (Table 11.2). The amount of change will be measured both as the number of ranges responding in each of the nine modes of change, and the total length of range extension or contraction resulting from the eight modes that include some range change.

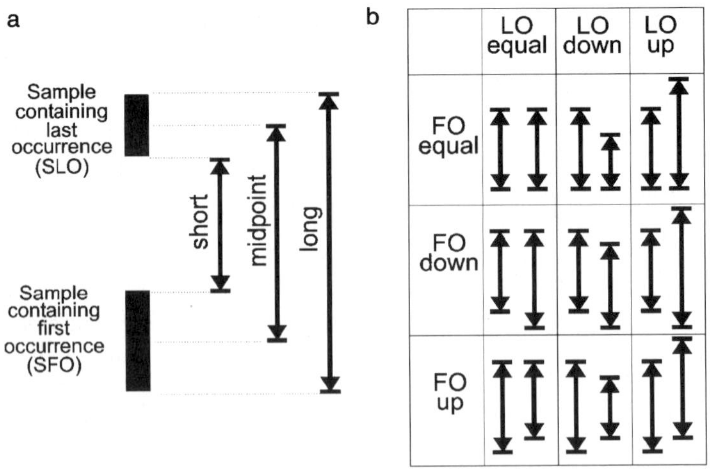

Fig. 11.3 Cartoons illustrating methods for calculating and comparing stratigraphic ranges. (**a**) Three different approaches to calculating stratigraphic ranges of taxa are possible if uncertainty in the age of samples is considered. (**b**). Comparisons of ranges results in a matrix with nine cells, each a combination of three possible responses in the upper limit of ranges and three modes of change in the lower limit of ranges

Table 11.2 Comparison of stratigraphic ages of reef-coral species in the OLDDR and NEWDR data sets, including total length of species ranges in each data set and a matrix showing the total length of change and number of species ranges (in parentheses) for each type of response

	Change in combined durations			
	OLDDR	NEWDR	Difference	Percent change
Total duration	1,004.785	910.485	94.3	9.39
	Changes in range endpoints			
	LO no change	FO down	FO up	Total
FO no change	0 (108)	0.10 (1)	3.2 (7)	3.3 (116)
FO down	0 (0)	0 (0)	0 (0)	0 (0)
FO up	69.15 (35)	2.1 (1)	49.45 (19)	120.7 (55)
Total	69.15 (143)	2.2 (2)	52.65 (26)	124 (171)

11.2.3 Calculating and Comparing Diversity and Taxonomic Turnover

To study the temporal pattern of change in regional reef-coral diversity during the Neogene, species richness and the number of first and last occurrences of species were counted within a series of stratigraphic bins. Selection of bin size and the position of boundaries between adjacent bins can have a marked effect of resulting estimates (Foote, 2000). For example, short pulses of extinction or origination may not be detectable if bin lengths are sufficiently long that the effects of the pulses are lost in long counts of background turnover. Similarly, inadvertently placing a bin boundary in the midst of an interval of accelerated turnover could have the effect of masking the turnover event as counts of extinction or origination are partitioned between adjacent bins. A related problem is determining whether or not a first or last occurrence event falls within a particular bin in cases where the age ranges assigned to a SFO or SLO cross more than one stratigraphic bin. We attempt to overcome potential pitfalls by selecting an arbitrary bin length of one myr and dividing bins between 11 Ma and Recent and then applying counting methods that are designed to be robust to bin selection. The median length of the pre-Late Pleistocene faunules included in the NEWDR data set is 0.6 myr, and the one-myr bin length was selected as a compromise so that most sample ages can be contained within a single bin but that bins were not too long to miss detecting relatively rapid pulses of taxonomic turnover.

Regional species richness was calculated within the one-myr bins using the range-through technique, so that species were inferred to be present in the region in all bins that fall between the bin containing the lower limit of the age assigned to the SFO and the bin containing the upper limit of the age range assigned to the SLO. A variety of alternative measures of within-bin richness can be estimated (Foote, 2000), but the range-through measure is adequate to document large-scale trends in regional richness because it is not strongly influenced by uneven sampling except near the edges of the total stratigraphic interval under consideration. We are confident that in the overwhelming majority of cases, the absence of species that are expected to occur in bins results from inadequate sampling rather than temporary regional extinction followed by re-invasion from outside the Caribbean. Besides, range-through richness is highly correlated with within-bin richness (Spearman rank correlation for the NEWDR data, $rho = 0.98$, $p < 0.001$).

The rate of reef-coral species turnover within the Caribbean region is estimated by counts of species first and last occurrences within the series of one-myr stratigraphic bins. However, many of the SFOs or SLOs have been assigned age ranges that cross more than one bin, so credit for each origination or extinction event is proportionally distributed to all crossed bins. For example, if a taxon last occurs in a sample that has been assigned a stratigraphic age that crosses two bins, then the total number of last occurrences in each of the crossed bins is incremented by one half. If three bins had been crossed by the SLO, then the totals for the three bins would have been increased by one third each. This approach is required because for the NEWDR data set, only 38 (36%) of the 105 pre-Late Pleistocene faunules have been assigned age ranges that fall within a single one-myr bin. Weighting numbers of first and last occurrence counts by the length of the stratigraphic range assigned to the SFOs and SLOs may result in smoothed turnover pattern, especially if stratigraphic resolution of samples is coarse relative to bin width. However, alternative approaches such as forcing all samples into a single bin can result in falsely pulsed turnover intervals.

Counts of taxa (richness or the numbers of first and last occurrences) are known to be strongly influenced by sampling completeness, so we compare these measures with the number of faunules that cross each of the 1 myr stratigraphic bins. The number of faunules provides a rough indication of sampling effort that can be used to identify visually likely biases in the observed pattern of faunal change. Gaps in sampling can create spurious peaks of origination following the gap or artificially inflated rates of extinction prior to the sampling gap (Koch, 1987). Turnover estimates are compiled from the distribution of stratigraphic ranges of taxa, so varying the ages assigned to a sample or excluding samples from the analysis can result in different turnover patterns. Removing or changing the interpreted ages of samples can also alter the distribution of sampling gaps in the record and remove or create artifacts in the turnover pattern that result from uneven sampling. These effects may obscure true patterns well above and below the actual gap. For example, a sampling gap in the Middle Miocene could obscure patterns in the late Pliocene by resulting in the overestimation of species origination during the Late Pliocene. These effects will be even stronger if there is a geographic partitioning of the regional species pool so that archaic assemblages persist in refugia after they have become extinct elsewhere in the region. Inclusion or exclusion of samples from these refugia may have marked effects of estimates of regional taxonomic turnover.

11.3 Results

11.3.1 Effects of a New Age Model for the Cibao Valley, Dominican Republic

The new age model contains significantly more precise age estimates for the faunules collected from sections exposed in the river valleys of the Rio Cana and Rio Gurabo (Fig. 11.4). The median length of old age assignments was 1.3 myr,

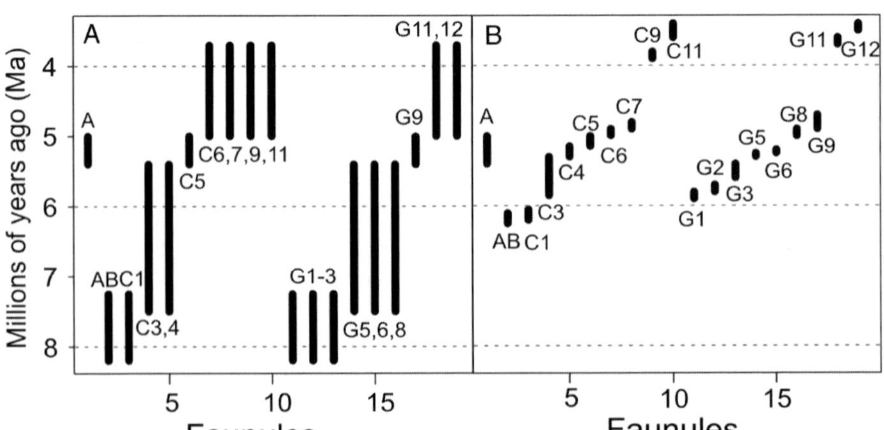

Fig. 11.4 Old and new age determinations for faunules from the Cibao Valley, Dominican Republic. (A) Plot of ages extracted from the age model of Saunders et al. (1986), and (B) new ages summarized in (McNeill et al., this volume). Faunule codes are A = Amina, AB = Arroyo Bellaco, C1 to C11 = Cana 1 to Cana 11, and G1 to G12 = Gurabo 1 to Gurabo 12

significantly longer than a median length for new age assignments of 0.1 myr (Kruskal-Wallis test of null hypothesis of no difference in duration, chi-squared =26.7, 1 d.f., $p < 0.0001$). As expected for stratigraphic data, the rank ordering of faunule ages remains consistent (Kruskal-Wallis test of null hypothesis that median ages of faunules are not different; chi-squared = 3.30, 1 d.f., $p = 0.0689$), although the median midpoint age for all faunules moved forward in time from 6.45 Ma to 5.2 Ma. This shift of older faunules from the Late Miocene into the Early Pliocene is the most important consequence of the new age model. Only the Arroyo Bellaco and Cana_1 faunules remain in the Late Miocene and these two faunules are assigned near the end of the Messinian (Appendix), all other faunules from the Rio Cana and all faunules from the Rio Gurabo are placed in the Early Pliocene in the NEWDR data set. Comparison of old and new stratigraphic ranges assigned to faunules from the Cibao Valley in the OLDDR and NEWDR data sets shows a net shortening of the total length of ranges for all faunules of 22% (from 98.37 myr to 76.37 myr; Table 11.3). The main mode or range change is for both lower and upper limits of age ranges to shift up in the record, with a total change of 40 myr, comprising 41% of the combined length of faunule age ranges in the OLDDR data. The relative magnitude of these changes is caused by the preponderance of data from the Cibao Valley in the Caribbean coral database as well as the generally poor stratigraphic resolution of Cibao Valley faunules in the OLDDR data (Fig. 11.2).

Examination of the distribution of stratigraphic ranges of taxa calculated with the OLDDR and NEWDR data sets shows that the main effects of the new age model for samples from the Cibao Valley is a shift of large numbers of first occurrences from intervals between 9 and 7 Ma to intervals between 7 and 5 Ma (Fig. 11.5). This change includes a reduction of 9.4 percent of the combined length of species ranges in the OLDDR data set (Table 11.2). Comparison of range sets indicates that changes were concentrated in upward shifts in age of first occurrences combined with either no change in the position of last occurrences or with an upward extension in the age of species last occurrence. A total of 55 species ranges (32%,

Table 11.3 Comparison of stratigraphic ages assigned to faunules from the Cibao Valley in the OLDDR and NEWDR data sets, including total length of ranges of faunules in each data set and a matrix showing the total length of change and number of faunules (in parentheses) for each type of response

	Change in combined durations			
	OLDDR	NEWDR	Difference	Percent change
Total duration	98.374	76.374	22	22.4

	Changes in range endpoints			
	LO no change	FO down	FO up	Total
FO no change	0 (106)	1.2 (1)	0 (0)	1.2 (107)
FO down	0 (0)	0 (0)	0 (0)	0 (0)
FO up	0.25 (1)	2.4 (2)	36.15 (14)	38.8 (17)
Total	0.25 (107)	3.6 (3)	36.15 (14)	40 (124)

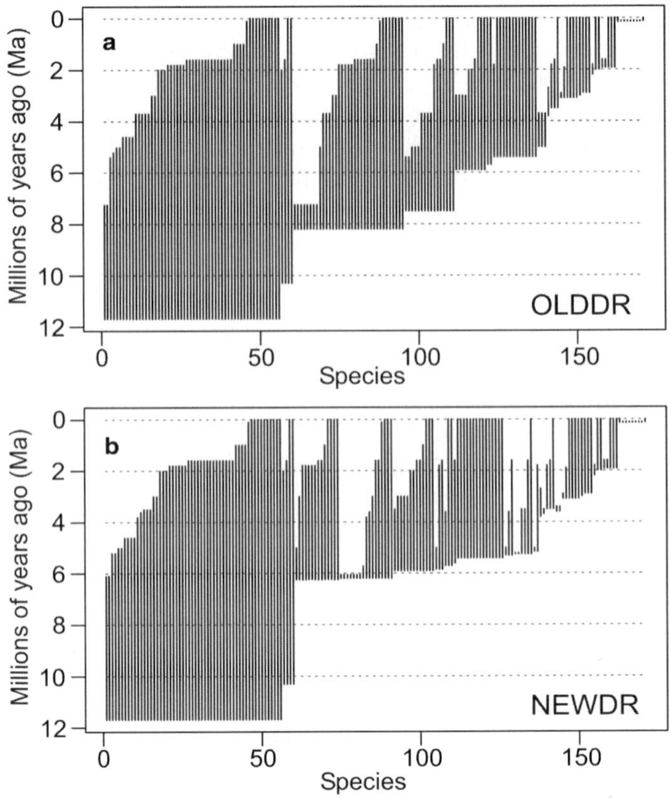

Fig. 11.5 Stratigraphic ranges of taxa estimated using OLDDR (**a**) and NEWDR (**b**) data. In each plot, the stratigraphic ranges are represented by vertical bars. Species are sorted by order of first and last occurrence

n = 171) responded with upward shifts of first occurrences, with a combined change of 120.7 myr of range, corresponding to 13.6% of the total length of ranges in the OLDDR data set (Table 11.2). Examination of stratigraphic distribution of faunules in the full data set (Fig. 11.2) clearly shows that the overall effect of the new age models is to expose a previously unrecognized sampling gap in the Late Miocene record of the Caribbean reef-coral fauna. With the old stratigraphic model for the Cibao Valley, there were 10 faunules crossing intervals between 8 and 6 Ma, and many of these faunules are characterized by large numbers of well-preserved specimens facilitating documentation of the contained faunas (Appendix). With the new age model, these faunules are shifted into the Early Pliocene leaving a large gap for most of the Late Miocene (Fig. 11.6A). There are collections from the Bahamas Drilling Project and Costa Rica that span the Late Miocene/Early Pliocene boundary (Budd and Manfrino, 2001; Budd et al., 1999), but the only remaining faunules in Late Miocene intervals older than 7 Ma are two small assemblages from

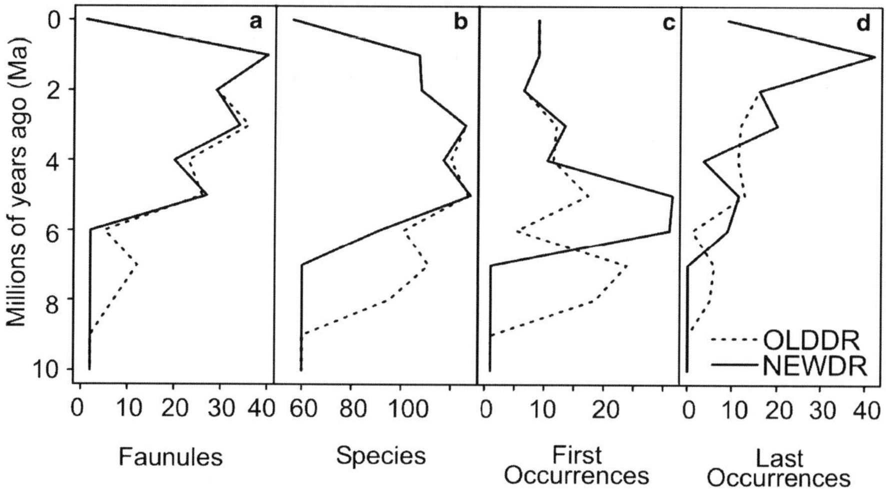

Fig. 11.6 A comparison of sampling coverage and temporal pattern of taxonomic turnover calculated with old and new age dates for collections from the Cibao Valley, Dominican Republic. In each panel, the solid line indicates the results of analysis of the NEWDR data set and results of an analysis of the DROLD data set are shown by the grey line. (**a**) Stratigraphic distribution of faunules is shown as the number of faunules in each stratigraphic bin. (**b**) Range-through species richness is lower in Late Miocene bins when the new age model is applied. (**c**) A Late Miocene pulse in first occurrences of species is transformed into an Early Miocene pulse if the new age model is used. (**d**) Updated age interpretations for collections from the Cibao Valley have little impact on the observed Plio-Pleistocene extinction pulse

the Seroe Domi Formation containing a total of 108 specimens. Our sampling is clearly too limited to document adequately the Late Miocene reef-coral diversity of the Caribbean region.

The upward shift of species first occurrences in the NEWDR data set relative to the OLDDR data set has a strong impact on estimates of species richness and taxonomic turnover. Using the OLDDR data set, regional species richness during the Late Miocene was calculated to increase from 60 to 95 species by the interval from 9 to 8 Ma and then to increase a second time during the Early Pliocene from just over 100 to a maximum of 127 (Fig. 11.6B), but using the NEWDR data set, range-through species richness remains at a level of 60 and increases rapidly during the last bin in the Late Miocene (7–6 Ma) and first bins of the Early Pliocene. In both cases, maximum richness of 127 species is calculated for the Late Miocene/Early Pliocene bin extending from 6 to 5 Ma. Decreased numbers of species in the Late Miocene is mirrored by counts of first occurrences. The large number of faunules from the Rio Cana and Rio Gurabo that were placed in the Late Miocene in the OLDDR data set created an apparent pulse or origination in the Late Miocene bins (9–8 Ma and 8–7 Ma; Fig. 11.6C). For the NEWDR data set, this pulse is shifted into the latest Late Miocene and Early Pliocene bins and is slightly more intense, with 31.5 species first occurring per million-year bin in the Early Pliocene using the new age models instead of a maximum of 23.8 species per myr using the OLDDR data set. In contrast to species richness and rates of first occurrence, the new age model for faunules does not substantially change the temporal pattern of species last occurrences from the Cibao Valley (Fig. 11.6D). The main extinction peak observed in the region occurs in the interval from 2 to 1 Ma, several millions of years beyond the range of faunules from the Cibao Valley.

The newly recognized Late Miocene sampling gap greatly reduces our confidence in patterns of species origination during the Late Miocene and Early Pliocene, but the Plio-Pleistocene extinction is robust to these new age interpretations.

11.3.2 Influence of Collections from the Cibao Valley, Dominican Republic

Some important aspects of the observed pattern of taxonomic turnover in the Caribbean reef fauna are sensitive to changes in age models for faunules collected from the Cibao Valley, Dominican Republic. This suggests that our understanding of Caribbean reef-coral evolution may be strongly influenced by material from the Dominican Republic. To test this hypothesis and to determine which aspects of the observed pattern are dependent on data from the Dominican Republic, we have compared stratigraphic ranges and diversity dynamics calculated including and excluding collections from the Cibao Valley. Examination of lists of species in the NEWDR and NODR data sets indicates that a total of 21 species are endemic to the Cibao Valley, including mainly taxa that are in open nomenclature pending formal taxonomic description. These include two species of *Leptoseris*, four species of *Montastraea*, three species of *Dichocoenia*, and one species each of *Agaricia*, *Pavona*, *Undaria*, *Favia*, and *Mussismilia*. Removing all Cibao Valley endemic species reduces the combined length of stratigraphic ranges of species by 11.6 myr or 1.3% of the combined length of ranges in the NEWDR data set (Table 11.4).

Table 11.4 Comparison of stratigraphic ages of reef-coral species in the NODR and NEWDR data sets, including changes in the number of species and total length of species ranges in each data set and a matrix showing the total length of change and number of species ranges (in parentheses) for each type of response

	Change in combined durations			
	OLDDR	NEWDR	Difference	Percent change
Total duration	790.785	910.485	−119.7	−15.14
Endemic species	0 (0)	11.6 (21)		

	Changes in range endpoints			
	LO no change	FO down	FO up	Total
FO no change	0 (108)	0 (0)	59.1 (12)	59.1 (120)
FO down	47.7 (29)	0 (0)	1.3 (1)	49 (30)
FO up	0 (0)	0 (0)	0 (0)	0 (0)
Total	47.7 (137)	0 (0)	60.4 (13)	108.1 (150)

Comparison of stratigraphic ranges of the 150 species that occur in both the NEWDR and NODR data sets suggests that the main effect of excluding faunules from the Cibao Valley is a shift of first occurrences of species above the bin extending from 7 to 6 Ma (Figs. 11.5B, 11.7A). This shift results from a decline in number of faunules from the bins extending from 6 to 4 Ma (Fig. 11.8A), in effect spreading the Late Miocene sampling gap further up into the Early Pliocene. Reduced numbers of faunules in the intervals from 6 to 4 Ma results in suppression of the Late Miocene/Early Pliocene origination peak, instead producing a low plateau of origination spanning most of the Pliocene (Fig. 11.8C). As for the impact of new ages for faunules from the Cibao Valley, excluding all Cibao Valley faunules has little effect on observed pattern of turnover (Fig. 11.8D).

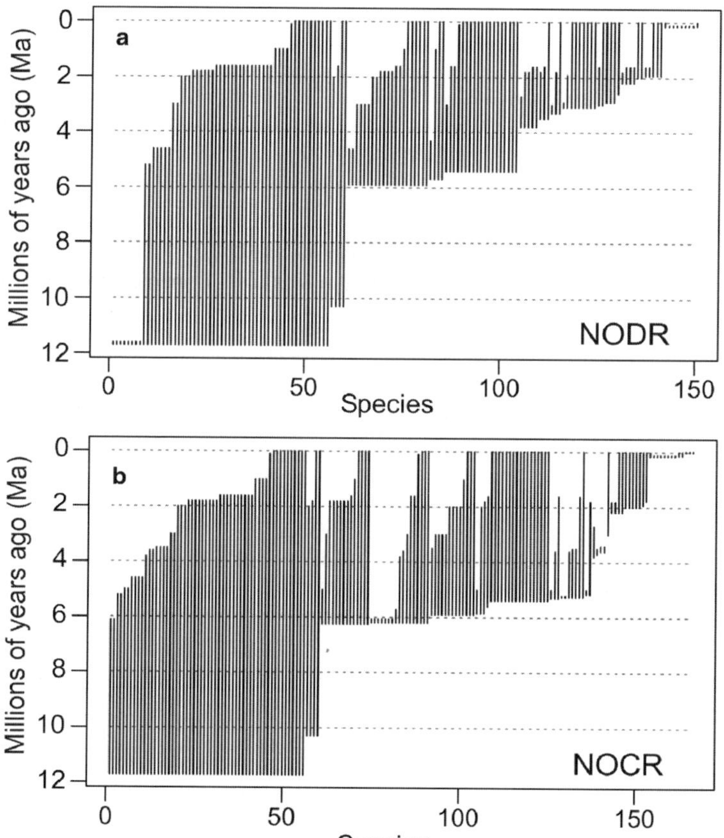

Fig. 11.7 Stratigraphic ranges of taxa estimated using NODR (**a**) and NOCR (**b**) data. In each plot, the stratigraphic ranges are represented by vertical bars. Species are sorted by order of first and last occurrence

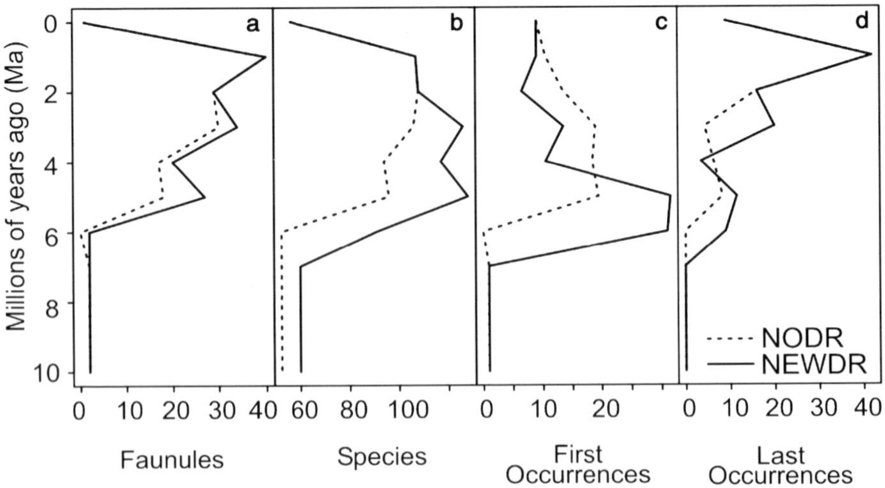

Fig. 11.8 A comparison of sampling coverage and temporal pattern of taxonomic turnover calculated including and excluding collections from the Cibao Valley, Dominican Republic. In each panel, the solid line indicates the results of analysis of the NEWDR data set and results of an analysis of the NODR data set are shown by the grey line. (**a**) Stratigraphic distribution of faunules is shown as the number of faunules in each stratigraphic bin, and excluding collections from the Cibao Valley decreases the number of faunules recovered from Late Miocene and Early Pliocene intervals. (**b**) Range-through species richness increases in the Late Miocene and Early Pliocene if collections from the Cibao Valley are included. (**c**) Collections from the Cibao Valley create a strong pulse of species first occurrence during bins from the latest Miocene and Early Pliocene, but additional data have little effect on the distribution of species last occurrences (**d**)

11.3.3 Influence of Collections from the Limon Basin, Costa Rica

These results suggest that estimates of origination during the Late Miocene and Pliocene are strongly influenced by data from the Dominican Republic, and that our knowledge of rates of origination in this interval is still not robust to change.

But what about estimates of species extinction and the Plio-Pleistocene extinction pulse? To determine whether or not the observed pattern of extinction was sensitive to the inclusion or exclusion of sets of collections, we compared stratigraphic ranges and patterns of taxonomic turnover for the NEWDR and the NOCR data sets. The stratigraphic ranges of collections from the Limon Basin of Costa Rica span the Pliocene (Fig. 11.2) and include key pre-extinction faunules from the Lomas del Mar coral reef trend (Budd et al., 1999).

Five species of reef-coral were endemic to the Limon Basin (Table 11.5), including the extant Eastern Pacific species *Porites colonensis*, two species of mussid genera currently in open nomenclatures (*Mycetophyllia* and *Isophyllastrea*) and three extinct species (*Diploria sarasotana, Archohelia limonensis*, and *Thysanus crassicostatus).* These five taxa have combined ranges of 3.96 myr, only 0.4 percent of the total length of species ranges in the NEWDR data set. Adding the collections from Costa Rica results in a two to three-fold increase in the number of Pliocene collections in the compilation (Fig. 11.9A), and as with the addition of collections from the Cibao Valley, the main effects of adding collections from the Limon Basin of Costa Rica is extension of species ranges both up and down in the section (Fig. 11.7B). However, the magnitude of changes is much less with the addition of samples from Costa Rica than was observed after addition of samples from the Dominican Republic (Table 11.5). A total of 25 species ranges were extended by the addition of collections from Costa Rica, with a total extension of 24.645 myr or only 2.7% of the combined length of ranges in the NEWDR compilation. This is 83.45 myr less total range change than was created by the addition of collections from the Cibao Valley.

The pattern of taxonomic turnover was also not altered strongly by the addition of collections from the Limon Basin of Costa Rica (Fig. 11.9). The main effect was a slight increase in range-through species richness in Pliocene stratigraphic bins (Fig. 11.9B), created by increased numbers of first occurrences in all Pliocene bins. If collections from Costa Rica are left out, the numbers of species inferred to have

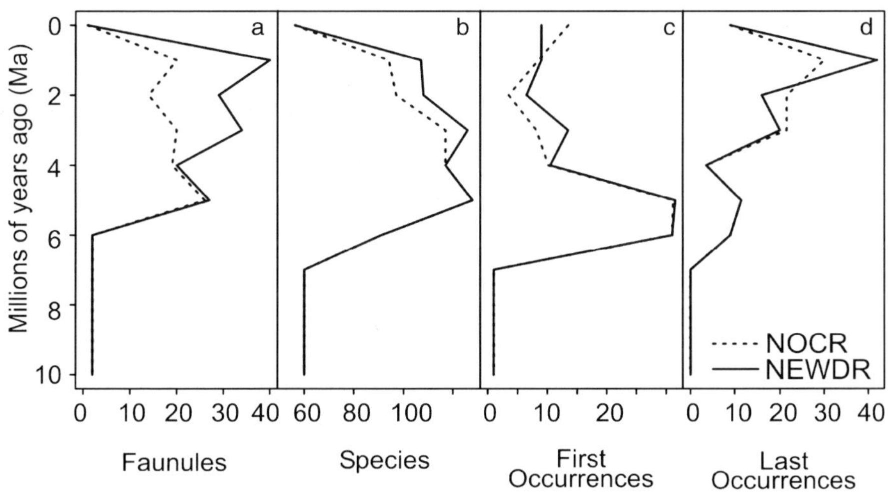

Fig. 11.9 The temporal pattern of sampling coverage and taxonomic turnover calculated including and excluding collections from the Limon Basin, Costa Rica shows decline in Valley, Dominican Republic. In each panel, the solid line indicates the results of analysis of the NEWDR data set and results of an analysis of the NOCR data set are shown by the grey line. (**a**) Culling samples from Costa Rica suppresses the number of Pliocene samples in the compilation, but this reduction in sample density does not have a significant effect on range-through species richness (**b**). Rates of species first occurrence (**c**) are slightly depressed in the Early Pliocene and slightly increased in the Pleistocene, and although the position of the Plio-Pleistocene extinction pulse is not altered (**d**), the estimated magnitude of the pulse declines as some of the species first occurrences are pushed back into bins extending from 4 to 2 Ma

Table 11.5 Comparison of stratigraphic ages of reef-coral species in the NOCR and NEWDR data sets, including changes in the number of species and total length of species ranges in each data set and a matrix showing the total length of change and number of species ranges (in parentheses) for each type of response

	Change in combined durations			
	OLDDR	NEWDR	Difference	Percent change
Total duration	881.88	910.485	−119.7	−15.14
Endemic species	0 (0)	3.96 (5)		

	Changes in range endpoints			
	LO no change	FO down	FO up	Total
FO no change	0 (141)	0 (0)	6.1 (10)	6.1 (151)
FO down	18.545 (15)	0 (0)	0 (0)	18.545 (15)
FO up	0 (0)	0 (0)	0 (0)	0 (0)
Total	18.545 (156)	0 (0)	6.1 (10)	24.645 (166)

originated during the past million years is slightly higher than if collections from Costa Rica are included (Fig. 11.9C). The position of the Plio-Pleistocene extinction pulse is insensitive to the inclusion or exclusion of collections from Costa Rica (Fig. 11.9D), but the estimated magnitude of the pulse is lower when collections from Costa Rica are excluded because the last occurrences of 10 species with their youngest records in Late Pliocene collections from Costa Rica are pushed back to collections from other geographic regions. The magnitude of the extinction during the bin from 2 to 1 Ma increases from 30 to 42 if collections from Costa Rica are considered in the analysis, an increase over two times greater than the increase expected by the inclusion of the five species only recovered from the Limon Basin. In summary, the reef-coral occurrences from Costa Rica do not have a significant effect on patterns of origination when compared with collections from the Cibao Valley. Costa Rica collections do contribute to the magnitude but not the position of the Plio-Pleistocene extinction pulse.

11.4 Implications

Maintaining compilations of paleontological occurrence data will require incorporating new interpretations including updated age models and new sets of taxon occurrences from old or new collecting localities. These revisions may alter the observed pattern of diversity change inferred from the available sample of the record. In general, adding new collections will result in extensions of stratigraphic ranges and may also result in the discovery of new species. All stratigraphic ranges underestimate the true durations of taxa, so it is not surprising that additional collecting is likely to extend the position of first occurrences back in time and extend the position of last occurrences forward in time. The full taxonomic effects of adding new collections remains to be determined for taxonomies constructed from morphospecies, because increased knowledge of the morphologic variation of species can result in either the merging of existing taxa as morphological gaps are filled by new material or by the erection of new taxa to incorporate new forms discovered in the new collections. In the analyses here, adding new collections always resulted in the addition of new species, but there were more species endemic to the Dominican Republic than were endemic to collections from the Limon Basin of Costa Rica.

Stratigraphic ranges of coral species were also changed by applying new age models or by excluding data from either the Cibao Valley or Costa Rica, but the range sets and observed pattern of taxonomic turnover in time were both more strongly influenced by changing age assignments or excluding collections from the Cibao Valley than by excluding collections from Costa Rica. Why are collections from the Cibao Valley more influential than collections from the Limon Basin? Many factors can increase the influence that sets of collections exert on interpretations of the temporal pattern of diversity dynamics. At the most basic level, collections resulting from more intensive sampling will on average include richer biotas and have greater influence on estimates of taxonomic turnover than will collections resulting from less intensive sampling. The relative intensity of sampling required to ensure adequate redundancy for estimating the true regional biota within a time interval depends on spatial heterogeneity among samples on local to regional scales as well as the rate of taxonomic turnover within the interval. In general, more collections are needed if each one is not capturing a large proportion of the regional biota either because they are incomplete samples of the local biota, or because samples do not cover the full range of spatial heterogeneity in the regional biota. More collections will also be required during intervals of rapid taxonomic turnover in order to ensure redundancy in the sample set. For the Caribbean reef-coral compilation, we suspect that the critical factor is the geographic distribution of collections. Shifting the ages of most of the faunules from the Cibao Valley from the Late Miocene to the Early Pliocene exposed a previously unrecognized Late Miocene sampling gap because there were few Late Miocene samples in other geographic regions. In contrast, excluding data from Costa Rica did not have as great an effect because of the more complete geographic coverage of the Pliocene record. Collections from the Pliocene were studied from rocks in the Bahamas, Curacao, the Dominican Republic, Jamaica, and Panama. The result of this redundant sampling is a robust pattern of taxonomic turnover for the late Pliocene.

The Late Miocene part of the record is much less complete and interpretations of the diversity dynamics from this interval is highly sensitive to the data compilation, for example the inclusion of collections from the Seroe Domi Formation of Curacao (Budd and Johnson, 1999) or the shift of age of most of the collections from the Cibao Valley from the Late Miocene to the Pliocene. It appears that more collections are required from the Late Miocene record to fully document patterns of change. But where might these data come from? In our current compilation we have attempted full coverage of the Caribbean region, but large Late Miocene faunules seem to be absent. The lack of Miocene reef-coral units might be related to Neogene changes in regional oceanography characteristic of the Late Oligocene to Pliocene interval in the Caribbean. Although reef building was apparently widespread during the Late Oligocene (Frost, 1977), there is evidence in the form of fossil distributions (Edinger and Risk, 1994) as well as paleoceanographic modeling (von der Heydt and Dijkstra, 2006) for increased regional surface productivity in the Miocene to Early Pliocene record of the Caribbean basin. Surface productivity apparently declined again during the Pliocene (O'dea et al., 2007) as a result of constriction of the Central American Seaway and decreasing influence of surface water flowing from the Eastern Pacific (Schneider and Schmittner, 2006). It is well known that significant carbonate production and extensive reef building are inhibited by increased surface productivity (Mutti and Hallock, 2003), and extensive Miocene accumulations of reef framework are rare (Johnson, K.G. and Jackson, J.B.C., unpublished data, 2007). However, regional range-through diversity is still relatively high through the Late Miocene even in the absence of large collections. Based on Early and Middle Miocene collections from Anguilla, Curacao, Panama, Puerto Rico, and Trinidad (see above), we would expect to discover more than 60 species from the Late Miocene of the Caribbean. This is a number equivalent to the extant diversity of the region. However, we would expect the ecological distribution of these Miocene species to be significantly different, as they are more likely to have been living in off-reef or other sites not considered typical for reef-building corals (Johnson et al., 1995).

Acknowledgments We thank T.A. Stemann for helping to collect and identify corals. R. Ginsburg arranged access to specimens and data from the Bahamas Drilling Program. Collections and curation support was provided by T. Adrain, J. Darrell, S. Cairns, H. Filkorn, J. Golden, and R. Panchaud. Funding was provided by the US National Science Foundation Grants EAR-9909485 and DBI-0237337 to KGJ, NSF EAR 0445789 to AFB, and NSF EAR 0446768 to DFM. This chapter benefited from discussions with A. McGowan and J. Todd, and was reviewed by J. Adrain. All analyses were performed using the R Statistical Programming Language version 2.41 (R Development Core Team, 2006).

Appendix

List of faunules included in the analysis, with numbers of taxa, occurrences and specimens indicated. Country names are indicated as B = Bahamas, CI = Cayman Islands, CR = Costa Rica, CU = Curacao, DR = Dominican Republic, J = Jamaica, and P = Panama. Age dates assigned to each faunule are indicated by intervals extending from Start to End values. New ages for collections from the DR are included in the columns labeled "New Start and "New End"

Faunule	Taxa	Occs	Specs.	Country	Start (Ma)	End (Ma)	New Start (Ma)	New End (Ma)
Pre-late miocene	56	56	56					
Bah_Pl	14	14	25	B	1.6	0.2		
Bah_LP	40	40	91	B	2.2	1.6		
Bah_LM	13	13	16	B	5.4	5.2		
Central Lagoon	31	31	34	CI	0.125	0.125		
E Reef	8	8	34	CI	0.125	0.125		
NW Lagoon	7	7	28	CI	0.125	0.125		
NW Patch	24	24	407	CI	0.125	0.125		
NW Reef	27	27	321	CI	0.125	0.125		

(continued)

Appendix (continued)

Faunule	Taxa	Occs	Specs.	Country	Start (Ma)	End (Ma)	New Start (Ma)	New End (Ma)
SW Lagoon	8	8	28	CI	0.125	0.125		
SW Patch	17	17	115	CI	0.125	0.125		
SW Reef	22	22	372	CI	0.125	0.125		
South Lagoon	9	9	41	CI	0.125	0.125		
WE Reef	9	9	30	CI	0.125	0.125		
Av.Barracuda-dorms	5	5	31	CR	1.93	1.6		
Bahia Portete	4	4	9	CR	1.93	1.6		
Corales overkill	36	36	36	CR	1.93	1.6		
Lomas del Mar	3	3	5	CR	1.93	1.6		
Lomas_e_1	22	22	56	CR	1.93	1.6		
Lomas_e_10	48	48	441	CR	1.93	1.6		
Lomas_e_4	32	32	131	CR	1.93	1.6		
Lomas_e_5	28	28	63	CR	1.93	1.6		
Lomas_e_7	25	25	44	CR	1.93	1.6		
Lomas_e_8	21	21	96	CR	1.93	1.6		
Lomas_e_9	31	31	51	CR	1.93	1.6		
Lomas_w_2	24	24	79	CR	1.93	1.6		
Lomas_w_3	12	12	49	CR	1.93	1.6		
Portete_1	34	34	118	CR	1.93	1.6		
Portete_2	19	19	49	CR	1.93	1.6		
Empalme_3	14	14	30	CR	2.9	1.9		
Pueblo Nuevo	6	6	13	CR	2.9	1.9		
St_Rosa_2	19	19	113	CR	2.9	1.9		
St_Rosa_3	24	24	66	CR	2.9	1.9		
St_Rosa_4	13	13	17	CR	2.9	1.9		
Buenos_1	7	7	58	CR	3.1	2.9		
Buenos_2	20	20	102	CR	3.1	2.9		
Buenos_3	18	18	62	CR	3.1	2.9		
Buenos_4	12	12	49	CR	3.1	2.9		
Buenos_6	19	19	50	CR	3.1	2.9		
Buenos_7	35	35	284	CR	3.1	2.9		
Buenos_8	13	13	30	CR	3.1	2.9		
Buenos_9	22	22	44	CR	3.1	2.9		
Chocolate_4	27	27	127	CR	3.1	2.9		
rt32- Chiquita	1	1	15	CR	3.1	2.9		
Q.Chocolate	6	6	46	CR	3.5	3.2		
Q_Choco	12	12	17	CR	3.5	3.2		
Q_Choco_1	4	4	4	CR	3.5	3.2		
Q_Choco_5	14	14	14	CR	3.5	3.2		
Brazo Seco	19	19	88	CR	5.7	4.3		
LP_Curacao	21	21	45	CU	0.125	0.125		
terr_l40	16	16	20	CU	0.2	0.1		
terr_m80	17	17	25	CU	0.6	0.5		
seacliff_17	18	18	58	CU	2.6	2		
seacliff_21	22	22	104	CU	2.6	2		
seacliff_23	20	20	147	CU	2.6	2		

(continued)

Appendix (continued)

Faunule	Taxa	Occs	Specs.	Country	Start (Ma)	End (Ma)	New Start (Ma)	New End (Ma)
seacliff_26	24	24	133	CU	2.6	2		
terr_h38	9	9	106	CU	2.6	2		
terr_hs39	8	8	52	CU	3	2.5		
ridges_31	18	18	92	CU	5.4	3		
ridges_32	22	22	98	CU	5.4	3		
ridges_33	15	15	46	CU	5.4	3		
ridges_59	13	13	65	CU	5.4	3		
ridges_60	26	26	98	CU	5.4	3		
ridges_63	25	25	73	CU	5.4	3		
ridges_65	17	17	33	CU	5.4	3		
ridges_70	25	25	70	CU	5.4	3		
ridges_71	24	24	100	CU	5.4	3		
ridges_74	24	24	96	CU	5.4	3		
ridges_77	25	25	50	CU	5.4	3		
salina_35	25	25	135	CU	5.9	4.6	5	4.9
salina_50	24	24	110	CU	5.9	4.6	4.9	4.8
salina_51	14	14	56	CU	5.9	4.6	3.9	3.8
salina_55	24	24	152	CU	5.9	4.6	3.9	3.8
salina_57	17	17	158	CU	5.9	4.6		
salinaE_53	18	18	84	CU	10.3	7.8		
salinaE_67	11	11	24	CU	10.3	7.8		
Boca Chica A	13	13	27.9	DR	0.125	0.125		
Boca Chica B	16	16	59.3	DR	0.125	0.125		
Boca Chica C	14	14	69	DR	0.125	0.125		
Cana_11	37	37	263	DR	5	3.7	3.6	3.4
Cana_6	18	18	38	DR	5	3.7	5	4.9
Cana_7	12	12	24	DR	5	3.7	4.9	4.8
Cana_9	18	18	91	DR	5	3.7	3.9	3.8
Gurabo_11	28	28	101	DR	5	3.7	3.7	3.6
Gurabo_12	31	31	74	DR	5	3.7	3.5	3.4
Amina_1	7	7	35	DR	5.4	5		
Cana_5	63	63	589	DR	5.4	5	5.15	5
Gurabo_9	16	16	393	DR	5.4	5	4.9	4.7
Cana_3	12	12	95	DR	7.5	5.4	5.85	5.3
Cana_4	14	14	59	DR	7.5	5.4	5.3	5.15
Gurabo_5	24	24	209	DR	7.5	5.4	5.3	5.25
Gurabo_6	57	57	1,050	DR	7.5	5.4	5.25	5.2
Gurabo_8	48	48	582	DR	7.5	5.4	5	4.9
Arroyo Bellaco	32	32	461	DR	8.2	7.25	6.25	6.1
Cana_1	43	43	271	DR	8.2	7.25	6.2	6.05
Gurabo_1	4	4	18	DR	8.2	7.25	5.9	5.8
Gurabo_2	20	20	76	DR	8.2	7.25	5.8	5.7
Gurabo_3	15	15	70	DR	8.2	7.25	5.6	5.4
Rio Bueno Harbor	1	1	1	J	0.125	0.125		
Falmouth	17	17	127	J	0.15	0.12		
DBML Qu.-terr2	3	3	3	J	1.8	1		

(continued)

Appendix (continued)

Faunule	Taxa	Occs	Specs.	Country	Start (Ma)	End (Ma)	New Start (Ma)	New End (Ma)
HopeGate_1	18	18	77	J	1.8	1		
HopeGate_2	17	17	90	J	1.8	1		
HopeGate_3	14	14	19	J	1.8	1		
HopeGate_4	13	13	72	J	1.8	1		
NavyIs_1	17	17	57	J	1.93	1.6		
NavyIs_2	22	22	71	J	1.93	1.6		
FollyPt_1	11	11	93	J	2	1.4		
FollyPt_2	17	17	62	J	2	1.4		
FollyPt_3	21	21	78	J	2	1.4		
HectorsR	12	12	55	J	2	1.4		
TropicW	12	12	86	J	2	1.4		
OldPera_2	26	26	153	J	2	1.8		
OldPera_1	29	29	240	J	3	2		
Bowden (C)-float	14	14	95	J	3.3	3		
Bowden(C) = ER180	10	10	17	J	3.3	3		
Bowden(A)-Shell Beds	14	14	55	J	3.8	2.7	5.25	5.2
Bowden(B)-Shell Beds	10	10	59	J	3.8	2.7	5	4.9
Swan Cay	6	6	6	P	1.77	0.78		
Fish_Hole	44	44	105	P	2.2	1.8		
Hill_Pt	18	18	59	P	2.2	1.8		
Paunch	14	14	27	P	2.2	1.8		
Ground Creek- west	2	2	23	P	3.6	1.806	4.9	4.7
TOTAL	171	2,436	12,756					

References

Adrain, J.M. and Westrop, S.R., 2000, An empirical assessment of taxic paleobiology, *Science*, **289**:110–112.

Alroy, J., 2000, Successive approximations of diversity curves: ten more years in the library, *Geology*, **28**:1023–1026.

Alroy, J. (ed.), 2007, *The Paleobiology Database*, world wide web site at http://pbdb.org, last checked June 2007.

Berger, W.H. and Jansen, E., 1994, Mid-Pleistocene climate shift: the Nansen connection, *Am. Geophys. Union Monogr.*, **84**:295–311.

Berggren, W.A., Kent, D.V., Flynn, J.J., and van Couvering, J.A., 1985, Cenozoic geochronology, *Geol. Soc. Am. Bull.*, **96**:1407–1418.

Berggren, W.A., Kent, D.V., Swisher, C.C., and Aubry, M.-P., 1995, A revised Cenozoic geochronology and chronostratigraphy, in: *Geochronology, Time Scales, and Global Stratigraphic Correlation* (W.A. Berggren, D.V. Kent, M.-P. Aubry, and J. Hardenbol, eds.), SEPM Special Publication 54, pp. 129–212.

Budd, A.F. and Johnson, K.G., 1999, Origination preceding extinction during late Cenozoic turnover of Caribbean reefs, *Paleobiology*, **25**:188–200.
Budd, A.F. and Manfrino, C.M., 2001, Coral assemblages and reef environments in the Bahamas Drilling Project cores, in: *Results of the Bahamas Drilling Project* (R.N. Ginsburg, ed.), *Society of Sedimentary Geology Special Publication*, number 70, Tulsa, OK, pp. 41–59.
Budd, A.F. and McNeill, D.F., 1998, Zooxanthellate scleractinian corals of the Bowden Shell Bed, southeast Jamaica, *Contrib. Tert. Quat. Geol.*, **35**:47–61.
Budd, A.F., Johnson, K.G., and Edwards, J.C., 1995, Caribbean reef coral diversity during the early to middle Miocene: an example from the Anguilla Formation. *Coral Reefs*, **14**: 109–117.
Budd, A.F., Johnson, K.G., and Stemann, T.A., 1993, Plio-Pleistocene extinctions and the origin of the modern Caribbean reef-coral fauna, in: *Global Aspects of Coral Reefs* (R.N. Ginsburg, ed.), Rosenstiel School of Marine and Atmospheric Science, University of Miami, Miami, FL, pp. H7–H13.
Budd, A.F., Stemann, T.A., and Johnson, K.G., 1994, Stratigraphic distributions of genera and species of Neogene to Recent Caribbean reef corals, *J. Paleontol.*, **68**:951–977.
Budd, A.F., Petersen, R.A., and McNeill, D.F., 1998, Stepwise faunal change during evolutionary turnover: a case study from the Neogene of Curacao, Netherlands Antilles, *Palaios*, **13**:170–188.
Budd, A.F., Johnson, K.G., Stemann, T.A., and Tompkins, B.H., 1999, Pliocene to Pleistocene reef coral assemblages in the Limon Group of Costa Rica, *Bull. Am. Paleontol.*, **357**:119–158.
Budd, A.F., Foster, C.T., Jr., Adrain, T., and Dawson, J.P., 2006, *Neogene Marine Biota of Tropical America*, world wide website at http://nmita.geology.uiowa.edu, last checked June 2007.
Chiappone, M., Sullivan, K.M., and Lott, C., 1996, Hermatypic scleractinian corals of the southeastern Bahamas: a comparison to Western Atlantic reef systems, *Carib. J. Sci.*, **23**:1–13.
Dustan, P. and Halas, J.C., 1987, Changes in the reef-coral community of Carysfort Reef, Key Largo, Florida: 1974–1982, *Coral Reefs*, **6**:91–96.
Edinger, E.N. and Risk, M.J., 1994, Oligocene-Miocene extinction and geographic restriction of Caribbean corals: roles of turbidity, temperature, and nutrients, *Palaios*, **9**:576–598.
Fenner, D., 1999, New observations on the stony coral (Scleractinia, Milleporidae, and Stylasteridae) species of Belize (Central America) and Cozumel (Mexico), *Bull. Mar. Sci.*, **64**:143–154.
Foote, M., 2000, Origination and extinction components of taxonomic diversity: general problems, in: *Deep time: Paleobiology's Perspective* (D.H. Erwin and S.L. Wing, eds.), *Paleobiology (Suppl.)*, **26**:74–102.
Frost, S.H., 1977, Cenozoic reef systems of Caribbean; prospects for paleoecologic synthesis, *Stud. Geol.*, **4**:93–110.
Frost, S.H., Harbour, J.L., Beach, D.K., Realini, M.J., and Harris, P.M., 1983, Oligocene reef tract development, southwestern Puerto Rico, *Sedimenta*, **9**:1–144.
Gilinsky, N.L. and Signor, P.W. (eds.), 1991, *Analytical Paleobiology, Short Courses in Paleontology*, Number 4, The Paleontological Society, Knoxville, TN, pp. 1–216.
Gradstein, F., Ogg, J., and Smith, A., 2004, *A Geologic Time Scale 2004*, Cambridge University Press, Cambridge.
Guzmán, H.M. and Guevara, C.A., 1998, Arrecifes coralinos de Bocas del Toro, Panamá: I. Distribución, estructura y estado de conservación de los arrecifes continentales de la Laguna de Chrirquí y la Bahía Almirante, *Revista de Biología Tropical*, **46**:601–622.
Guzmán, H.M. and Holst, I., 1994, Inventario biológico y estado actual de los arrecifes coralinos a ambos lados del Canal de Panamá, *Revista de Biología Tropical*, **42**:493–514.
Holst, I. and Guzmán, H.M., 1993, Lista de corals hermatípicos (Anthozoa: Scleractinia; Hydrozoa: Milleporina) a ambos labos del istmo de Panamá, *Revista Biologia Tropical*, **41**:871–875.
Hunter, I.G. and Jones, B., 1996, Coral associations of the Pleistocene Ironshore Formation, Grand Cayman, *Coral Reefs*, **15**:249–267.

Jackson, J.B.C. and Johnson, K.G., 2000, Life in the last few million years, in: *Deep Time: Paleobiology's Perspective* (D.H. Erwin and S.L. Wing, eds.), *Paleobiology (Suppl.)*, **26**:221–235.

Johnson, K.G., 2001, Middle Miocene recovery of Caribbean reef corals; new data from the Tamana Formation, Trinidad, *J. Paleontol.*, **75**:513–526.

Johnson, K.G., 2007, Reef-coral diversity from the Late Oligocene Antigua Formation and temporal variation of local diversity on Caribbean Cenozoic Reefs, in: *Proceedings of the Ninth International Symposium on Fossil Cnidaria and Porifera* (B. Hubmann and W.E. Piller, eds.), Österr. Akad. Wiss., Schriftenreihe der Erdwissenschaftlichen Kommissionen, Graz, pp. 471–491.

Johnson, K.G. and Kirby, M.X., 2006, The Emperador Limestone rediscovered: early Miocene corals from the Culebra Formation, Panama, *J. Paleontol.*, **80**:283–293.

Johnson, K.G. and McCormick, T., 1999, The quantitative description of biotic change using palaeontological databases, in: *Numerical Palaebiology* (D. Harper, ed.), Wiley, Chichester, pp. 227–247.

Johnson, K.G., Budd, A.F., and Stemann, T.A., 1995, Extinction selectivity and ecology of Neogene Caribbean reef corals, *Paleobiology*, **21**:52–73.

Klaus, J.S. and Budd, A.F., 2003, Comparison of Caribbean coral reef communities before and after Plio-Pleistocene faunal turnover: analyses of two Dominican Republic reef sequences, *Palaios*, **18**:3–21.

Koch, C.F., 1987, Prediction of sample size effects on the measured temporal and geographic distribution patterns of species, *Paleobiology*, **13**:100–107.

Marshall, C.R., 1994, Confidence intervals on stratigraphic ranges: partial relaxation of the assumption of randomly distributed fossil horizons, *Paleobiology*, **20**:459–469.

Maslin, M.A. and Ridgewell, A.J., 2005, Mid-Pleistocene revolution and the 'eccentricity myth', in: *Early-Middle Pleistocene Transitions: The Land-Ocean Evidence* (M.J. Head and P.L. Gibbard, eds.), *Special Publications of the Geological Society*, London, pp. 19–34.

McNeill, D.F., Klaus, J.S., Evans, C.C., Budd, A.F., and Maier, K.L., This volume, An overview of the regional geology and stratigraphy of the Neogene deposits of the Cibao Valley, Dominican Republic, in: *Evolutionary Stasis and Change in the Dominican Republic Neogene*.

Mutti, M. and Hallock, P., 2003, Carbonate systems along nutrient and temperature gradients: some sedimentological and geochemical constraints, *Int. J. Earth Sci.*, **92**:465–475.

Overpeck, J.T., Otto-Bliesner, B.L., Miller, G.H., Muhs, D.R., Alley, R.B., and Kiehl, J.T., 2006, Paleoclimatic evidence for future ice-sheet instability and rapid sea-level rise, *Science*, **311**:1747–1750.

O'Dea, A., Jackson, J.B.C., Fortunato, H., Smith, J.T., D'Croz, L., Kenneth G. Johnson, and Todd, J.A., et al., 2007, Environmental change preceded Caribbean extinction by 2 million years, *Proc. Natl. Acad. Sci. USA*, **104**:5501–5506.

R Development Core Team, 2006. *R: a language and environment for statistical computing*. R Foundation for Statistical Computing, world wide web site at http::/R-project.org, last checked June 2007.

Raup, D.M., 1972, Taxonomic Diversity during the Phanerozoic, *Science*, **177**:1065–1071.

Saunders, J.B., Jung, P., and Biju-Duval, B., 1986, Neogene paleontology in the northern Dominican Republic, 1. Field surveys, lithology, environment, and age, *Bull. Am. Paleontol.*, **89**:1–79.

Schneider, B. and Schmittner, A., 2006, Simulating the impact of Panamanian seaway closure on ocean circulation, marine productivity, and nutrient cycling, *Earth Planet. Sci. Lett.*, **246**:367–380.

Sepkoski, J.J., Jr., 1982, A compendium of fossil marine animal families, *Milwaukee Publ. Mus. Contrib. Biol. Geol.*, **51**:1–125.

Sepkoski, J.J., Jr., 2002, A compendium of fossil marine animal genera, *Bull. Am. Paleontol.*, **363**:1–560.

Stemann, T.A., 2004, Reef corals of the White Limestone Group of Jamaica, *Cainozoic Res.*, **3**:83–107.

Tomascik, T. and Sander, F., 1987, Effects of eutrophication on reef-building corals. II. Structure of scleractinian coral communities on fringing reefs, Barbados, West Indies, *Mar. Biol.*, **94**:53–75.

Valentine, J.W., 1969, Patterns of taxonomic and ecological structure of the shelf benthos during Phanerozoic time, *Palaeontology*, **12**:684–709.

von der Heydt, A. and Dijkstra, H.A., 2006, Effect of ocean gateways on the global ocean circulation in the late Oligocene and early Miocene, *Paleoceanography*, **21**:PA1011.

Weisbord, N.E., 1971, Corals from the Chipola and jackson Bluff Formations of Florida, *Florida Bur. Geol. Geol. Bull.*, **53**:1–100.

Weisbord, N.E., 1973, New and little-known corals from the Tampa Formation of Florida, *Florida Bur. Geol. Geol.. Bull.*, **56**:1–147.

Wells, J.W. and Lang, J.C., 1973, Systematic list of Jamaican shallow-water Scleractinia, *Bull. Mar. Sci.*, **23**:55–58.

Chapter 12
Science Education and the Dominican Republic Project

Ross H. Nehm[1,2], Jupiter Luna[1], and Ann F. Budd[3]

Contents

12.1 Introduction

Two recent developments – the retirement of the architects of the Dominican Republic Project (DRP) and the decline of the Naturhistorisches Museum Basel as a DRP research center – have provided an opportunity to re-envision sources of human capital for the DR project while expanding the diversity of individuals involved in geoscience research. In response to this opportunity, efforts are underway to engage new teams of students in geoscience research, especially those from underrepresented racial and ethnic groups. Of particular concern to some has been the persistent lack of Dominican and Dominican-American involvement in this research project over the past 30 years. The continued under-representation of minority groups in geoscience research (National Science Foundation, 1996), coupled with the erosion of science literacy in the United States (National Research

[1] School of Education, The City College C.U.N.Y., Convent Avenue at 138th Street, New York, NY, USA. Email: jupiter.luna@gmail.com

[2] The Ohio State University, Columbus, OH, USA. Email: nehm.1@osu.edu

[3] Department of Geoscience, University of Iowa, Iowa City, IA. Email: ann-budd@uiowa.edu

R.H. Nehm, A.F. Budd (eds.) *Evolutionary Stasis and Change in the Dominican Republic Neogene,*

Council, 2007), has caused many of the agencies that fund palaeobiological research, particularly the National Science Foundation, to encourage geoscientists to integrate basic research efforts with educational outreach activities. We briefly review two education efforts that have occurred in parallel to the empirical research discussed in this volume: Dominican American science education in the United States and international educational outreach and development in the Dominican Republic.

The challenge of providing quality education to students of diverse cultures, races, and classes is no longer restricted to particular geographic regions or urban centers (Garcia, 1999). In the past 30 years, dramatic demographic shifts have changed the face of school children in the United States. Nationwide, 42% of public school students are from minority groups, nearly a 100% increase since the 1970s (Dillon, 2007). In the American West, Whites now comprise a minority of public school students (~46%). Although the American Midwest remains the whitest region of the nation, minorities now comprise 26% of Midwestern school students. The most pronounced growth in minority enrollment has been among Latino students: from 1972 to 2005 this student population has increased from 6% to 20% nationwide (Dillon, 2007).

Programs promoting educational opportunities for minorities in the sciences over the past 25 years, in response to the significant growth in minority students in the United States, have not produced concomitant minority representation in science careers. In 1999, for example, only 3.4% of employed scientists and engineers were Latino. Particular disciplines, such as the geosciences, are characterized by even lower participation rates (National Science Foundation, 1999). A brief look at the state of secondary science education in the United States partially explains these patterns.

International comparisons of American student performance in science and math reveal alarmingly low scores relative to other industrialized nations and largely static scores among all racial and ethnic groups over the past 30 years (National Center for Education Statistics, 1996; Schmidt, McNight, and Raizen, 1997; O'Sullivan et al., 2003). At every grade level, White students are characterized by higher science content and inquiry skills than Black and Latino students (Rakow, 1985). Black and Latino student performance in math and science have remained well below those of White students. As Lee and Luykx (2006) point out, White and Asian American 8th grade math and science performance is very similar to that of African American and Latino 12th graders. Importantly, student attitudes toward science do not appear to parallel performance measures. Rakow (1985) and Kahle (1982), for example, found that minority student positive attitudes toward science were comparable to, or in some cases greater than, White students. Additionally, science career choice aspirations were similar among racial and ethnic groups (Lee and Luykx, 2006). Significant work is clearly needed to narrow the persistent educational performance gap between White and underrepresented minorities. The proceeding pages will focus on Dominican Americans, a large but greatly neglected minority group in the United States.

12.2 Science Education in the Dominican-American Community

12.2.1 Overview

Dominican-Americans are a growing population within the United States in general and New York City (NYC) in particular (Fig. 12.1). In NYC, for example, Dominican Americans comprise the largest Latino subgroup with a population greater than 500,000 (Torres-Saillant and Hernandez, 1998). Dominican Americans have also comprised the largest share of students entering the NYC public school system since the 1980s. Despite high aspirations (see Section 2.3), Dominican-American students fare worse than all other major ethnic subgroups in terms of educational attainment. Instructional Region 10, for example, the area with the highest concentration of Dominican children, has one of New York State's worst records in terms of performance testing (Torres-Saillant and Hernandez, 1998:87). Significantly less than half of all Dominican American students graduate from high school (National Research Council, 2004). Between 1996 and 1999, only 5% of Dominican Americans graduated from college (Leavitt, 2001:52). Only a fraction of these college graduates majored in the sciences. Clearly, there is a profound need for quality education in the Dominican American community.

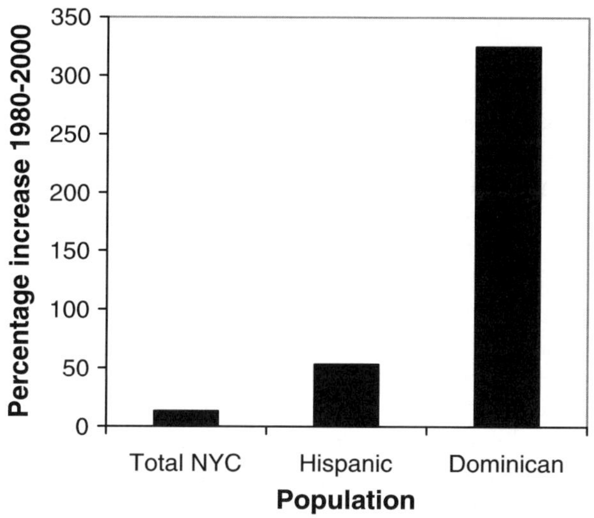

Fig. 12.1 Growth in New York City's Dominican American population from 1980 to 2000 in percentages compared to overall population growth and that of the overall Latino population. Data from Queens College Department of Sociology

The Dominican American community in NYC is unique in that it may be considered to include a large number of "transnational villagers" (*sensu* Leavitt, 2001). "Transnationalism" refers to a pattern of living and working in the United States for most of the year but keeping strong social, economic, and political ties to one's homeland. This transnational connection extends beyond politics and economics, however; it includes familial relationships, religion, and culture (Leavitt, 2001). The DR thus remains a central part of many Dominican American students' lives; indeed, many Dominican Americans have been known to "mythologize" the DR as a "paradise" (Gray, 2001:198). Many Dominican American students' lives thus involve living in two very different worlds. This aspect of the community has important implications for science education efforts (see Section 2.3).

The DRP science education program was developed with these factors in mind. Specific science education goals included: (1) broadening the participation of minority undergraduate students and teachers in Dominican Republic Project field and laboratory research; (2) involving minority K-12 students in Dominican Republic Project laboratory research; (3) developing innovative and culturally relevant K-12 science curricula for schools with primarily Dominican American students; and (4) providing professional development for teachers who work with Dominican American students.

12.2.2 Sense of Place

"Sense Of Place" (SOP) has recently emerged as an important theoretical framework for exploring a broad array of issues in contemporary education. These issues include minority science education (Riggs and Riggs, 2003), rural education (Howley et al., 1996), 'at-risk' education (Bailey and Stegelin, 2003), environmental education (Sanger, 1997), urban education (Chrispeels and Rivero, 2001; Barton, 2002) and curriculum design (Nehm, 2004a) to name a few. As a consequence of the diverse knowledge domains in which SOP has been employed, its meaning varies greatly. Nevertheless, all definitions appear to encompass one's personal connection with a particular spatial location. Sanger (1997) for example, who has discussed SOP within the context of environmental education, defined SOP as "an experientially based intimacy with the natural processes, community, and history of one's place." Likewise, Howley et al. (1996) discussed SOP in the context of rural youth and their magnitudes of attachment to family and the local community.

Within the field of science education considerable research has focused on how increasing one's SOP can be used to enhance teaching and learning outcomes (Sanger, 1997, and references therein). Sanger (1997) has discussed how contemporary science education considers students' relationships with their "lived" environments as marginal, uninteresting, and unimportant. The purported marginalization of SOP in education has been argued to be tied to students' alienation from their natural environment and alienation from science itself (Sanger, 1997). For these reasons, SOP has been receiving increased attention as a part of science education research.

For example, the increasing autonomy of native peoples over their lands, and the central role of ‘place’ in indigenous cultural traditions, has made an understanding of SOP central to effective educational reforms for native peoples in the United States and elsewhere (Riggs and Riggs, 2003; Robinson and Hughes, 1999). Likewise, the recognition that Euro-Americans and indigenous peoples often hold starkly different conceptualizations of natural processes, community, and the history of one’s place in nature have served as a fruitful starting point for challenging the universality of “scientific knowledge” (Riggs and Riggs, 2003, see also Haraway, 1991). Overall, SOP is an important theoretical framework for exploring fundamental aspects of science education including how science relates to students’ lived experiences.

Within the field of urban minority science education, SOP has recently emerged as an important framework for locating knowledge resources in immigrant youth. Nehm (2004a) for example, has argued that SOP can be employed as a construct for locating knowledge resources in Dominican-American students in New York City. Specifically, geological, meteorological, geographical, biological, and ecological knowledge about the Dominican Republic, which Dominican-American immigrant students harbor to varying extents, can be “mined” during science instruction to generate improved understandings of scientific patterns and processes (e.g., climate, earthquakes, etc.). Science curricula can be constructed in ways that prompt students to contrast natural history components of the Dominican Republic with their new environment in New York City, thereby making use of their personally experienced environments and ecologies.

Curriculum that integrates SOP is potentially important because culturally congruent curricula and culturally-aware teachers have been demonstrated to offset the well-documented disengagement in learning science among many urban and minority student populations (Bouillion and Gomez, 2001). Decreasing student marginalization can break the cumulative effects of disengagement, which culminate in low graduation rates (Oakes, 1990). Student interest and success in secondary science is necessary for increasing the likelihood of science career choice later in life. Many minority students unfortunately do not have access to inquiry-based, culturally relevant science curricula in their schools, are alienated from science, continue to under-perform in science classes, and avoid undergraduate science majors (Torres-Saillant and Hernandez, 1998). These problems are exacerbated by teachers who have not been prepared to teach diverse populations effectively. Thus, in order to improve the quality of science education in the Dominican American community, considerable educational reform is necessary.

The DRP science education efforts with Dominican-American students have progressed along four related lines: (1) Funds of knowledge research relating to SOP in secondary students (Nehm and Luna, 2006); (2) development of curricula and resources relating to the DRP (Nehm and Budd, 2006); (3) teacher professional development using the knowledge from (1); and (4) involvement of Dominican American middle, high, and college students and teachers in DRP research projects (e.g., Jarrett et al., 2004; Harvey et al., 2004; Nehm and Luna, 2006; Rivera et al., this volume).

12.2.3 *Funds of Knowledge Research*

'Funds of knowledge' research has a longstanding tradition within the field of minority science education (reviewed in Gonzalez et al., 2005). The need to connect the science content taught in schools with students' cultures and home communities requires knowledge about students' backgrounds, experiences, and interests (Bouillion and Gomez, 2001; National Research Council, 2004). Simply put, "students enjoy learning more, and they learn better, when topics are personally interesting and related to their lives" (National Research Council, 2004). The DR funds of knowledge research program, which set out to identify and value typically unrecognized cultural and cognitive capital harboured by Dominican American students, was instigated in order to generate knowledge for the production of culturally relevant science curricula and to enhance science teacher professional development (see Section 2.5).

Despite being one of the largest ethnic groups in New York City public schools, no research had focused on funds of knowledge in Dominican American students. Related research has provided a general outline of funds of knowledge in New York City Latino (specifically Puerto Rican) households (Mercado, 2005). Our work began with a pilot study which involved developing and refining a paper and pencil instrument for documenting (i) funds of natural history knowledge, (ii) transnational experiences, and (iii) attitudes toward schooling and science. Interviews with Dominican American students (in English and Spanish) supplemented the paper and pencil instrument and were used to explore possible sources of knowledge, refine instrument questions, and explore general attitudes towards schooling and science. In 2006, the revised and validated instrument was administered to seventy-three Dominican American high school students from New York City public schools. All of the voluntary participants completed the anonymous paper and pencil instrument sufficiently for subsequent analyses.

The average age of the students in the sample was 16 years (minimum 15 years, maximum 19 years). Males comprised 43.8% of the sample and females comprised 56.2% of the sample. The majority of students (68.5%) were born in the United States whereas 31.5% were born in the Dominican Republic (DR). A minority (11%) of the students had lived in the United States for 5 years or less. The majority of students (52.1%) never lived outside of the US, but 100% had travelled to the DR. On average, students in the sample had travelled to the DR 5.6 times in their lives (minimum 1 visit, maximum more than 10 visits). When students were asked where they went when they visited the DR, the majority reported that they spent most of their time in both the city and the country (58.9%). In contrast, 28.8% reported spending most of their time in the city and 12.3% reported spending most of their time in the country (Fig. 12.2).

Nearly all of the students reported that their parents spoke Spanish at home (95.9%; n = 73). Approximately 45% of the students' parents had not completed high school. Nevertheless, the students sampled had high educational aspirations: 80.8% reported that they expected to attend college. Additionally, the career aspirations

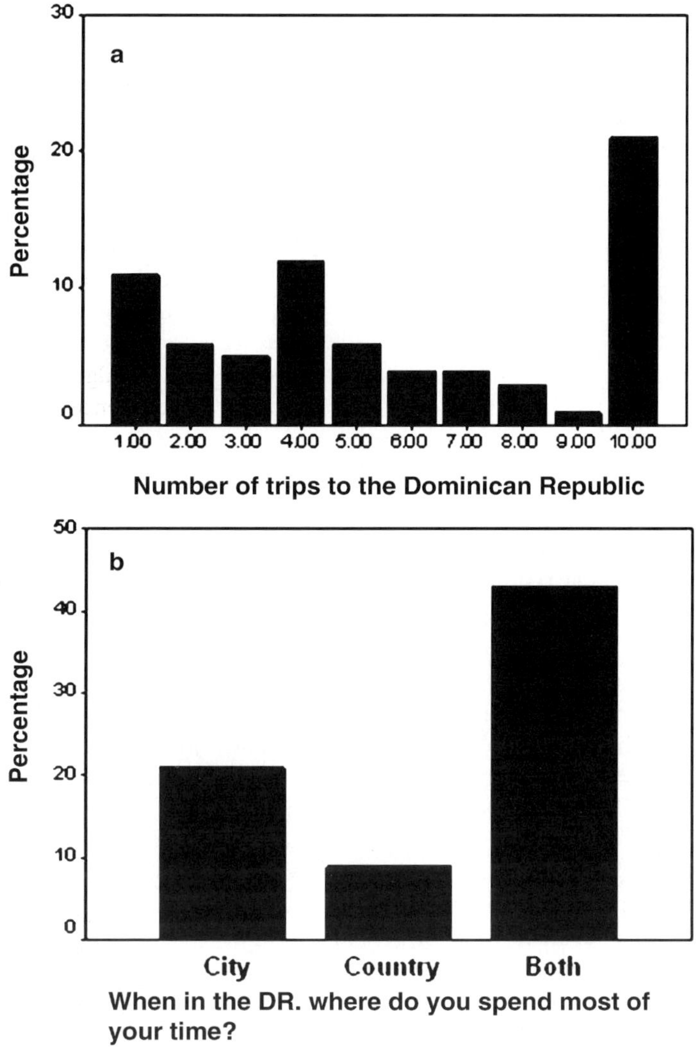

Fig. 12.2 **a**. Number of trips to the DR by Dominican American students (n = 73). Note that 10 trips = 10 or more trips. **b**. Where student time is spent while in the DR, Data from Dominican American public school secondary students from New York City

of the students included many high-skilled occupations (e.g., lawyer, doctor, teacher, etc.; Table 12.1). Finally, the vast majority of students (96%) reported that their parents would be very happy if they went on to attend college.

Six Likert-scale questions (on a five-point scale, where 1 is "not at all" and 5 is "very much") were used to assess student educational experiences relating to science: (1) How much Dominican-American students liked the science classes that they took in middle or high School; (2) how important science was to their lives;

Table 12.1 Career aspirations of Dominican American public school secondary students from New York City in order of abundance (n = 73)

Career aspiration	Rank	Career aspiration	Rank
Lawyer	1	Fashion designer	14
Doctor	2	Hospital management	15
Police officer	3	Hospital work	16
Medical assistant	4	Hotel manager	17
Nurse	5	International attorney	18
Teacher	6	Law firm	19
Chef	7	Law-related	20
Child psychologist	8	Scientist	21
Computer technician	9	Sociologist	22
Dancer	10	Sports agent	23
Designer	11	Technologist	24
Detective	12	US Navy Seal	25
Engineer	13	Veterinarian	26

(3) whether their teachers ever talked about jobs relating to science; (4) whether their teachers encouraged them to study/pursue science as a career or job; (5) whether they thought that aspects of Dominican culture, Dominican scientists, or Dominican society belonged in a science class; and (6) whether any aspects of their everyday or home life were included in their science classes.

As a group, students reported liking science "a fair bit" (Mean = 3.37, sd = 1.15) and viewed science as "somewhat important" (Mean = 3.33, sd = 1.27). Most students reported that their teachers talked about science jobs very little or not at all (Mean 2.35, sd = 1.28) and generally did not encourage students to pursue science (Mean 2.01; sd = 1.26) (Fig. 12.3). Dominican-American students also reported that some aspects of their everyday or home life were included in their science classes (Mean = 2.61, sd = 1.29). Finally, when asked whether they thought that aspects of Dominican culture, scientists, or society belonged in a science class they responded with an average value of 3.4 (sd = 1.3).

As expected, Dominican-American students harbored a wide range of knowledge relating to the natural history of the Dominican Republic (e.g., land animals, land plants, and coastal life), although the amount of knowledge varied significantly among students. Table 12.2 provides a list of the general topics spontaneously mentioned by the students in our study. Many students did not seem to be aware of indigenous species from the Dominican Republic. In contrast, many students mentioned common domesticated animals and plants related to agricultural production (e.g., cows and rice). While awareness of differences in weather and climate between the DR and New York were noted by nearly all students, knowledge of the geological history of both regions appeared to be low.

Overall, our study revealed that Dominican-American students do harbor significant funds of natural history knowledge. Such natural history knowledge appears to be primarily derived from student experiences in the Dominican Republic, although some knowledge of New York natural history was noted. The funds of

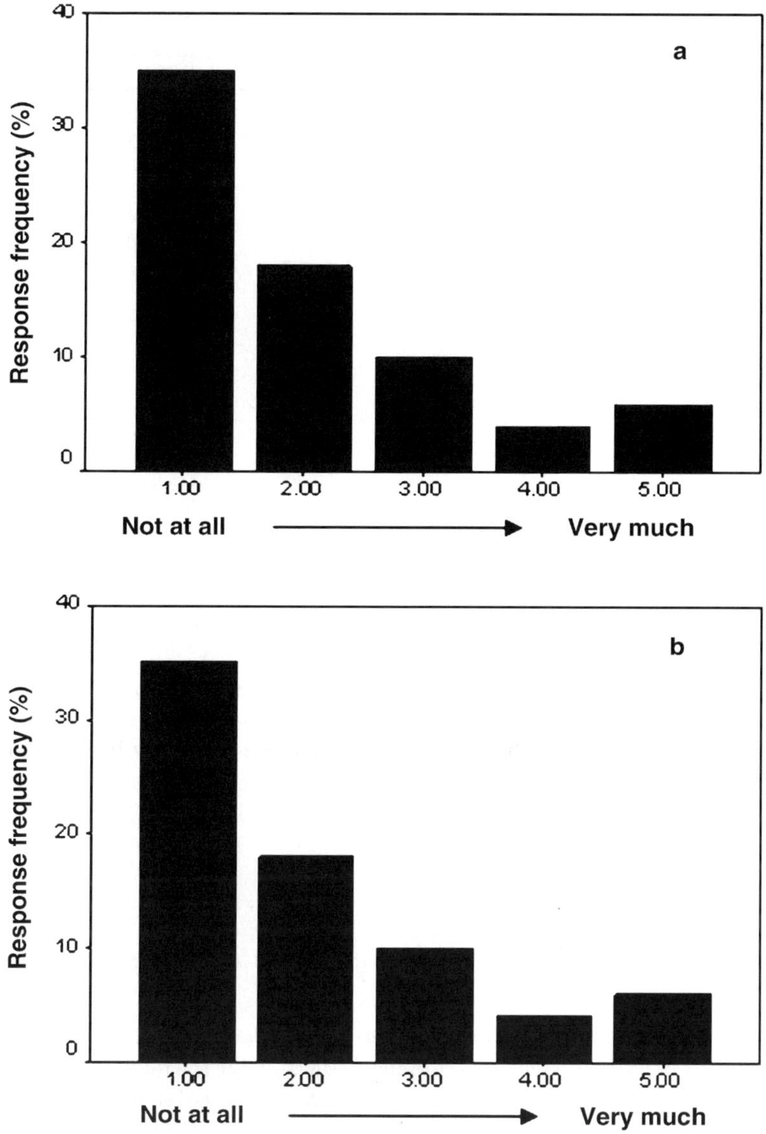

Fig. 12.3 **a.** "Have your teachers ever talked about jobs related to science?" **b.** "Have your teachers encouraged you to study/pursue science as a career or job?" Data from Dominican American public school secondary students from New York City (n = 73)

knowledge that we documented in our sample of Dominican American students provide a first step in building a resource bank that can be used to build curricula that more explicitly connect students' lived experiences with school science topics (Barba and Reynolds, 2003).

Table 12.2 New York City Dominican American public secondary students' observations in the DR from most to least common (n = 73 students responding)

DR Coastal life	Rank	DR Land animals	Rank	DR Plants/Crops	Rank
Fish	1	Cows	1	Plantains	1
Crabs	2	Mosquitoes	2	Sugar cane	2
Coconut trees	3	Dogs	3	Mango	3
Fruit trees	4	Horses	4	Rice	4
Palm trees	5	Pigs	5	Banana	5
Seaweeds	6	Chickens	6	Avocado	6
Farm animals	7	Bees	7	Yucca	7
Birds	8	Lizards	8	Com	8
Starfish	9	Birds	9	Oranges	9
Sugar cane	10	Butterflies	10	Coffee	10
Grapes	11	Donkeys/Mules	11	Yams	11
Almendra	12	Snakes	12	Beans	12
Fish ("rare")	13	Goats	13	Limonsilla	13
Coral reef	14	Roosters	14	Tomatoes	14
Corals	15	Crickets	15	Coconut	15
Ducks	16	Sheep	16	Tamarind	16
Dolphins	17	Matat de coco	17	Batata	17
Turtles	18	Ormigas	18	Pineapple	18
Tadpoles	19	Frogs	19	Potato	19
Plantains	20	Lice	20	Carambola	20
Beach bugs	21	Ants	21	Strawberries	21
Jellyfish	22	Turtles	22	Lettuce	22
Ticks	23	Rats	23	Eggplant	23
Sea shells	24	Spiders	24	Tobacco	24
Sea gulls	25	Flies	25	Grapes	25
Whales	26				
Sharks	27				
Mosquitoes	28				

12.2.4 Curricula and Teaching Resources Relating to the DRP

The low rate of Latino school success in general has prompted several reform efforts, most notably the Hispanic Dropout Project (HDP), which gathered data on Latino school experiences and proposed recommendations for improving the educational experiences of Latino students (ERIC, 2000). HDP and several other studies (Barba and Reynolds, 1998; Barba, 1995) agree on the broad attributes that curricula for Latino students in general should share: (1) Science curricula should be connected to the "real world" (i.e., the students' homes and communities); (2) Curricula should be built around culturally familiar contexts; (3) Science curricula should be built upon concrete experiential knowledge; (4) Latino role models should be incorporated into curricula; (5) Students should be able to make use of their personal experiences while learning; and (6) Mutual assistance and socialization should be incorporated in the lesson activities because Latino culture values collaborative approaches to achievement. Curricula that share these attributes are expected to decrease the need for Latino students to part with their culture, traditions, and home knowledge while in schools. This is significant because this issue has been implicated as one major reason why Latinos are reluctant to be involved in the sciences (Barba and Reynolds, 1998; Barba, 1995).

In response to the HDP and numerous other studies and reform documents, the National Science Foundation (1998) advocated the development of more culturally relevant curriculum materials for the nation's growing minority student population. Unfortunately, recent research indicates that the curriculum materials used in US classrooms "are not culturally relevant to non-mainstream students" and "cultural diversity is not adequately represented in textbooks and materials" (Lee and Luykx, 2007:177). The few cases in which such culturally responsive materials have been developed and evaluated suggest that they do indeed produce improved science learning experiences. Specifically, culturally-relevant curricula have been shown to be associated with significantly higher achievement scores and significantly more positive attitudes towards science than traditional curriculum materials (Matthews and Smith, 1994; Bouillion and Gomez, 2001). Much more work is clearly needed to develop new culturally congruent curricula and explore their efficacy for underrepresented students' science achievement.

A goal of the DRP education efforts therefore was to help provide resources that could be used to interweave Dominican American funds of knowledge and science interests into middle and high school science curricula. The purpose of such efforts was to assemble resources that could be used to explicitly acknowledge the experiences, interests, and backgrounds of Dominican American public school students while covering national and state-mandated science curriculum benchmarks (e.g., plate tectonics, evolution, biodiversity). The DRP website (Nehm, 2004b) was developed to provide access to historical information, maps, scientific images, video, and curriculum materials relating to the Dominican Republic (see also Nehm and Budd, 2006). Because many science teachers have great difficulty envisioning how standard curriculum topics (e.g., plate tectonics) could be connected to Dominican American students' personal experiences (see also "Science Teacher Professional Development" below), we briefly outline several simple approaches relating to earthquakes, biodiversity, and field studies.

The standard earth science curriculum topics of earthquakes and plate tectonics can be connected culturally *via* an exploration of the great Dominican earthquake of 1532, which destroyed the city of Santiago (the Dominican Republic's second largest city) and necessitated its relocation nearby the old city limits. This topic provides important opportunities for teachers to make connections among history, science, and the DR. More contemporary examples could also be used. For example, a lesson could be embedded in the context of students asking their relatives if they remember the more recent devastating earthquake of August, 1946 (Lynch and Bodle, 1948). In class, students could share stories of what their relatives remember from the event and explore why the Dominican Republic has larger and more frequent earthquakes than New York, for example. These types of lessons can facilitate student engagement in earth sciences on a more personal level by involving their families and relatives in student learning.

Technology and the internet should be part of every science subject, and the online database NMITA (Neogene Marine Biota of Tropical America, nmita.geology.uiowa.edu, see Budd et al., this volume) provides a useful resource for lessons that connect the concepts of extinction, systematics, evolution, and animal biology to the DR and the Caribbean region. This website contains photos, maps, and detailed information about Dominican invertebrates (e.g., corals, molluscs, and arthropods). Student activities related to this website could be combined with actual specimens of Dominican invertebrate fossils for more hands-on activities relating to abstract curriculum topics such as extinction and evolution (Nehm and Budd, 2006).

One challenge that every science teacher faces is how to create "entry points" into curriculum topics for students. The topic of environmental change might become more interesting to students if they were asked to view photos or videos and collaboratively explain how a spectacular fossil marine coral reef come to be located on land in the Cibao Valley of the DR in an area now known for tobacco and rice production. Such simple examples can provide ways to make standard science topics more closely connected to students' experiences, backgrounds, and cultures.

Field studies of nature are an important part of learning about biology and earth science. In New York City, for example, many publicly accessible beaches contain an amazing diversity of marine animals that leave shells that can be collected, surveyed, or photographed in the field (An exhibit on New York's molluscs at the American Museum of Natural History can be used to help identify species). These shells provide useful data for comparisons with the Caribbean faunas of the Dominican Republic. Many students spend part of their summers in the DR (see above) and can make surveys, assemble small collections, or take photos of common shells and contrast them with the marine molluscs of New York City. Activities can be developed that have students hypothesize why the shells from these two regions are of different shapes, sizes, and diversities. These specimens could provide hands-on resources for many other activities relating to biodiversity, classification, biomes, geography, and ecology. In summary, one goal of the DRP education efforts was to produce materials for teachers to assist in the modification of standard curriculum topics to make them less foreign to student populations that have historically been underrepresented in the sciences.

12.2.5 Science Teacher Professional Development

Practicing science teachers often lack sufficient preparation to work effectively with the nation's growing minority student population (Lee and Luykx, 2006). More problematic are data indicating that many mainstream teachers who have participated in teacher preparation and professional development programs in multicultural education resist ideological change and display feelings of disbelief and defensiveness relating to multicultural perspectives of teaching and learning (Lee and Luykx, 2006:105; Bryan and Atwater, 2002). Problematic worldviews, such as the notion that minority students are less capable than White students, that inequality is 'a given' in education, or that student racial, ethnic, and cultural differences have little significance for science teaching, are also widespread (Bryan and Atwater, 2002). Additionally, many science teachers find exploring their students backgrounds, and in so doing crossing "cultural borders", to be threatening "because accepting a new set of norms or beliefs could imply that something was wrong with their original beliefs" (Lee and Luykx, 2006:106). Overall, there is a great need to better prepare mainstream science teachers through fostering the skills and belief systems necessary to work more effectively with minority students.

The cultural, socioeconomic, and geographic differences between Dominican American students in New York City (NYC) and their teachers are great. Science teachers who work with Dominican American students in NYC are predominantly White, middle class, from suburban backgrounds, and enrolled in so-called 'emergency' credentialing programs such as the NYC Teaching Fellows program. Most of these teachers are transplants from other regions of the country and came to NYC in response to educational and financial incentives designed to ameliorate the shortage of science and math teachers in urban schools. Consequently, the vast majority of these teachers have little familiarity with the social and cultural backgrounds of the students they teach, and the students have little in common with the teachers who teach them. The lack of published research on Dominican American student backgrounds, interests, and funds of knowledge (see above) hinders the professional development of the teachers who have an interest in bridging cultural differences with their students.

A goal of the DRP science education program was to integrate knowledge of Dominican American transnationalism, 'sense of place,' and funds of knowledge into a science teacher preparation program enrolling mostly NYC Teaching Fellows. Three specific goals included: (1) Raising science teacher's awareness that Dominican American students have funds of science knowledge; (2) educating science teachers to recognize that funds of knowledge that are different from their own also have value; and (3) preparing science teachers to utilize student experiences and knowledge in their classrooms in order to produce more culturally congruent instruction and help students connect their lives to schooling (see also National Research Council, 2004).

Science teacher attitudes and beliefs are known to greatly influence the effectiveness of professional development efforts (Bryan and Atwater, 2002; Jones and Carter, 2007). The majority of science teachers involved in our teacher education program ($n \sim 100$) maintained a transmission model of science teaching that was generally stable throughout the short (<2 years) program. In brief, the transmission model, which is well known to be at odds with how students actually learn (NRC, 2001), assumes that students' minds are like empty vessels and can be filled with science knowledge *via* traditional pedagogical methods such as lecturing. This model of teaching and learning ignores or downplays the importance of prior student knowledge and misconceptions, views personal and cultural background as largely irrelevant to science learning, and largely rejects a constructivist view of knowledge acquisition. Ironically, teachers' prior beliefs about science learning significantly hindered their ability to recognize the importance of their students' prior knowledge and experiences about science. Similar results have been widely reported in the literature (reviewed in Jones and Carter, 2007).

Most of the science teachers had great difficulty negotiating the foreign cultural, socioeconomic, and linguistic backgrounds of their students. Science teachers commonly remarked that their Dominican American students "didn't know anything", despite efforts to explicitly demonstrate that teacher and student funds of knowledge were non-overlapping but both were nevertheless of value. Little cross-cultural exchange occurred, because student funds of knowledge were generally not valued by teachers. These ideas formed parts of a larger "deficit model" of Dominican American students (see Delpit, 1995). Teachers thus excused themselves as a cause of student failure; it was a result of *student* deficiencies. Similar patterns have been observed in other mainstream teachers working with Latino secondary students (Espinoza-Herold, 2003).

These factors collectively made it difficult for the science teachers working with Dominican American students to recognize the importance of their students' background knowledge and experiences in creating classroom learning environments that were less alienating to their students. Consequently, many teachers did not make an effort to learn about students' funds of science knowledge and cultural backgrounds or use them to modify science instruction. Lucas, Henze, and Donato (1990) have shown that Latino student success has occurred when teachers value students' language and culture and have high expectations of minority students. Thus, student failure is not just about students; it also depends largely on teachers (Espinoza-Herold, 2003).

Unfortunately, the experiences of the teachers involved in this teacher preparation program were similar to those of many other white, middle class, suburban teachers who began teaching minority students in urban schools: teachers and students remained in two separate worlds and considerable resistance characterized attempts to bridge these worlds (Rodriguez, 1998; Bryan and Atwater, 2002; Espinoza-Herold, 2003; Lee and Luykx, 2006). This component of the project illustrated an important lesson: reform documents, innovative curriculum materials, and knowledge of student backgrounds are necessary but not sufficient for producing change. Much more work is clearly needed to help teachers ameliorate their deficit models of minority students and better navigate cultural borders.

12.2.6 Student Participation in DRP Research Projects

In the past 30 years, it is concerning that no Dominicans or Dominican Americans have been involved in DR project research. Since instigating DRP research in 2002 at the City University of New York (CUNY), which enrolls more Dominican American students than any other university in the United States, many Dominican American students have expressed interest in working on geoscience research projects relating to the Dominican Republic. Outreach efforts with local middle and high schools, talks in science courses, and websites were important for making students aware of research opportunities at the college. In the past 5 years, 24 students (eight middle school, three high school, ten undergraduate, and three science teachers) have worked on laboratory and/or field research projects lasting from 1–3 years (see also Rivera et al., this volume) (Fig. 12.4). All of these research projects have involved presentations at local, regional, or national science fairs, science competitions, and scientific conferences such as the Geological Society of America. One high school student team was a Siemens-Westinghouse national semi-finalist and an undergraduate student team was awarded best team research at the American Institute of Biological Sciences conference on biodiversity in 2006. Thus, Dominican American students do have an interest in geoscience research and, if given the opportunity, have the potential to make major contributions to research relating to the Dominican Republic Neogene. We hope that future DRP research will continue to engage Dominican American students in research.

Fig. 12.4 Student and teacher participation in Dominican Republic Project research from 2002 to 2007

12.3 International Educational Outreach Efforts in the Dominican Republic

In addition to the work with Dominican American students in New York City, other efforts have focused on undergraduates in the Dominican Republic (Fig. 12.5). As part of an ongoing effort to develop partnerships with Dominican institutions, we (Budd, McNeill, Klaus) conducted two NSF-funded educational workshops for undergraduate students at the Universidad Autónoma de Santo Domingo (UASD): (1) Paleoecology and Sedimentology of Ancient Coral Reefs in the Dominican Republic, March 16–17, 2006, in Santo Domingo; and (2) Paleoecology of Ancient Coral Reefs and Sample Curation, January 9–10, 2007 in Valverde Mao. The first workshop was attended by 17 students from four different departments (biology, geology, geography, engineering), and the second by 14 students.

The goal of the first workshop (based in Santo Domingo) was to show how studies of fossil reef systems, thousands to millions of years old, are relevant to addressing modern-day issues in reef conservation. During a full day of lectures at UASD, we introduced students to past extinctions and faunal turnover events on coral reefs and their relationship to changes in climate and tectonics. We focused on community change on Caribbean reefs during an episode of biotic turnover that occurred between 6 and 1 million years ago in association with the closure of the Central American isthmus and the onset of Northern Hemisphere glaciation. Laboratory exercises taught students how to identify skeletons of Holocene corals, and how different assemblages of corals are associated with different reef environments. A fieldtrip to the rich and exquisitely-preserved fossil Holocene reefs of the

Fig. 12.5 Field studies with undergraduate students from the Universidad Autónoma de Santo Domingo in the Dominican Republic

Enriquillo Valley provided a hands-on example of how fossil coral assemblages can be used to interpret ancient reef environments in the fossil record. One outcome of the workshop was increased interest by faculty and students in developing the natural history collections at the UASD, and making these collections available to university faculty and students and the Dominican public.

The second workshop (based in Valverde Mao) had two objectives: (1) To describe our ongoing multidisciplinary research project on the Mio-Pliocene fossil reefs of the Cibao Valley; and (2) to train students and researchers in collection care and management, including preventive conservation, collection organization, and data preservation and management. We emphasized how museum collections contribute to understanding biodiversity, and the importance of sharing collection information with researchers and institutions around the world and participating in international museum initiatives. This workshop consisted of two evenings of lectures and exercises identifying Mio-Pliocene corals, and a full-day fieldtrip to Arroyo Bellaco. During the fieldtrip, students collected a total of 117 coral specimens, and were taught how to take field notes and how to pack, label, and properly document field collections (especially stratigraphic and locality data). As in the previous workshop, we discussed how fossil coral assemblages can be used to interpret ancient reef environments in the fossil record. On our return to Santo Domingo, we showed students how to wash and curate the collected specimens, and enter the associated information into a specimen database. The materials were deposited at a UASD biological collections facility. We continue to work with faculty at the university to develop an online catalogue for the facility, which conforms with international museum standards. Workshop guidebooks, along with numerous photos of the two workshops, can be accessed at the Neogene Marine Biota of Tropical America (NMITA) web page http://nmita.geology.uiowa.edu/DRworkshop/uasd-mar2006.htm.

12.4 Conclusions

For the past 30 years, the Dominican Republic Project has focused exclusively on science research and has not involved Dominican American or Dominican students or scientists. Significant growth in minority students in the United States, including Dominican Americans, has not been accompanied by increasing representation in science careers, especially those in the geosciences. In this chapter we reviewed new DRP science education projects with Dominican-American students that have involved: (1) funds of knowledge research with Dominican American secondary students; (2) development of curricula and resources relating to the DRP; (3) science teacher professional development; (4) involvement of Dominican American middle school, high school, and college students and teachers in DRP research projects; and (5) international outreach and development activities. Although preliminary and small in scope, we hope that this work will be of use to other scientists working to improve the quality of science teaching and learning in Dominican American as well as other minority groups.

Acknowledgments We thank Francisco Geraldes (UASD) and Haydee Dominguez Tejo (UASD), Ramona Hernandez (Dominican Studies Institute at CUNY), NSF CAREER to RHN, NSF EAR 0445789 to AFB, and NSF EAR 0446768 to DFM. Many students and teachers at CUNY provided important insights into minority science education. Reviews by and discussions with Andrew Ratner, Andrea Gay, and Brian Baldwin helped to improve the manuscript.

References

Bailey, B. and Stegelin, D.A., 2003, Creating a sense of place: anchoring at-risk students within K-12 classrooms, *J. At-Risk Issues*, **9**:17–26.

Barba, R.H., 1995, *Science in the Multicultural Classroom: A Guide to Teaching and Learning*, Allyn & Bacon, Boston, MA.

Barba, R.H. and Reynolds, K.E., 1998, Towards an equitable learning environment in science for Hispanic students, pp. 925–939 in Fraser, B.J. and Tobin, K.G. International Handbook of Science Education, Kluwer: Dordrecht.

Barba, R.H. and Reynolds, K.E., 2003, Toward an equitable learning environment in science for Latino students, in: *International Handbook of Science Education* (Fraser, B.J. and Tobin, K.G., eds.), Kluwer, London, pp. 925–940

Barton, A.C., 2002, Urban science education studies: a commitment to equity, social justice and a sense of place, *Studies Sci. Ed.*, **38**:1–37.

Bouillion L.M. and Gomez, L.M., 2001, Connecting school and community with science learning: real world problems and school-community partnerships as contextual scaffolds, *J. Res. Sci. Teach.*, **38**:878–898.

Bryan, L.A. and Atwater, M.M., 2002, Teacher beliefs and cultural models: a challenge for science teacher preparation programs, *Sci. Ed.*, **86**:821–839.

Chrispeels, J.H. and Rivero, E., 2001, Engaging Latino families for student success: how parent education can reshape parents' sense of place in the education of their children, *Peabody J. Ed.*, **76**:119–169.

Delpit, L., 1995, *Other People's Children: Cultural Conflict in the Classroom*, New Press, New York.

Dillon, S., 2007, US data show rapid minority growth in school rolls, *The New York Times*, June 1, p. A21.

Espinoza-Herold, M., 2003, *Issues in Latino Education: Race, School Culture, and Politics of Academic Success,* Allyn & Bacon, Boston, MA.

Garcia, E.E., 1999, *Student Cultural Diversity: Understanding and Meeting the Challenge*, 2nd ed., Houghton Mifflin, Boston, MA.

Gonzalez, N., Moll, L.C., and Amanti, C. (eds.), 2005, *Funds of Knowledge: Theorizing Practices in Households, Communities, and Classrooms*, Lawrence Erlbaum, Mahwah, NJ.

Gray, D.M., 2001, *High Literacy and Ethnic Identity: Dominican American Schooling in Transition*, Rowman & Littlefield, New York.

Haraway, D.J., 1991, Situated knowledges: the science question in feminism and the privilege of partial perspective, in: *Simians, Cyborgs, and Women: the Reinvention of Nature*, Routledge, New York, pp. 183–201.

Harvey, M., Nehm, R.H., and Jarrett, N., 2004, Exploring the relationships among mollusk community change and the tempo and mode of speciation in the upper Gurabo Formation, Neogene, Dominican Republic, *Geol. Soc. Am., Abstr. Prog.*, **36**:366.

Howley, C.B., Harmon, H.L., and Leopold, G.D., 1996, Rural scholars or bright rednecks? Aspirations for a sense of place among rural youth in appalachia, *J. Res. Rural Ed.,* **12**:150–160.

Jarrett, N., Nehm, R.H., Hidalgo, Y., and Cabrera, I., 2004, Quantifying the effects of sampling method and effort on estimates of species richness and relative abundance of Neogene benthic marine mollusks, Cibao Valley, Dominican Republic, *Geol. Soc. Am., Abstr. Prog.*, **36**:366–367.

Jones, M.G. and Carter, G., 2007, Science teacher attitudes and beliefs, in: *Handbook of Research on Science Education* (Abell, S.K. and Lederman, N.G., eds.), Lawrence Erlbaum, Mahwah, N.J., pp. 1067–1104.

Kahle, J.B., 1982, Can positive minority attitudes lead to achievement gains in science? Analysis of the 1977 National Assessment of Educational Progress, attitudes toward science, *Sci. Ed.*, **66**:539–546.

Leavitt, P., 2001, *Transnational Villagers*. University of California Press, California.

Lee, O. and Luykx, A., 2007, Science education and student diversity: race/ethnicity, language, culture, and socioeconomic status, in: *Handbook of Research on Science Education* (Abell, S.K. and Lederman, N.G., eds.), Lawrence Erlbaum, Mahwah, NJ, pp. 171–197.

Lee, O. and Luykx, A., 2006, *Science Education and Student Diversity: Synthesis and Research Agenda,* Cambridge University Press, New York.

ERIC Clearinghouse on Urban Education, 2000, School practices to promote the achievement of Latino students, Digest 133, March.

Lucas, T., Henze, R., and Donato, R., 1990, Promoting the success of language-minority students: an exploratory study of six high schools, *Harvard Ed. Rev.*, **60**:315–337.

Lynch, J.J. and Bodle, R.R., 1948, The Dominican earthquakes of August 1946, *Bull. Seismol. Soc. Am.*, **38**:1–18.

Matthews, C.E. and Smith, W.S., 1994, Native American related materials in elementary science instruction, *J. Res. Sci. Teach*, **31**:363–380.

Mercado, C.I., 2005, Reflections on the study of households in New York City and Long Island: a different route, a common destination, in: *Funds of Knowledge: Theorizing Practices in Households, Communities, and Classrooms* (Gonzalez, N., Moll, L.C. and Amanti, C., eds.), Lawrence Erlbaum, Mahwah, NJ, pp. 233–255.

National Center for Education Statistics, 1996, *Pursuing Excellence: A Study of US Eighth Grade Mathematics and Science Teaching, Learning, Curriculum, and Achievement in International Context,* US Department of Education, Office of Educational Research and Improvement, Washington, DC.

National Research Council, 2001, *How People Learn*, National Academy Press, Washington, DC.

National Research Council, 2004, *Engaging Schools*, National Academy Press, Washington, DC.

National Research Council, 2007, *Rising Above the Gathering Storm*, National Academy Press, Washington, DC.

National Science Foundation, 1998, *Infusing Equity in Systemic Reform: An Implementation Scheme*, National Science Foundation, Washington, DC.

National Science Foundation, 1999, *Doctorate Recipients from United States Universities: Summary Report 1998*. Arlington, VA.

National Science Foundation Directorate for Education and Human Resources, 1996, *Review Of Instructional Materials For Middle School Science*, National Science Foundation, Arlington, VA.

Nehm, R.H., 2004a, Integrating the Dominican Republic Project with new approaches to multicultural urban science education, *Geol. Soc. Am. Abstr. Prog.,* **36**:13.

Nehm, R.H., 2004b, The Dominican Republic Project, available online at: www.dominicanrepublicproject.org.

Nehm, R.H., 2005, The effects of a marine and coastal resource education project on urban science teachers' curricula, pedagogical practices, and 'sense of place', *Proceedings of the OCEANS 2005 conference*, IEEE publishers, Washington, DC, pp. 1–9.

Nehm, R.H. and Budd, A.F., 2006, Missing 'links' in bioinformatics education: expanding students' conceptions of bioinformatics using a biodiversity database of living and fossil reef corals, *Am. Biol. Teach.*, **68**:e91–e97.

Nehm, R.H. and Luna, J., 2006, Dominican-American transnationalism, sense of place, and cultural capital: their roles in science teacher preparation and multicultural curriculum reform, *Association of Science Teacher Educators International Conference Proceedings* #12015, pp. 1–11.

Oakes, J., 1990, *Multiplying Inequalities: The Effects of Race, Social Class, and Tracking on Opportunities to Learn Mathematics and Science*, Rand Corporation, Santa Monica, CA.

O'Sullivan, C.Y., Lauko, M.A., Grigg, W.S., Qian, J., and Zhang, J., 2003, *The Nation's Report Card: Science 2000*, US Department of Education, Institute of Education Sciences, Washington, DC.

Rakow, S.J., 1985, Minority students in science: perspectives from the 1981–1982 national assessment in science, *Urban Ed.*, **20**:103–113.

Riggs, E.M. and Riggs, D.M., 2003, Cross-Cultural Education of Geoscience Professionals: The Conferences of the Indigenous Earth Sciences Project, *J. Geosci. Ed.,* **51**:527–535.

Robinson, C. and Hughes, P., 1999, Creating a Sense of place: indigenous peoples in vocational education and training, *National Centre for Vocational Education Research*, Leabrook, Australia.

Rodriguez, A., 1998, Strategies for counter-resistance: toward sociotransformative constructivism and learning to teach science for diversity and for understanding, *J. Res. Sci. Teach*, **35**:589–622.

Sanger, M., 1997, Sense of place and education, *J. Environ. Ed.,* **29**:4–8.

Schmidt, W.H., McNight, C.C., and Raizen, S.A., 1997, *A Splintered Vision: An Investigation of US Science and Mathematics Education,* Kluwer, Dordrecht, The Netherlands.

Torres-Saillant, S. and Hernandez, R., 1998, *The Dominican Americans*, Greenwood Press, New York.

Chapter 13
The Neogene Marine Biota of Tropical America ("NMITA") Database: Integrating Data from the Dominican Republic Project

Ann F. Budd[1], Tiffany S. Adrain[1], Juw Won Park[2], James S. Klaus[3], and Kenneth G. Johnson[4]

Contents

13.1 Introduction

The NMITA database (http://nmita.geology.uiowa.edu) serves as a central data repository for the Dominican Republic Project and members of the team have been actively contributing data as the project progresses. NMITA was originally designed as a taxonomic database, the main purpose of which is to document the taxa that have been identified in paleontological collections from the Neogene of Tropical America as well as the taxonomic concepts and morphologic features that have been used in making identifications. NMITA thereby provides the taxonomic foundation for spatial and temporal analyses of biodiversity. NMITA is similar to a taxonomic monograph in that it contains high quality images and

[1] Department of Geoscience, University of Iowa, 121 Trowbridge Hall, Iowa City, IA 52242, USA. Email: ann-budd@uiowa.edu, tiffany-adrain@uiowa.edu

[2] Department of Computer Science, Information Technology Services, 2860–65 UCC, University of Iowa, Iowa City, IA 52242, USA. Email: jwpark@cs.uiowa.edu

[3] Department of Geological Sciences, University of Miami, 43 Cox Science Building, Coral Gables, FL, 3133. Email: jsklaus@rsmas.miami.edu

[4] Department of Palaeontology, Natural History Museum, Cromwell Road, London SW7 5BD, United Kingdom. Email: K.Johnson@nhm.ac.uk

R.H. Nehm, A.F. Budd (eds.) *Evolutionary Stasis and Change in the Dominican Republic Neogene*,

information about taxa (i.e., authorship, synonyms, type specimens) as well as morphologic characters used in their identification and stratigraphic and geographic occurrence data. The taxa in NMITA currently consist of: ~225 scleractinian coral species (zooxanthellate & azooxanthellate), ~330 bryozoan species (cheilostome & cyclostome), ~300 bivalve genera/subgenera, ~50 gastropod species (columbellid, muricid, marginellid), ~100 benthic foraminifera species, ~100 ostracode species, and ~230 teleost and elasmobranch fish species. Information has been contributed by specialists on the systematics of these groups, who are based at many different academic and museum institutions in the Americas and Europe (see http://nmita.geology.uiowa.edu/nmtstaff.htm for complete listing). In addition to the DR Project, the database also serves the Panama Paleontology Project led by J.B.C. Jackson and A.G. Coates of the Smithsonian Tropical Research Institute, as well as other team-based paleontological collection projects that use rigorous sampling and geologic-age dating protocols.

The taxa that are treated in NMITA are determined by querying specimen databases, including museum catalogs, which document the collections made by team projects such as the DR project (Fig. 13.1). In the case of zooxanthellate corals, we have developed the Cenozoic Coral Database (CCD) to capture specimen information, because the specimens are located in many different museums [including the Natural History Museum in Basel, Switzerland (NMB), the Paleontology Repository of the University of Iowa (SUI), among others]. Biodiversity analyses are performed using a third type of database ("ANALYSIS"), also separate from but overlapping with NMITA, which focuses on selected subsets of occurrences and is in a format amenable to performing computations and statistical analyses. In the case of zooxanthellate corals, this third type of database is also derived from queries of CCD and has been filtered to remove incomplete records. Different formats are used depending on the scientific question that is being addressed.

NMITA focuses on middle and lower taxonomic levels within the Linnean hierarchy (e.g., families, genera, and species) and was originally based on a single classification system for each taxonomic group. This system was determined by the specialist(s) who contributed images and information for that particular taxonomic group. Similarly, geologic age dates for occurrences in NMITA were based on a single system determined by integrating the most up-to-date source(s) available. Recently, for zooxanthellate corals, we have implemented a strategy to incorporate several different classification systems in NMITA, as well as a system for recording several different identifications of each specimen and several different age interpretations for each stratigraphic unit (or "faunule") in CCD. The purpose of this chapter is to briefly describe the strategies we are using to implement multiple specimen identifications and age interpretations as part of the DR Project, as well as to share our data with others in the scientific community. Here we emphasize the structure and content of the database, rather than the web pages that are dynamically created by queries of the database (see Budd et al., 2001). These new implementations have been developed using the corals in NMITA, so we focus our discussion on this group.

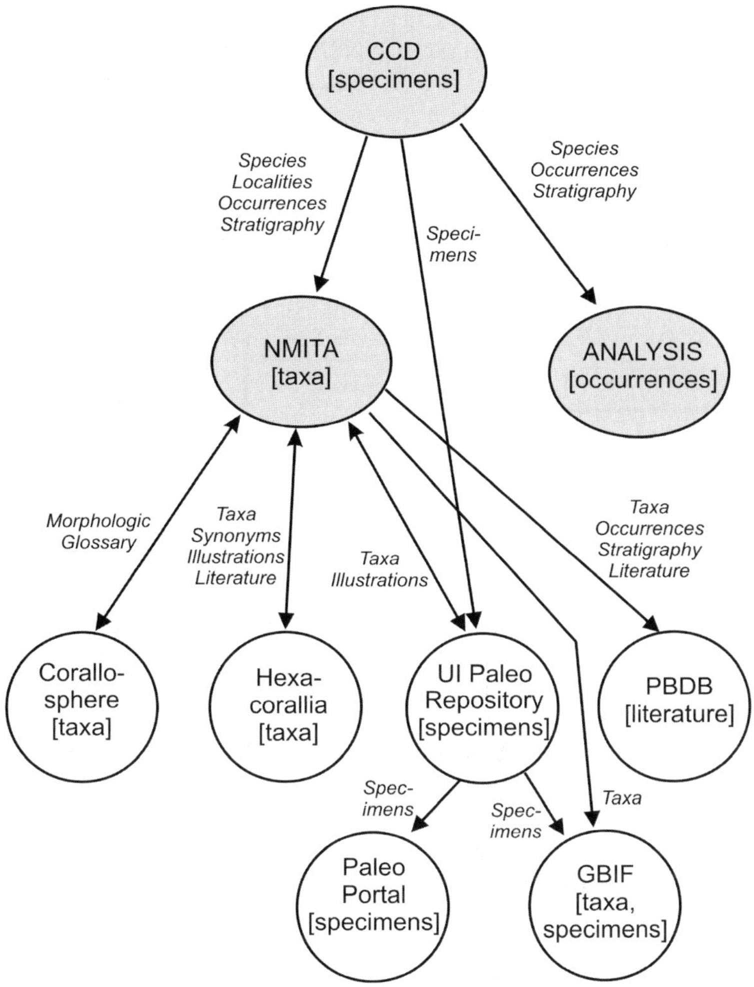

Fig. 13.1 Diagram showing the databases (circles and ellipses) and flow of data (arrows) associated with zooxanthellate corals in the DR Project. Terms in italics indicate the types of data being shared. Unshaded circles represent databases administered elsewhere in the scientific community

13.2 Overview of NMITA Database Structure

NMITA is a relational database consisting of 17 tables (spreadsheets with rows and columns of data) and >100 fields (data columns; Fig. 13.2). It currently contains data for >1,300 taxa, including >3,800 images. The database currently runs on an Oracle 9i server managed by the University of Iowa ITS database group. Applications are developed on a separate application server by University of Iowa ITS programmers, and involve: (1) NMITA Web Applications implemented using JAVA-JSP, and (2) NMITA Web Services implemented using APACHE-AXIS.

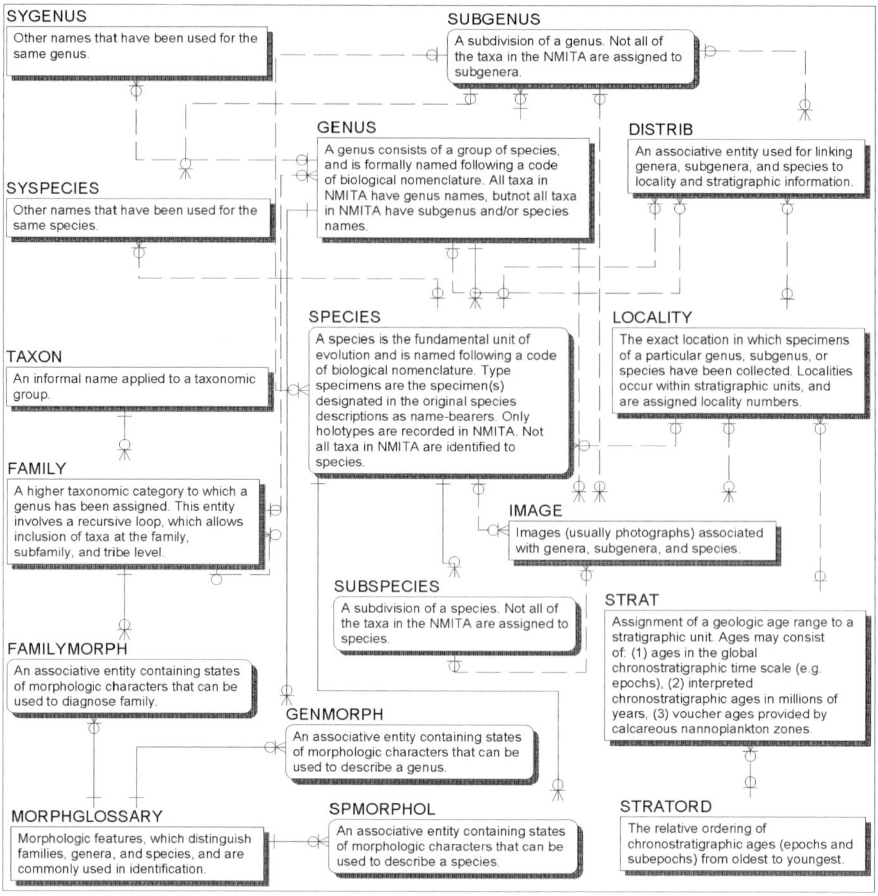

Fig. 13.2 NMITA database schema showing the tables (entities) and their definitions. A complete data dictionary is available at nmita.geology.uiowa.edu/NMITAdatadict/ NMITAdatadict.htm

The database can be subdivided into four primary subject areas: (1) taxonomy, (2) synonymy, (3) morphology, and (4) locality. In addition, there is a separate table for images, as well as one for bibliographic information input using EndNote computer software. The database model or schema was originally patterned after the 1993 Association of Systematic Collections (ASC) Information Model, but has been extensively modified and simplified over the past few years in order to facilitate data import and to minimize the need for extensive programming as well as expertise in database design and management. Not only have many ASC fields been dropped, but many complex relationships have been collapsed by combining tables. In the taxonomy area, as in the ASC model, the table for families involves a recursive loop (i.e., a loop within a loop), which allows for the inclusion of subfamily or tribe information if available. In the recursive loop, taxon names (families, subfamilies, and tribes), taxonomic ranks, and the next higher taxon to which a given taxon belongs are entered in separate columns. However, tables for genera, subgenera, species, and subspecies are arranged hierarchically. The fields for these tables include author/date information and type specimen or type species information. Synonyms can be provided at the genus and species levels. Multiple classification systems have been implemented by adding an additional field for classification system to the genus table (for genera within families, subfamilies, and tribes) and to the species table (for species within genera and subgenera).

At the heart of the morphology subject area is a morphologic glossary which defines and illustrates a standard set of characters and states for each higher level taxonomic group. These characters are used to generate morphologic descriptions of families, genera, and species in telegraphic style. They also form the basis of NIT (NMITA Identification Tools, see http://nmita.geology.uiowa.edu/idkeys.htm), which allows users to search for taxa that possess specific combinations of character states. NIT is highly flexible in that users may select one or more characters and one or more states for each character (Fig. 13.3). NIT is patterned after the Pollyclave program distributed by the University of Toronto, which is written in DELTA format (http://prod.library.utoronto.ca:8090/polyclave/); however, it is implemented using JAVA-JSP queries of the NMITA Oracle database. In corals, the morphologic glossary will share data with the Corallosphere database, which contains diagnostic morphologic information for all valid genera of scleractinian corals (fossil and Recent) and will be used to create the next edition of the Scleractinian Volume of the Treatise on Invertebrate Paleontology published by the University of Kansas and the Geological Society of America.

The locality subject area contains information about the geographic and stratigraphic occurrences of genera, subgenera, and species. Localities are georeferenced and grouped into stratigraphic units (sometimes termed "faunules", see Johnson et al., this volume), which are each associated with interpretations of geologic ages. Searches of occurrence data have been implemented using clickable maps and columns. The maps have been developed by importing scanned topographic maps into ARC/Info, registering them to the same coordinate system, making them clickable, and linking them to the Oracle database. Clicking on a locality number on a map or stratigraphic column initiates a search, which results in a list of all taxa that have been recorded at that locality. For users who have received special permission from the NMITA Database Coordinator, searches are also available, which output spreadsheets listing occurrences of a given taxon or occurrences of taxa found at a given locality (or stratigraphic unit, or range of geologic ages).

13.3 Multiple Specimen Identifications

Unlike the Paleobiology Database (pbdb.org), which is based on the published literature, NMITA is ultimately based on specimens that have recently been collected as part of team collecting projects, and the taxa and stratigraphic units that are contained in NMITA are based on specimen data. Literature citations have recently

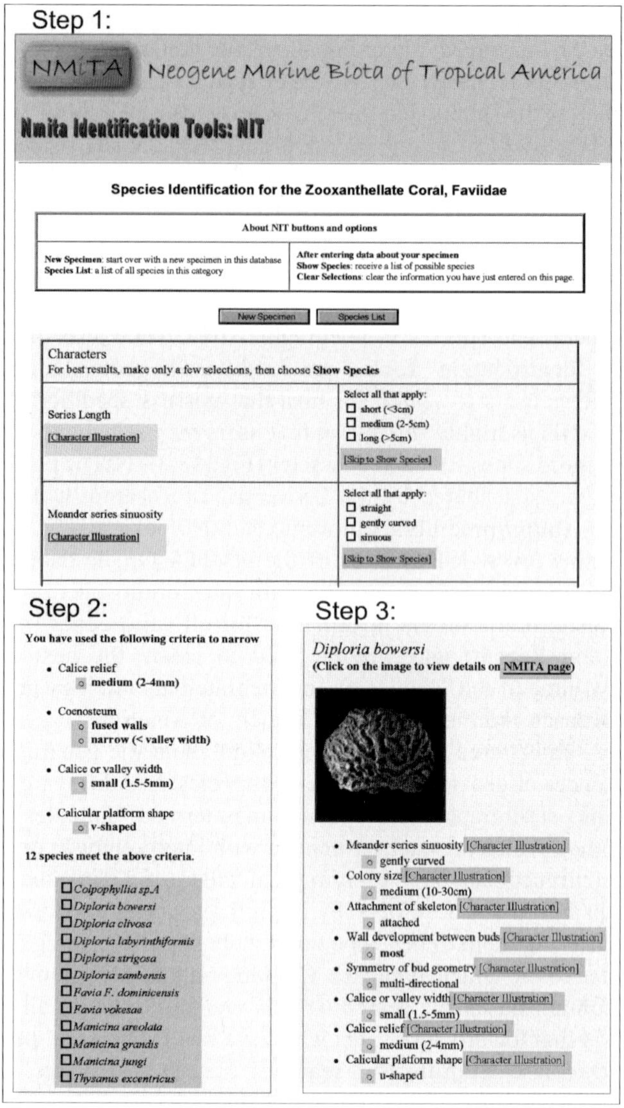

Fig. 13.3 NMITA Identification Tools (NIT) involving searches of morphologic characters. Step 1: the user selects one or more states for morphologic characters selected from a list of 25 characters. Step 2: a list of species having those character states is returned. Step 3: the user may then display all of the states for each of the species in the list and refine the search

been added to NMITA to facilitate sharing data with PBDB, but they do not play a central role in the functioning of the database. In corals, specimen data is captured in the Cenozoic Coral Database (CCD), a MS Access database consisting of five tables that is available for downloading from the NMITA website with special

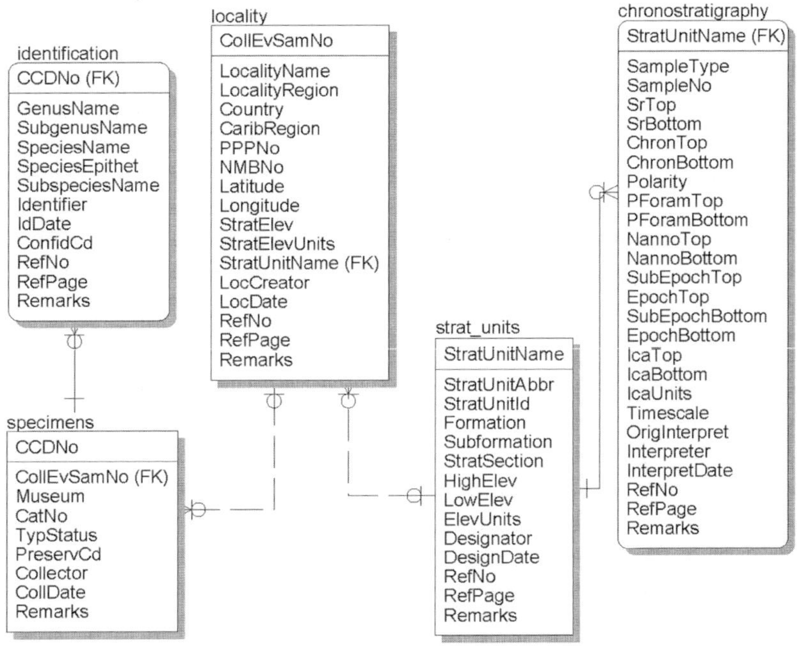

Fig. 13.4 CCD database schema showing the tables (entities) and their attributes. This database provided the basic data contained in NMITA as well as ANALYSIS (filtered data used for performing analyses of biodiversity through geologic time, see Johnson et al., this volume)

permission from the NMITA Database Coordinator (Fig. 13.4). After returning from the field and processing the samples, each specimen is assigned a unique CCD number (specimens table) and associated locality information is entered in the locality table, primarily using field notes. The specimens are then identified using the online identification keys in NMITA, and initial identifications are entered in the identification table. Subsequent identifications, made for example using morphometric analyses (see Budd et al., this volume, and Schultz and Budd, this volume), are also entered into the same table as they become available, thereby allowing a complete history of identifications for each specimen to be recorded.

13.4 Multiple Age Interpretations

CCD also contains tables for recording stratigraphic data, which are used in determining the stratigraphic ranges of individual taxa (Budd et al., this volume; Schultz and Budd, this volume; Beck and Budd, this volume) and studying patterns of biodiversity through geologic time (Johnson et al., this volume). Each of the collected stratigraphic sequences is first subdivided into stratigraphic units (sometimes termed "faunules") based on lithology. For example, in the DR Project, we have used the 21 lithostratigraphic units originally defined by Budd et al. (1996) and used in Klaus et al. (this volume) and Johnson et al. (this volume).

Geologic ages are interpreted for each stratigraphic unit by integrating data based on strontium isotope analyses, paleomagnetics, and/or microfossils using multiple time scales (see McNeill et al., this volume). As with specimen identifications, many different interpretations can be entered for any given stratigraphic unit, allowing a complete history of interpretations for each unit to be recorded. As described by Johnson et al. (this volume), this information facilitates comparison of patterns of biodiversity through time using different age interpretations.

13.5 Data Sharing

Ongoing efforts at data sharing focus on the University of Iowa Paleontology Repository, which houses many of the coral specimens displayed on the NMITA website as well as the coral specimens collected as part of the DR Project and cited in this book. Each colony is given a unique identity, or catalog number. This number is written on the specimen and recorded in a specimen database along with all available information about the specimen. The UI Paleontology Repository uses Specify Biodiversity Collections Software, a free database system available from the Specify Software Project, University of Kansas (http://www.specifysoftware.org/Specify). Specify is a Windows software platform for managing biodiversity collection data, with HTML web and XML DiGIR interfaces, fast Google-like database searching, customizable data entry forms, and a report writing system. The application manages specimen data and information about collections transactions. The Specify Web Interface runs on a Microsoft IIS or Apache server, and allows on-line access for searching and viewing data. Specimen images and NMITA web pages citing UI Paleontology Repository specimens can be recorded as a file network path or URL respectively in each relevant specimen record. The information is displayed with on-line search results as a link that displays the image or takes the user to the relevant NMITA page.

The Specify DiGIR Interface provides collection data as XML-structured data records compatible with discipline-based virtual communities such as the Paleontology Portal (http://www.paleoportal.org/). It also supports connections to warehouses of collection information like that of the Global Biodiversity Information Facility (http://www.gbif.org/). PaleoPortal provides a search interface that is capable of returning results from the collections databases of several institutions. We employ the DiGIR protocol and a custom "PaleoPortal" data schema to make this possible.

In addition to sharing data with the University of Iowa Paleontology Repository, XML protocols are also being developed in NMITA to share taxonomic and occurrence data with the database "Hexacorallians of the World", which treats the systematics of the six living cnidarian orders of Subclass Hexacorallia (hercules.kgs.ku.edu/hexacoral/anemone2/ index.cfm), and also with the Paleobiology Database (Fig. 13.1). One example involving occurrence data is provided as an Appendix.

Acknowledgments We thank Arnie Miller for comments. Current database efforts are supported by a collaborative grant from the National Science Foundation (EAR 0445789 to A.F. Budd, and EAR 0446768 to D.F. McNeill).

Appendix

Examples of XML protocols available to provide occurrence data

First, we provide a link to display all locality numbers ("collecting event sample numbers") in XML format. For example,

< coll_ev_sam_no > **AB93–05** < /coll_ev_sam_no >
< coll_ev_sam_no > **AB93–06** < /coll_ev_sam_no >
< coll_ev_sam_no > **AB93–21** < /coll_ev_sam_no >
< coll_ev_sam_no > **AB93–22** < /coll_ev_sam_no >
< coll_ev_sam_no > **AB93–23** < /coll_ev_sam_no >
< coll_ev_sam_no > **AB93–30** < /coll_ev_sam_no >
< coll_ev_sam_no > **AB93–31** < /coll_ev_sam_no >
< coll_ev_sam_no > **AB93–32** < /coll_ev_sam_no >
< coll_ev_sam_no > **AB93–35** < /coll_ev_sam_no >
< coll_ev_sam_no > **AB93–36** < /coll_ev_sam_no >
< coll_ev_sam_no > **AB93–37** < /coll_ev_sam_no >
< coll_ev_sam_no > **AB93–38** < /coll_ev_sam_no >
< coll_ev_sam_no > **AB93–39** < /coll_ev_sam_no >

Then using one of the locality numbers, a user can manually type or run an automated program to retrieve detailed information, again in XML format. For example, for locality = AB95–21,

< coll_ev_sam_no > **AB95–21** < /coll_ev_sam_no >
< strat_unit_name > **seacliff_21** < /strat_unit_name >
< locality_name > **St. Michiel seacliff** < /locality_name >
< country_name > **Curacao** < /country_name >
< latitude > **12.148619** < /latitude >
< longitude > **-68.999375** < /longitude >
< formation_name > **Seroe Domi Formation** < /formation_name >
< sub_epoch_name_bottom > **Late** < /sub_epoch_name_bottom >
< epoch_name_bottom > **Pliocene** < /epoch_name_bottom >
< sub_epoch_name_top > **Early** < /sub_epoch_name_top >
< epoch_name_top > **Pleistocene** < /epoch_name_top >
< ica_bottom > **2.5** < /ica_bottom >
< ica_top > **1** < /ica_top >
< ica_units > **Ma** < /ica_units >

- < distribution >
- < occurrence >
< genus_name > **Acropora** < /genus_name >
< species_name > **cervicornis** < /species_name >

< genus_name > **Agaricia** < /genus_name >
< species_name > **undata** < /species_name >
< genus_name > **Stephanocoenia** < /genus_name >
< species_name > **intersepta** < /species_name >
< genus_name > **Caulastraea** < /genus_name >
< species_name > **portoricensis** < /species_name >
< genus_name > **Colpophyllia** < /genus_name >
< species_name > **natans** < /species_name >
< genus_name > **Diploria** < /genus_name >
< species_name > **clivosa** < /species_name >
< genus_name > **Diploria** < /genus_name >
< species_name > **strigosa** < /species_name >
etc etc
< /occurrence >
< /distribution >

References

Budd, A.F., Foster, C.T. Jr., Dawson, J.P., and Johnson, K.G., 2001, The Neogene Marine Biota of Tropical America ("NMITA") Database: accounting for biodiversity in paleontology, *J. Paleontol.*,**75**:743–751.

Budd. A.F., Johnson, K.G., and Stemann, T.A., 1996, Plio-Pleistocene turnover and extinction in the Caribbean reef-coral fauna, in: *Evolution and Environment in Tropical America* (J.B.C. Jackson, A.F. Budd, and A.G. Coates, eds.), University of Chicago Press, Chicago, pp. 168–204.

Johnson, K.G., Budd, A.F., Klaus, J.S., and McNeill, D.F., This volume, The impact of fossil from the northern Dominican Republic on origination estimates for Miocene and Pliocene Caribbean reef corals, in: *Evolutionary Stasis and Change in the Dominican Republic Neogene*.

Klaus, J.S., Budd, A.F., and McNeill, D.F., This volume, Assessing community change in Miocene to Pliocene Coral Assemblages of the northern Dominican Republic, in: *Evolutionary Stasis and Change in the Dominican Republic Neogene*.

McNeill, D.F., Klaus, J.S., Evans, C.C., Budd, A.F., and Maier, K.L., This volume, An overview of the regional geology and stratigraphy of the Neogene deposits of the Cibao Valley, Dominican Republic, in: *Evolutionary Stasis and Change in the Dominican Republic Neogene*.

Index

Topics in Geobiology

Series editors

Neil H. Landman,
American Museum of Natural History, New York, landman@amnh.org
Peter Harries,
Department of Geology, University of South Florida, USA,
harries@shell.cas.usf.edu

Current volumes in this series

Volume 30: **Evolutionary Stasis and Change in the Dominican Republic Neogene**
Ross H. Nehm and Ann F. Budd
Hardbound, ISBN 978-1-4020-8241-6, 2008

Volume 29: **Biogeography, Time and Place: Distributions, Barriers and Islands**
Willem Renema
Hardbound, ISBN 978-1-4020-6373-2, 2007

Volume 28: **Paleopalynology: Second Edition**
Alfred Traverse
Hardbound, ISBN 978-1-4020-5609-3, 2007

Volume 27: **Neoproterozoic Geobiology and Paleobiology**
Shuhai Xiao and Alan J. Kaufman
Hardbound, ISBN 978-1-4020-5201-9, October 2006

Volume 26: **First Floridians and Last Mastodons: The Page-Ladson Site in the Aucilla River**
S. David Webb
Hardbound, ISBN 978-1-4020-4325-3, October 2006

Volume 25: **Carbon in the Geobiosphere – Earth's Outer Shell**
Fred T. Mackenzie and Abraham Lerman
Hardbound, ISBN 978-1-4020-4044-3, September 2006

Volume 24: **Studies on Mexican Paleontology**
Francisco J. Vega, Torrey G. Nyborg, María del Carmen Perrilliat, Marison Montellano-Ballesteros, Sergio R.S. Cevallos-Ferriz and Sara A. Quiroz-Barroso
Hardbound, ISBN 978-1-4020-3882-2, February 2006

Volume 23: **Applied Stratigraphy**
Eduardo A. M. Koutsoukos
Hardbound, ISBN 978-1-4020-2632-4, January 2005

Volume 22: **The Geobiology and Ecology of *Metasequoia***
Ben A. LePage, Christopher J. Williams and Hong Yang
Hardbound, ISBN 978-1-4020-2631-7, March 2005

Volume 21: **High-Resolution Approaches in Stratigraphic Paleontology**
Peter J. Harries
Hardbound, ISBN 978-1-4020-1443-7, September 2003

Volume 20: **Predator-Prey Interactions in the Fossil Record**
Patricia H. Kelley, Michal Kowalewski, Thor A. Hansen
Hardbound, ISBN 978-0-306-47489-7, January 2003

Volume 19: **Fossils, Phylogeny, and Form**
Jonathan M. Adrain, Gregory D. Edgecombe, Bruce S. Lieberman
Hardbound, ISBN 978-0-306-46721-9, January 2002

Volume 18: **Eocene Biodiversity**
Gregg F. Gunnell
Hardbound, ISBN 978-0-306-46528-4, September 2001